Basic and Intermediate

Solid Edge ST5

Modeling, Drafting, and Assemblies

A Project Oriented Workbook

By:
Stephen M. Samuel PE

**DESIGN
VISIONARIES**

**Superior Vision Yields
Optimal Products**

ISBN 978-1-935951056

Published by:
Design Visionaries
7034 Calcaterra Drive
San Jose, CA 95120
info@designviz.com
www.designviz.com
Local Phone: (408) 997 6323
Fax: (408) 997 9667

Printed in the United States of America
Published September 2013

**DESIGN
VISIONARIES**

Superior Vision Yields
Optimal Products

Dedication

We dedicate this book to all the men and women who see things differently. Just like JFK. said, 'there are those who see things as they are and say why, then there are others who see things as they could be and ask why not'. Thank you. You take us, sometimes kicking and screaming, from point A to point B.

About the Author:

Stephen M. Samuel PE, Founder and President of Design Visionaries, has over 20 years' experience in developing and using high-end CAD tools and mentoring its users. During a ten-year career at Pratt & Whitney Aircraft, he was responsible for implementing advanced CAD/CAM technology in a design/manufacturing environment. He has trained thousands of engineers in Unigraphics, written self-paced courses in UG Advanced Modeling and Best Practices, and performed design work for numerous Fortune 500 companies. Stephen is the author of distinctive publications on Nastran, UGNX CAD, Sold Edge, SolidWorks, and Teamcenter Engineering PLM. Stephen holds several US patents and enjoys a life of creativity and intellectual challenge in the city of San Jose, CA. He happily shares his life with three amazing children, his wife and an 81 year old powerhouse of a mother that lives in a home right next door.

Acknowledgements

We would like to thank the following people for their tireless efforts. Without the contributions from each of you this book would be a mere shadow of what it has become:

Landon Ritchie and Jenine Ruiz.

Special thanks to Landon Ritchie for designing the cover.

What readers have to say about our previous books:

Bryan McDonald, Product Design Manager at Apple Computers
"Bottom line: this is an excellent book. If anyone wants to learn how to use Unigraphics quickly, efficiently and practically, this is the way to do it."

Fred Dyen, Director of St. Louis University's Aviation Maintenance Institute (AMI)
"Practical Unigraphics NX2 Modeling for Engineers was extremely effective and much better than [other] textbooks. I would highly recommend this book to other professors and students alike."

Dr. Pat Spicer, Professor at Western Illinois University
"The UG NX2 textbook is well organized. Its tutorial style of learning is easy for students to utilize. The practice exercises are essential. From my experience in teaching students to use UG software, I have found that this is the best textbook currently on the market for teaching UG NX2."

Preface

Dear reader,

Thank you for purchasing our Solid Edge Workbook. We have taken great pride over writing the content in the book and we hope that you find huge benefit from using it. Design Visionaries is an engineering consulting firm that performs many design projects great and small, including industrial design, product design and engineering analysis. Our customers entrust us with the design of medical devices, aerospace components, heavy machinery, consumer products, etc.

The methods outlined in this book go beyond an academic use of the software. They are tricks of the trade that come from thousands of hours of actual use of the software to design some of the most difficult products in the world. In addition, Design Visionaries offers world class on-site training which enables us to develop and evolve our training material so that they provide the maximum benefit.

Please enjoy this text, and we invite you to log on to our website – www.designviz.com where you can download all the completed part files of this book, additional free material, and some extra goodies.

Thank you,

Stephen Samuel
July 20, 2013

Exercise Guide:

Ex.	Name	Description	Picture
i	UI Guide (Pg - 19 -)	A general look at the Solid Edge User Interface	
1	Configuring Defaults (Pg- 24 -)	How to change default templates and units as well as how to change units on a per file basis	
2	Switching and Viewing Windows (Pg - 31 -)	How to view and arrange multiple windows for maximum productivity	
3	How to make a basic sketch (Pg - 33 -)	How to use the rectangle tool and placing dimensions	
4	More Sketching (Pg - 39 -)	Using the Draw tool / circle tool and creating basic dimensions	
5	Constraints: Horizontal, Vertical and how to create midpoint constraints (Pg - 44 -)	How to use horizontal and vertical constraints to define your sketch. And how to create midpoint constraints	

6	Constraints – Equal (Pg - 47 -)	How to use the Equal Constraints	
7	Constraints - parallel and Concentric (Pg - 50 -)	How to use the concentric constraint	
8	Constraints – Concentric and Trim Tool (Pg - 55 -)	How to use the concentric constraint and trimming lines	
9	Constraints – Collinear (Pg - 58 -)	How to use the collinear constraint	
10	Constraints – Tangent (Pg - 60 -)	How to use the tangent constraint	
11	Constraints - Connect and Angle (Pg - 63 -)	How to use the connect constraints and angle dimension tool	
12	Creating Relationships in a sketch. (Pg - 66 -)	How to relate your constraints using formulas	

13	Extruding (Pg - 68 -)	How to extrude	
14	More Extruding (Pg - 74 -)	Extruding and learning how to use the arc tool	
15	Creating Holes (Pg - 81 -)	How to use the hole feature and its options	
16	More holes (Pg - 88 -)	Creating a hole on a curved surface	
17	Cutting (Pg - 92 -)	How to use the Cut feature	
18	Revolving (Pg - 98 -)	How to use the revolve tool	

19	Revolved Cut with angled datum planes (Pg - 101 -)	How to use angled datum planes to create an angled hole and revolving a cut.	
20	Mirror (Pg - 111 -)	How to mirror a body	
21	Draft, Round and Chamfer (Pg - 116 -)	How to use the ellipse tool with a draft and adding chamfers and rounds to a body.	
22	Sweep along a Path (Pg - 121 -)	How to sweep a section along a path to create a solid body.	
23	Thin Wall (Pg - 125 -)	How to give a solid body a uniform wall thickness by removing material.	
24	Mounting Bosses (Pg - 130 -)	How to create a boss	
25	Loft and Web networks (Pg - 133 -)	Using SE's great feature techniques to create web features on a nice lofted surface.	
26	Vent Tool (Pg - 138 -)	How to use the vent tool	

27	Rectangular Hole Pattern by Grouping Faces (Pg - 145 -)	How to create a group of faces then pattern them.	
28	Circular Hole pattern (Pg - 149 -)	How to create a circular pattern of holes	
29	Curved Pattern (Pg - 156 -)	How to create a pattern of holes along a path	
30	Helical Cut Neural (Pg - 163 -)	How to use the helical cut feature along with circular pattern to create a neural shape.	
31	Text Tool, Wrapping and Normal Cut (Pg - 173 -)	How to create a decal on a body exploring new features such as wrapping and the normal cut.	
32	Blue surface through 3 sketches (Pg - 179 -)	How to create a basic Blue Surface	
33	Blue surface with guide curves (Pg - 182 -)	Techniques on how to gain more control over a blue surface by using guide curves	
34	Changing Surfaces into a solid (Pg - 187 -)	Using the stitch tool to turn sheet bodies into a solid.	

35	Extrude Surfaces that are tangentially connected. (Pg - 191 -)	How to create curvature continuous surfaces and turning them into a solid.	
36	Surfacing, Trim, Swept, extend and split (Pg - 198 -)	Exploring several surfacing tools to create and edit a sheet body.	
37	Offset Surfaces (Pg - 204 -)	Using the revolved surface tool and using offset face.	
38	Replace Face (Pg- 208 -)	How to use the traditional part files 'synchronous' modeling technique – replace face.	
39	Copying Surfaces (Pg- 214 -)	How to copy a surface in SE	
40	Boolean and Curve Projection (Pg - 220 -)	How to use tool bodies to cut solids	
41	Adding Threads (Pg - 223 -)	How to create a thread on a cylinder	

42	Working in a synchronous part file (Pg - 230 -)	Quick guide to creating and modifying synchronous part files	
43	Live Sections and Live Rules (Pg - 238 -)	Explores how Live Sections and Live Rules can be utilized to aid in the Synchronous Modeling process.	
44	Adding Ordered Features to Synchronous Models (Pg - 247 -)	Demonstrates "Hybrid Modeling", or modeling in both Synchronous and Ordered mode within a single part.	
45	Editing Imported Parts using SolidEdge (Pg - 255 -)	How to utilize Synchronous Mode to edit a part not native to SolidEdge.	
46	Fish Food for Thought (Pg - 268 -)	Brings together the skills you've learned so far to create an organic, geometrically complicated model.	
47	Assembly 1 (Pg- 288 -)	How to put together a simple assembly - bolt and nut	

48	Assembly 2 (Pg - 293 -)	Exploring more assembly techniques with several different components.	
49	Assembly 3 (Pg - 303 -)	Assembling a complicated assembly	
50	The Gear Relationship (Pg - 310 -)	How to create gears in SE	
51	Assembly 4 "Do Nothing" (Pg - 317 -)	Importing external geometry into SE and creating a kinematic assembly	
52	Cam Tool "Hurdy Gurdy" (Pg - 327 -)	Creating another kinematic assembly	
53	Part Families and Alternate Assemblies (Pg - 336 -)	Creating families of parts and using alternate assemblies to display each part of the family	
54	Annotations (Pg - 347 -)	How to add dimensions and annotations to your 3D model	

55	Drafting (Pg - 353 -)	Beginners guide to drafting - including sections and detailed views	
56	Drafting (Broken View) (Pg - 361 -)	How to create a broken view in drafting mode	
57	Drafting Assemblies (Pg - 364 -)	How to create exploded views and a parts list	
58	Tabs and Flanges (Pg - 372 -)	How to use the basic Sheet Metal features Tab and Flange to create Sheet Metal models	
59	Contoured Flanges (Pg - 375 -)	How to create Contoured Flanges in a Sheet Metal model	
60	Bending Sheet Metal (Pg - 377 -)	How to bend Sheet Metal along a linear sketch	
61	Unbending and Re-Bending (Pg - 380 -)	How to unbend, modify, and then re-bend a Sheet Metal model	

62	2-Bend Corners (Pg - 384 -)	Applying corner treatments to 2-Bend Corners	
63	Dimples, Drawn Cutouts, and Etching (Pg - 386 -)	How to add dimple, drawn cutout, and etch features to Sheet Metal models	
64	Hems and Jogs (Pg - 392 -)	Applying Hems and Jogs to a Sheet Metal model	
65	Gussets (Pg - 395 -)	How to add Gussets to a Sheet Metal Model	
66	Louvers (Pg - 397 -)	How to add Louvers to a Sheet Metal model	
67	Converting Solid Parts to Sheet Metal (Pg - 402 -)	How to convert a model from a solid part to Sheet Metal	
68	Modifying Sheet Metal Models (Pg - 407 -)	Explores the Bend Angle and Bend Radius tools.	

(i) Guide to the Solid Edge User Interface

In the following few pages we will take a look at the Solid Edge user interface and discover the fastest way to navigate around the SE environment.

The following figure displays the screen discovered when first opening Solid Edge. From this screen it is possible to open existing Solid Edge files, create new SE files, complete some helpful SE tutorials, and get linked to support from GTAC (the Siemens help service for SE, NX, and the rest of their fantastic suite of software programs).

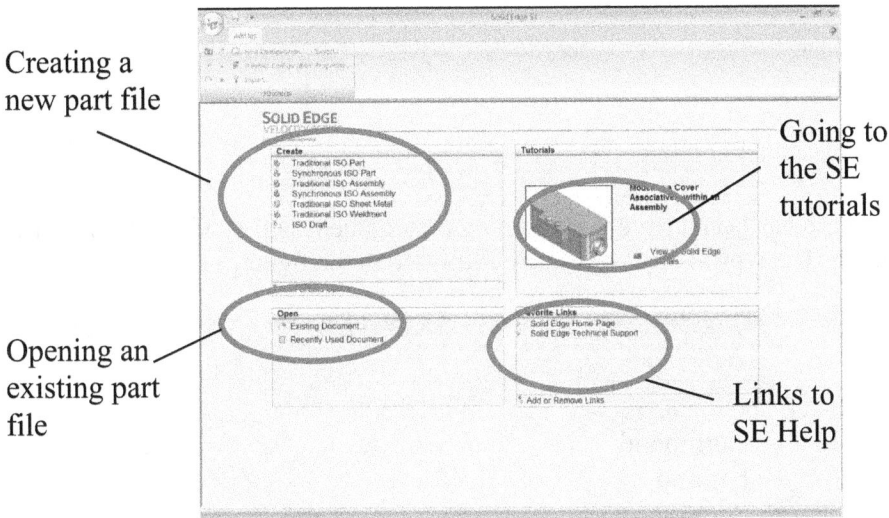

Creating a new part file

Going to the SE tutorials

Opening an existing part file

Links to SE Help

Click on the SE symbol in the top left corner. From this button is it possible to open or create parts and change system settings.

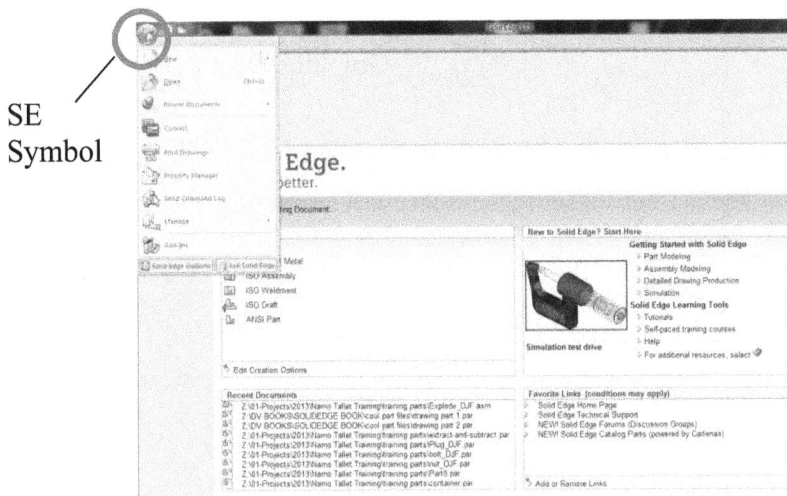

SE Symbol

Create a new part file and look at what's inside the modeling environment of SE. To create a new part file click on the **SE** symbol, select **New** and **ANSI Part.** *Note: Most of the parts in this book will be created using the ANSI Part, unless otherwise stated.*

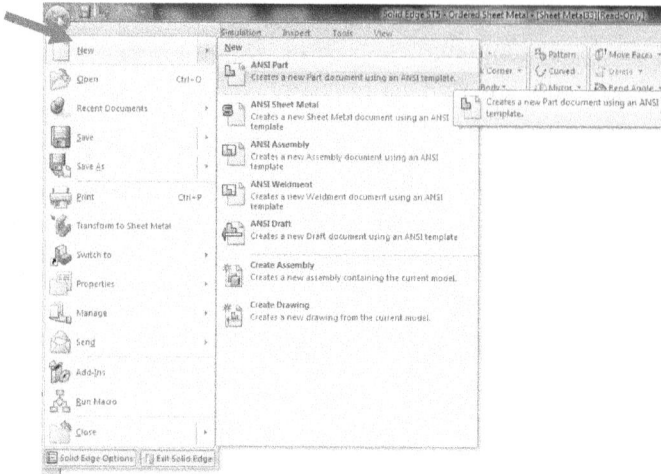

Once the new part file has been create the modeling environment will be viewed, as shown in the following figure. The typical areas of use are identified in the figure.

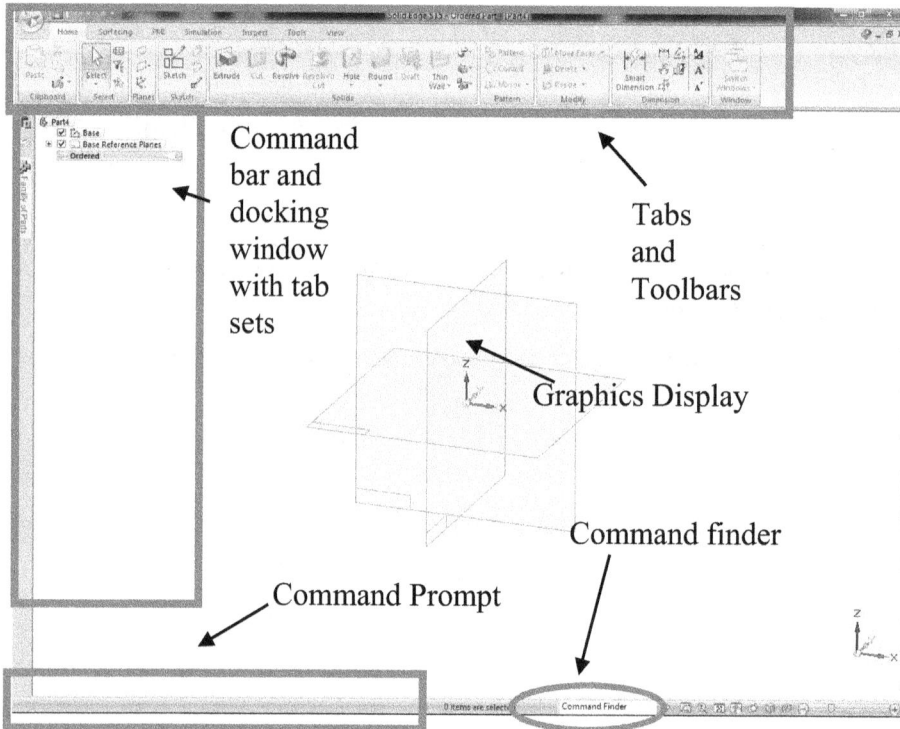

The 'Tabs and toolbars' area of the screen is where most of the features and tools that are found in SE. This includes the Sketcher, Extrude tool, Revolve tool, Surfacing features, and all the inspection tools. The Home tab is shown in the following figure. *Explore the other tabs to see all the amazing features, most of which, we will be covering in this book.*

The **Command Bar** comes to life when modeling. By default this appears on the left side of the screen. The Command Bar guides the user through the steps of feature creation. It provides all the different options that are available to the user and identifies what information each step of the process is looking for. For further assistance the **Prompt Bar,** at the foot of the screen, provides the user with more information about what the operation is specifically asking for. The command bar for the extrude command is shown below. *Notice: This bar may also be horizontal, depending on how it is configured in Solid Edge Options.*

The **Docking Window** is one of the places where the command bar can be positioned. The Docking Window is where all the 'bars' are located, including the path finder, feature library, layers etc. The following figure shows the Pathfinder bar populated with the feature history of a simple model. *Note: Any 'bar' in the Docking Window can be moved by dragging it to a different location on the screen.*

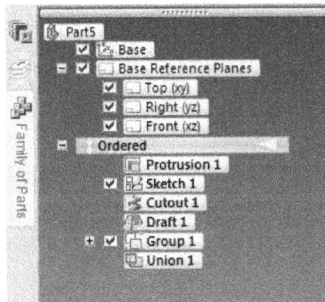

The **Command Finder** is a great tool to help any new SE user find the tools they are looking for. Say the user found him/herself with a new job as an engineer at Acme Inc. Its day one and they have been given some modeling to do on Solid Edge. Well, guess what? They have been working on 'ProE' for the last 10 years and have never touched SE. Well - it's their lucky day. The new user can simply type in names of features found in 'ProE' or any other CAD software and the command finder will look for the equivalent feature in Solid Edge. It will even highlight where the features exist on the toolbars.

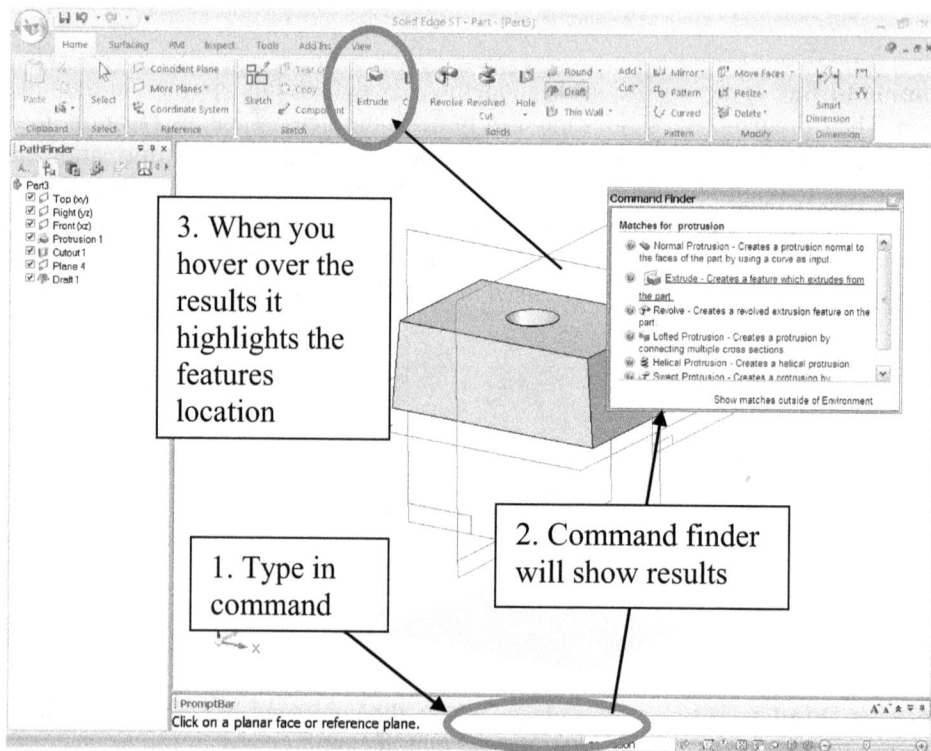

3. When you hover over the results it highlights the features location

2. Command finder will show results

1. Type in command

The **Graphics Display** is where the user can create, modify and view geometry. Two hands are better than one. By using a 3D connexion space ball, or any similar product, that can quickly rotate, zoom, pan around the modeling environment. This will leave the mouse free to click on the detail that you require. If a Space Ball is not available then the viewing tools found under the 'View' tab, shown in the following figure, will help navigation of the graphics display.

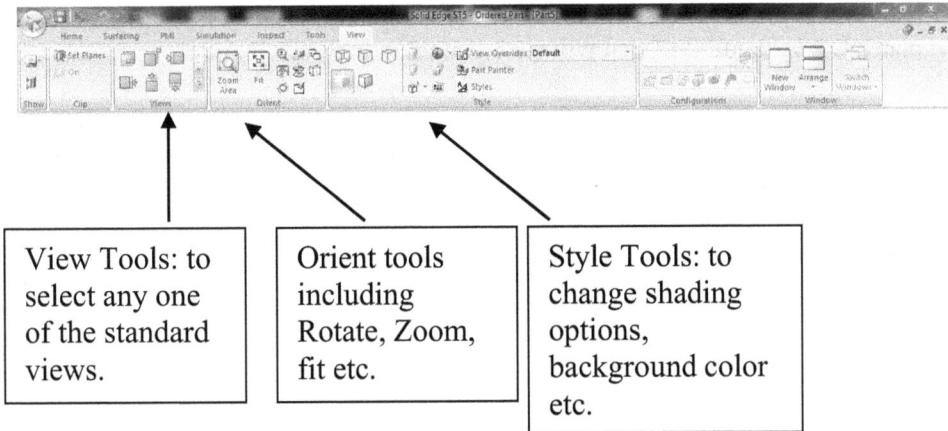

| View Tools: to select any one of the standard views. | Orient tools including Rotate, Zoom, fit etc. | Style Tools: to change shading options, background color etc. |

The 'Orient Tools' ' help the user navigate around the geometry, zooming in and out of detailed areas and rotating the model around to view different angles. The 'Views tools' will orient the model to the standard top, bottom, right views including the ISO view and a few others. The 'Style toolbar' includes such tools as the Part Painter which changes the color of the model. And the View tool will change the color of the background (among other things). It is also possible to change the view from 'shaded with edges' to 'shaded with no edges' or to multiple wireframe options.

Exercise 1: Configuring Defaults

All exercises in this book will be created in inches unless otherwise stated. For this reason, we will be using **ANSI templates** as our default. (ANSI templates have inches as the default unit while ISO and ANSI (mm) templates have millimeters as the default unit) Upon installation, you can choose the default templates you want to use. If you chose ANSI, congratulations, you're done! If not, you'll have to manually switch templates. Regardless, the ability to add new templates is a valuable skill that will save you time.

Just to get you familiar with your default templates, you reach them by clicking the Solid Edge icon in the top left corner of the screen and then hovering over the arrow next to "New". Our goal is for these to be **ANSI templates**.

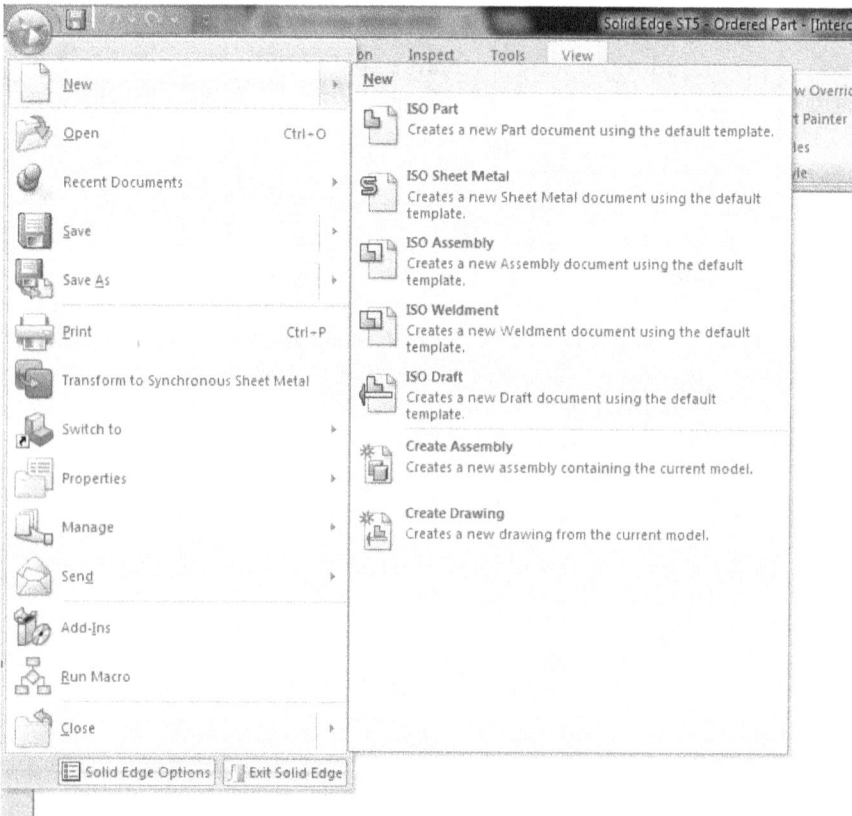

Note: You can open a new file through your defaults, or you can click "New" and browse for the template you want.

To change these templates, you click the "Solid Edge Options" button at the bottom of the menu shown in the picture above. (the window that pops up is a great resource for changing settings in Solid Edge, so feel free to explore it and personalize Solid Edge to your liking) Click "Helpers" as shown below.

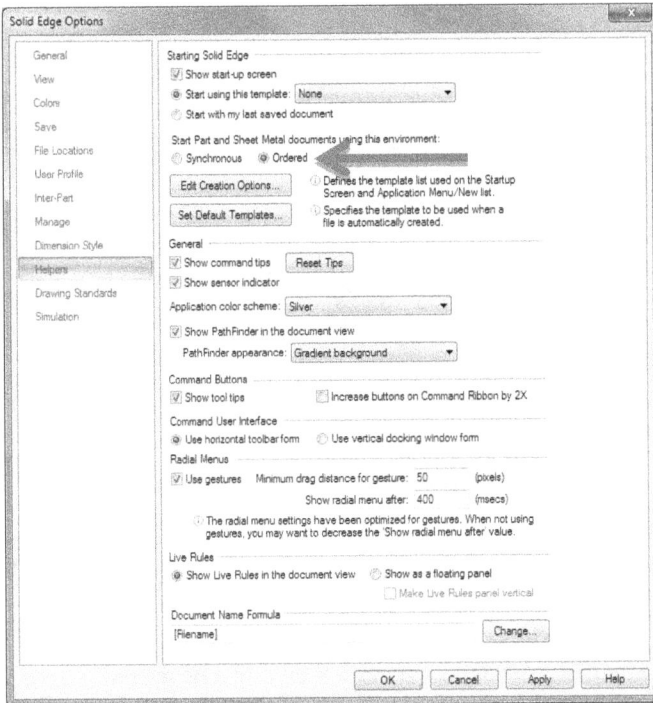

First of all, click the "Ordered" bubble as shown above. Unless otherwise stated, we will be using Ordered Modeling for all exercises. This setting is in no way final, it's simply the default when you open a file. You can switch between Ordered and Synchronous modeling at any time by right clicking anywhere in the design window and clicking "Transition to Ordered/Synchronous" at the bottom of the drop down menu.

Now that we've taken care of that, back to changing the default templates. Click "Edit Creation Options" just below the "Ordered" bubble mentioned previously. You should get the following menu.

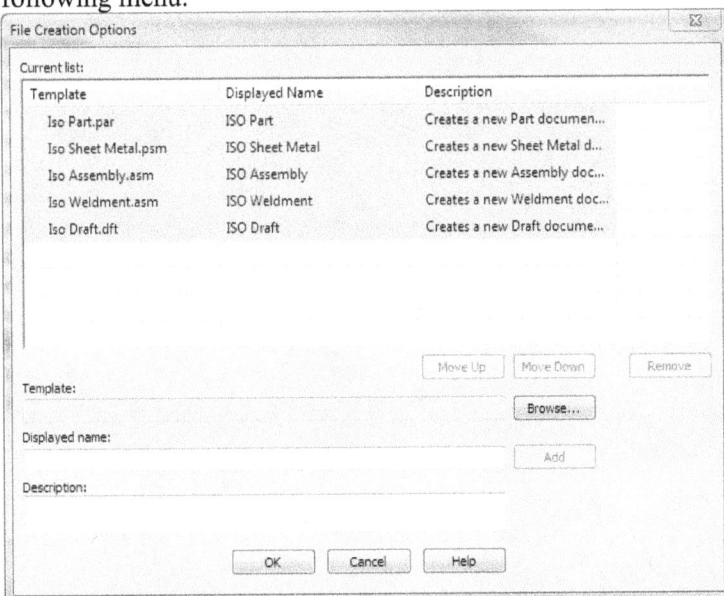

Click "Browse" and then choose "More". From here you can choose and input your ANSI templates.

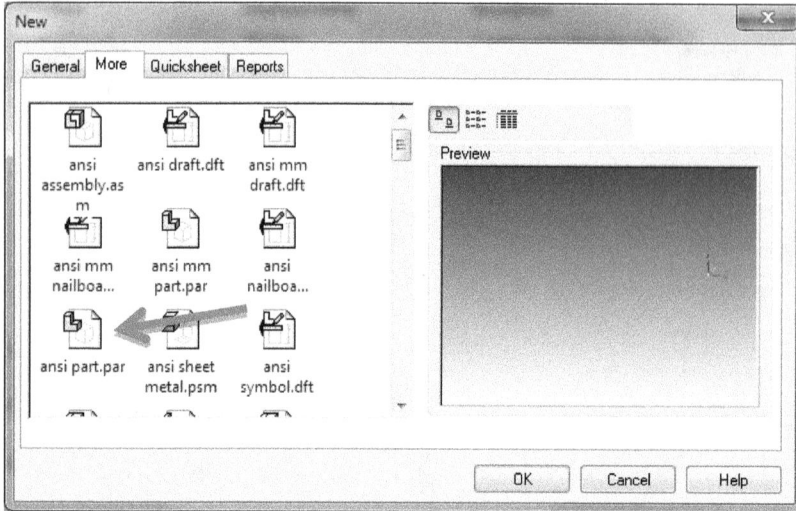

Click **ansi part.par** and hit OK. Input the name "ANSI Part" and then add a description that will help you remember its functionality.

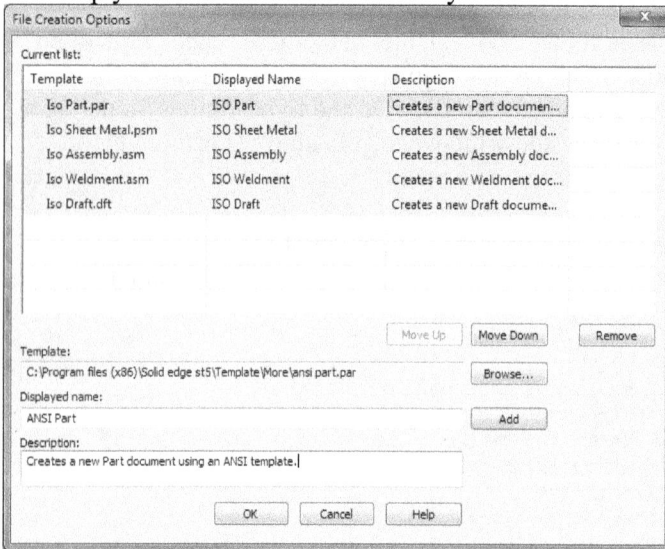

Hit "Add" and then move it to the top using the "Move Up" button. Now that you have an ANSI Part template, you can feel free to remove the ISO Part template.

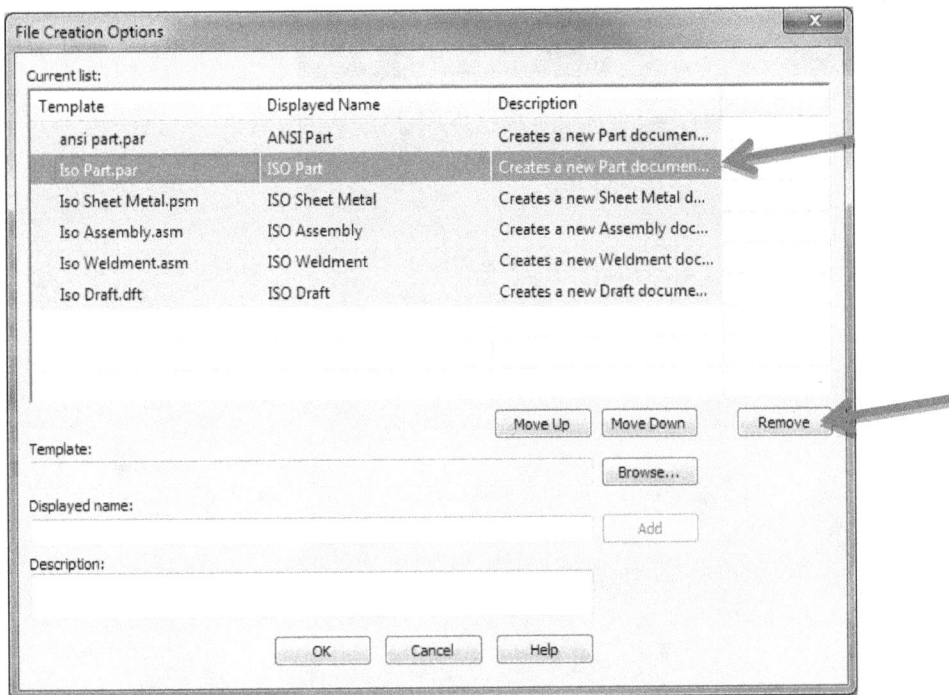

Do this with the rest of the ISO templates until they have all been replaced with ANSI templates. When you are finished, click "OK" to exit the "File Creation Options" window and then click "Apply" before exiting the Solid Edge Options window. Now go back to the Solid Edge menu and hover over the arrow on the "New" tab. Your menu should now look like the following one.

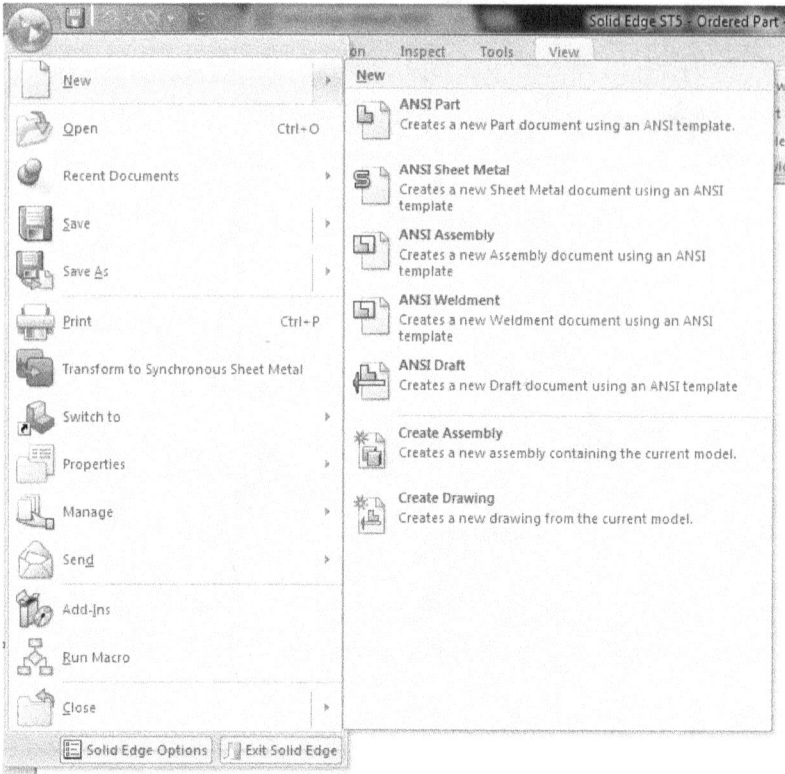

You've just set all of your defaults to ANSI, and therefore inches, but what if you want to change to millimeters within a specific file you're working on? This can be done in a few simple steps:

Click the Solid Edge logo in the top right corner of the page and on the properties tab, click "File Properties".

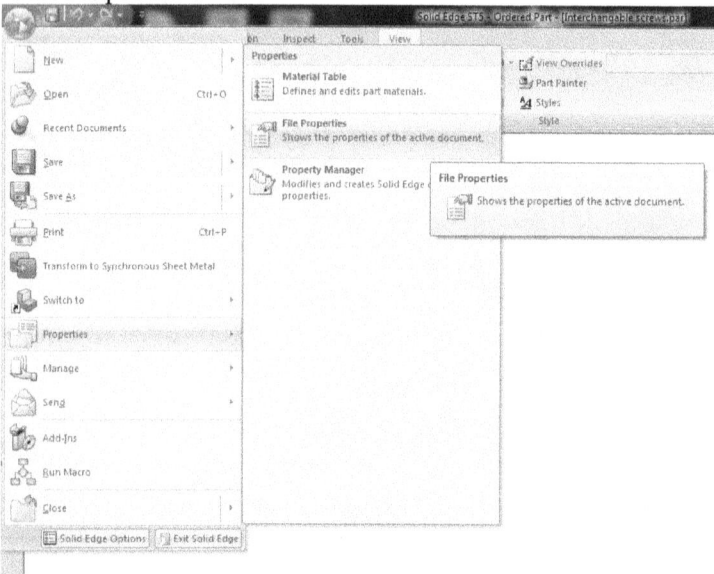

You should get the following window. From here you go to the "Units" tab and choose "mm" in the Length and Area readouts. Click OK.

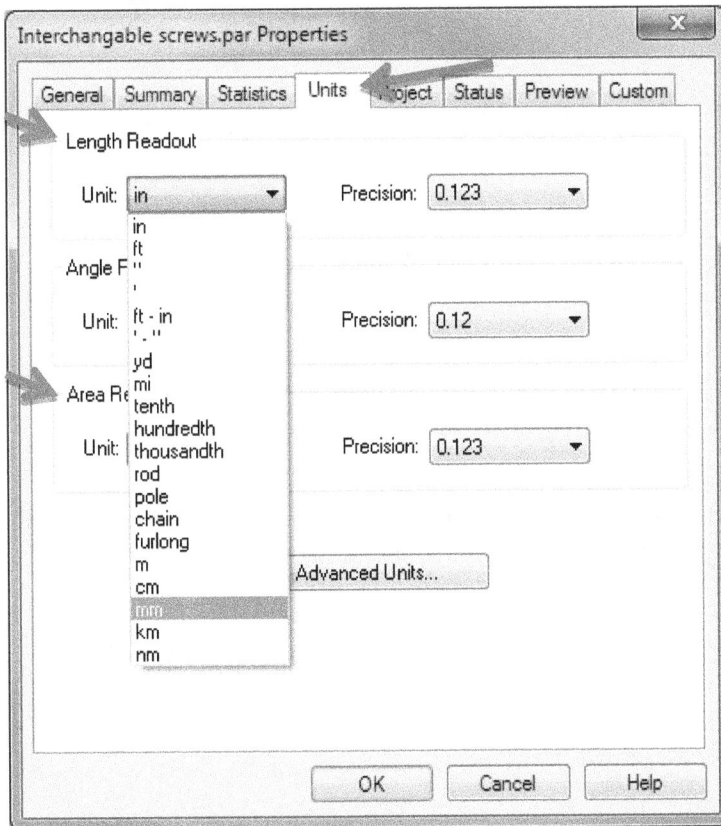

Changing the **File Properties** only changes the actual unit value. When a dimension is created it may still appear to be in the default units of the file. (inches if you're using an ANSI file)

To change the **Dimension Style,** click the **Solid Edge** symbol and click on the **Solid Edge Options** button, just as you did previously to change the default templates. When you get the Options window, go the **Dimension Style** tab and choose the style that corresponds with the units chosen in the previous step (ANSI mm).

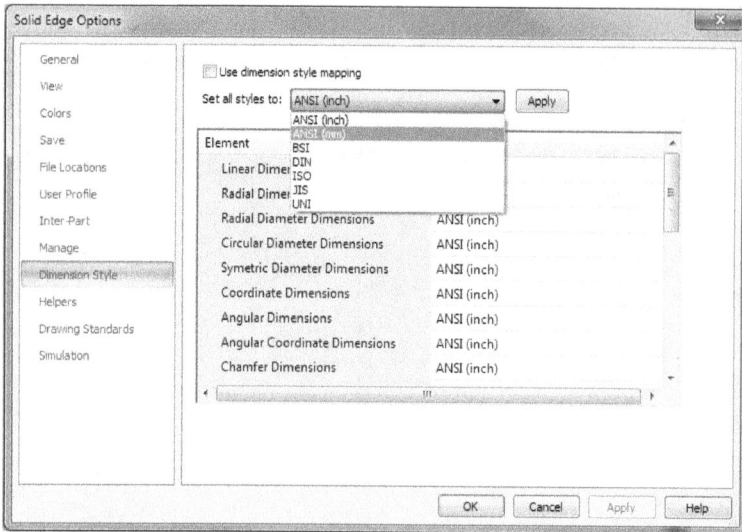

WARNING: After the previous steps it may come to the point, when sketching, where dimensions are being created and showing up in inches (not the mm expected to be shown). If this is the predicament you are in then simply click on the 'Smart Dimension' tool whilst in sketch mode and change the Format to mm, as shown in the following figure. *Notice that dimensions that existed prior to this change will remain in inches. Unfortunately those dimensions will have to redone if you want the old dimensions to match the new format.*

Exercise Complete

Exercise 2: Switching and Viewing Windows

Under the **View** tab in the Window options box it is possible to switch the window being viewed by choosing an open part in the Switch Windows dropdown menu. This is shown in the following figure.

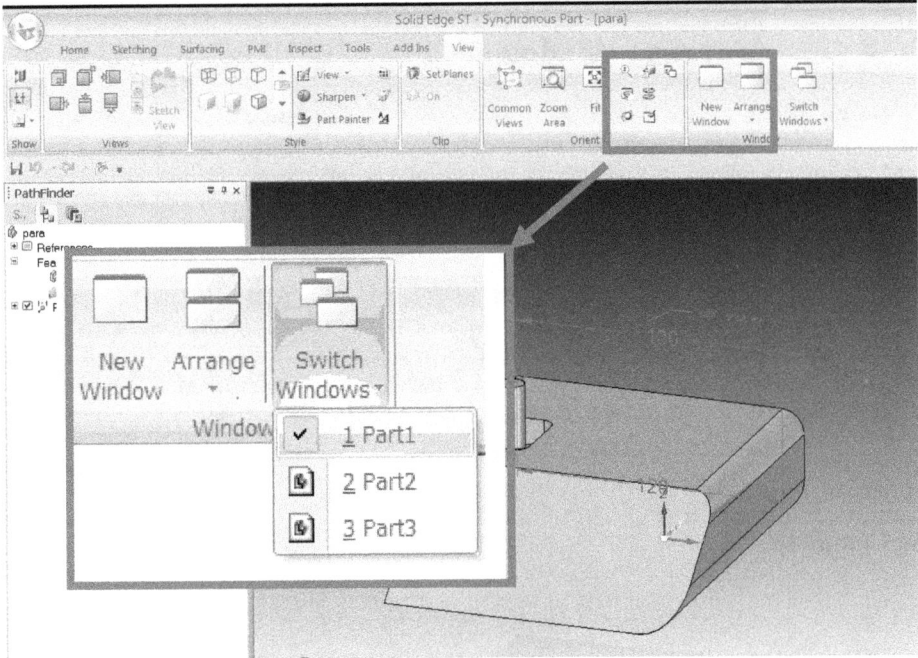

Functionality is also available to show all the open windows using the Arrange tool, as shown below.

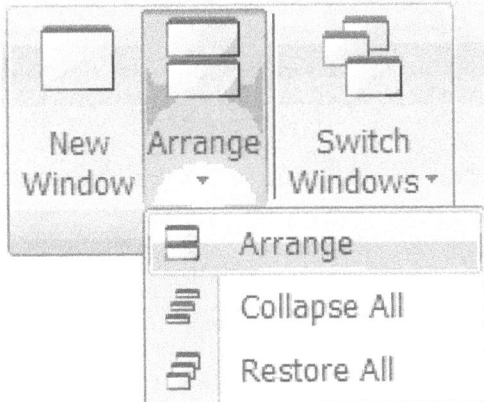

When selecting the **Arrange** tool, an options window should appear like the one below.

This lets the user choose how the windows will be displayed. Find an arrangement that suits your needs.

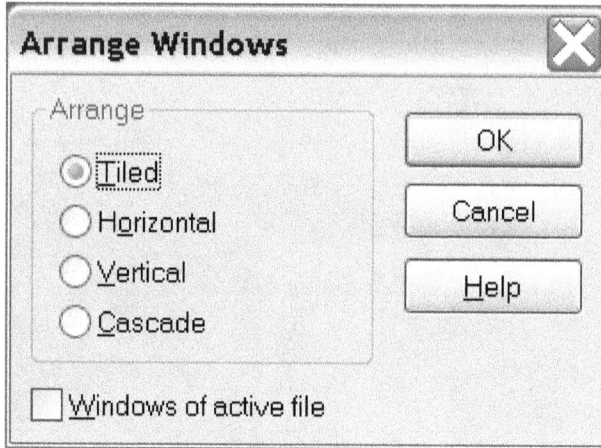

Arrange Windows

Arrange
- ◉ Tiled
- ○ Horizontal
- ○ Vertical
- ○ Cascade

☐ Windows of active file

OK

Cancel

Help

Exercise Complete

Exercise 3: How to make a Basic Sketch

Click **Solid Edge** symbol in top left corner of the screen and select **New / ANSI Part,** as shown in the following figure. *Note: all part files will be made in the traditional mode unless specified otherwise.*

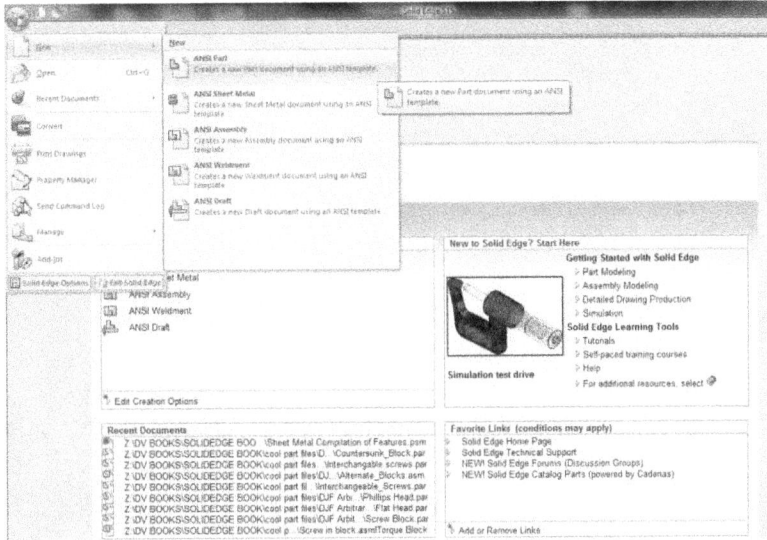

The modeling environment will now be displayed. This is where geometry can be created as shown below.

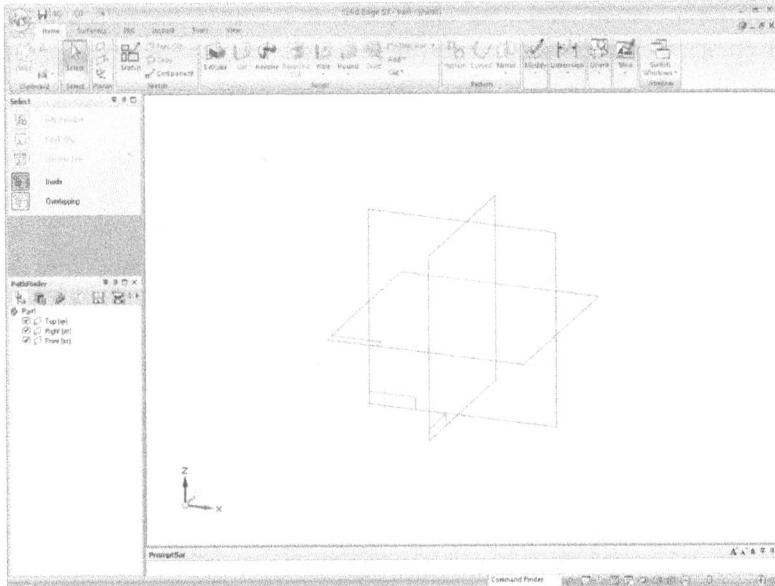

On the **Home** tab in the Sketch toolbar select **Sketch**, as shown above. *Notice: the three planes*

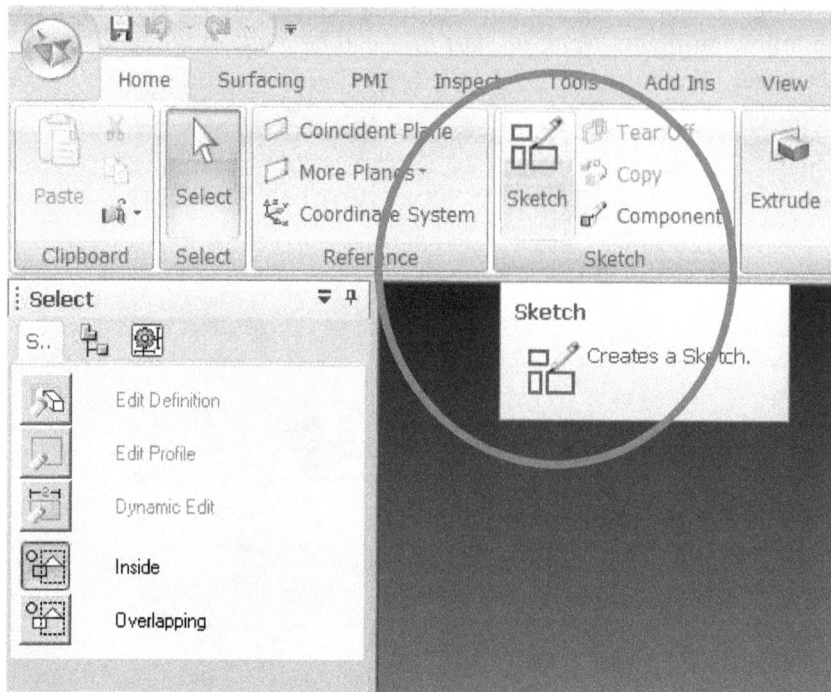

There are 3 default datum planes available for selection in the graphics display however there are other options for selection on the bar on the left, as shown in the following figure. For this exercise, click on one of the existing planes, in the graphics display, to begin sketching. *Note: The sketch plane will now be viewed.*

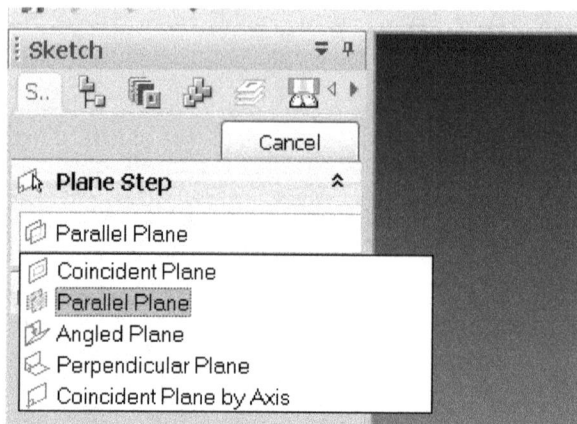

After a plane is selected the sketcher will orient the view to make life easier for the user. To begin sketching select a tool from the **Draw** options box. For this example select the **Rectangle** tool, circled in the following figure.

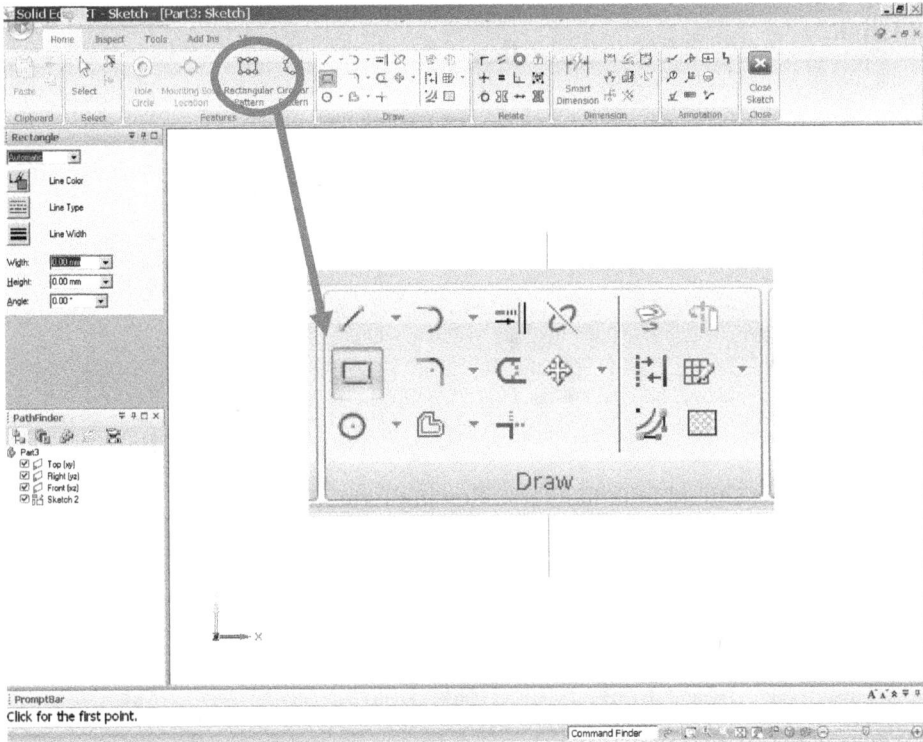

Note: We changed our background to white for clarity. To change the color of the background select the View Tab in the menu / Select View in the Style options box and change the background.

Click on the screen to place the top left corner of the rectangle as shown below. Place the 2nd point by clicking at the top right corner of rectangle and 3rd by clicking in bottom right corner. *Note: the lines will snap to vertical and horizontal positions.*

First Click

Second Click

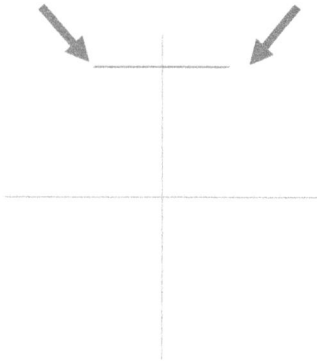

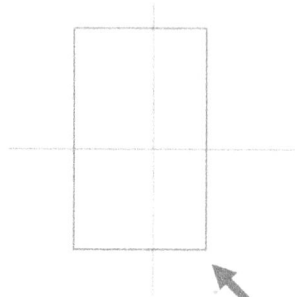

Third Click

The finished rectangle should look like the following figure.

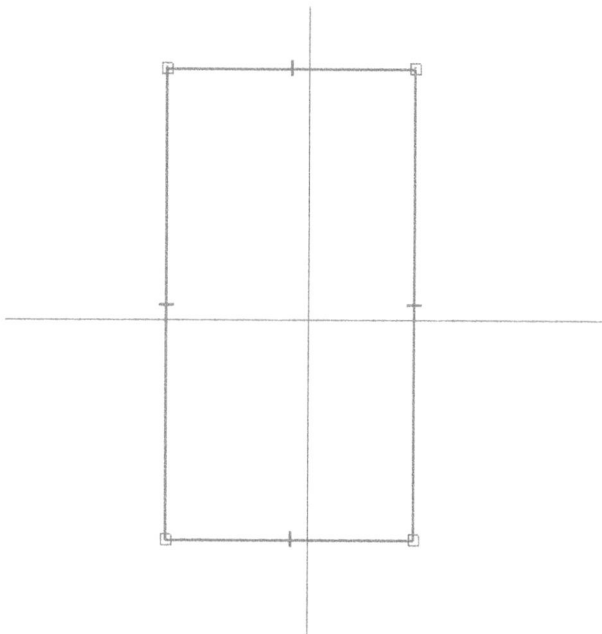

To change the size of the rectangle click on the **Smart Dimension** tool in the Dimension options box.

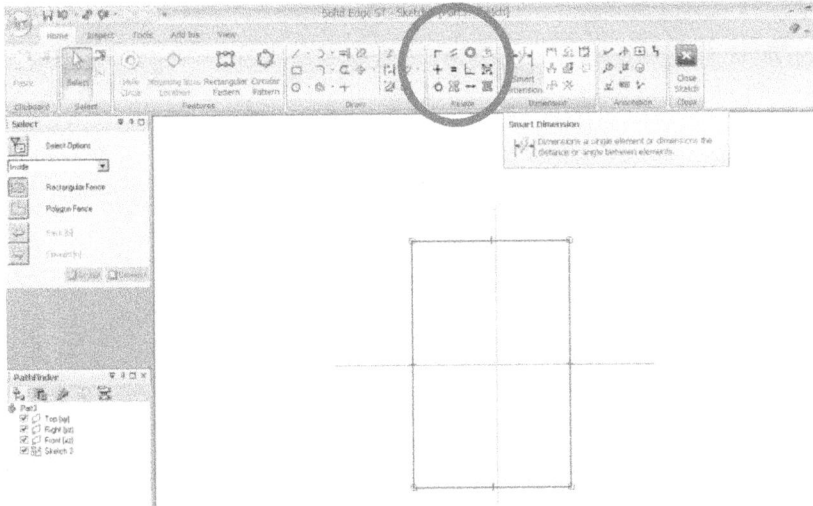

1st click on the top line of the rectangle and click a second time to place the dimension in a suitable location. To change the value, use the box labeled linear in the menu on the left. Type in a value of 1in and hit enter to **apply**.

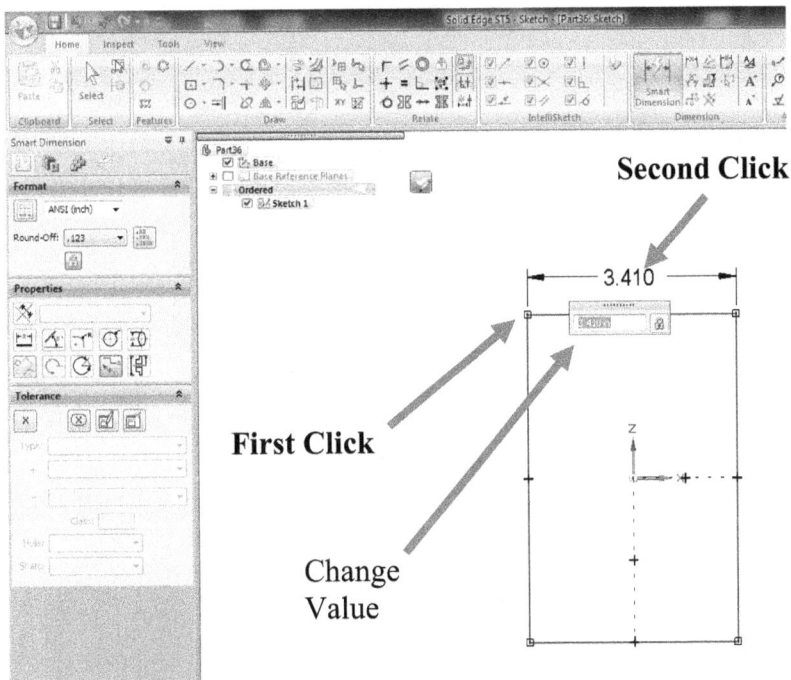

For this exercise change the horizontal dimension to 1 and the vertical dimension to 1.618, as shown in the following figure.

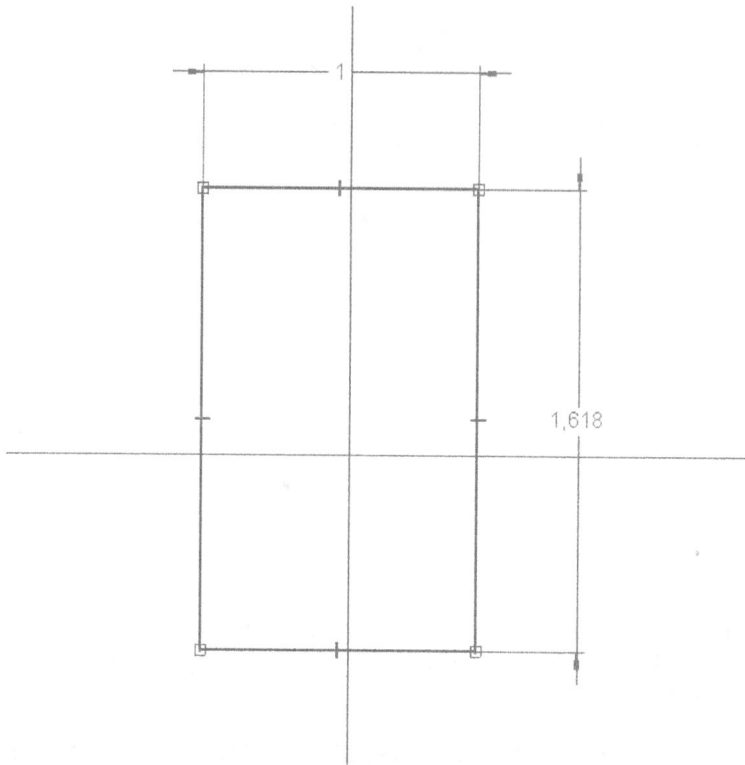

Exercise Complete

Exercise 4: More Sketching

Create a **new part file**. Using the **line** tool located in the Draw options box, draw the figure below.

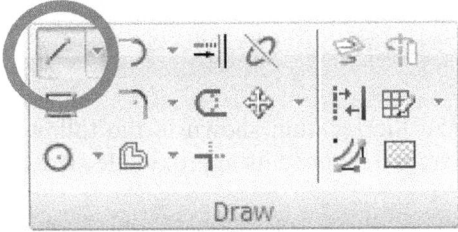

Note: The line tool will create a continuous chain of lines if end points are continued to be clicked.

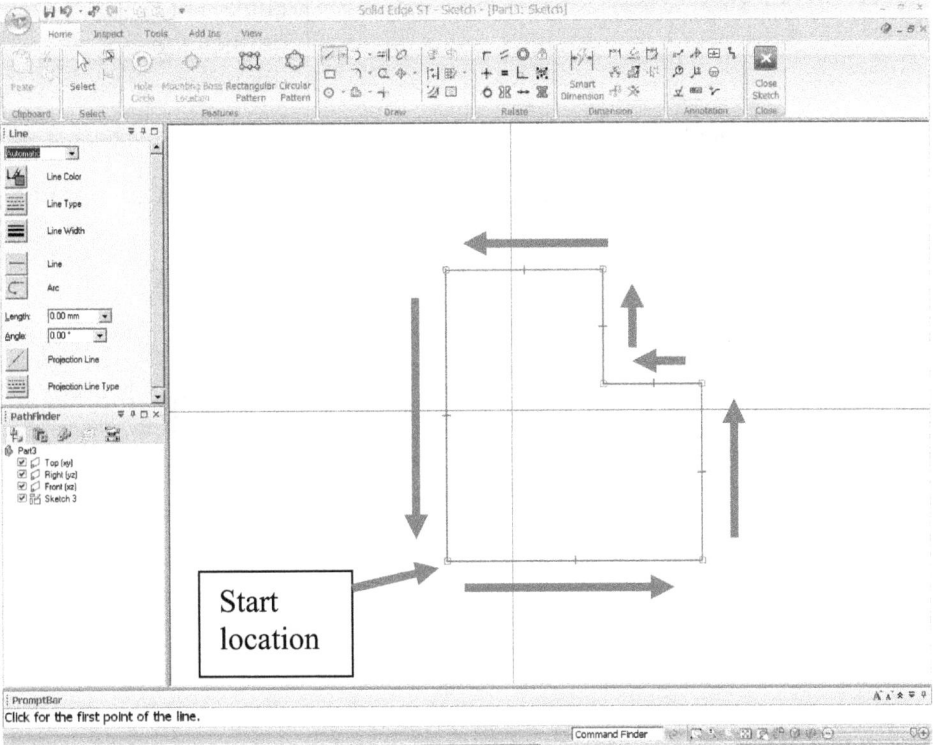

Select the **circle** tool, shown below.

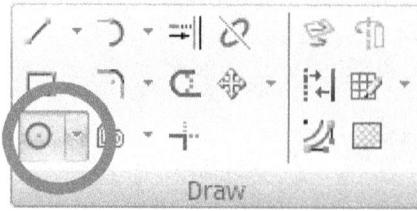

Draw

Click on the screen roughly in the location shown in the following figure. Move the cursor away from the center of the circle to choose its approximate size and click to confirm.

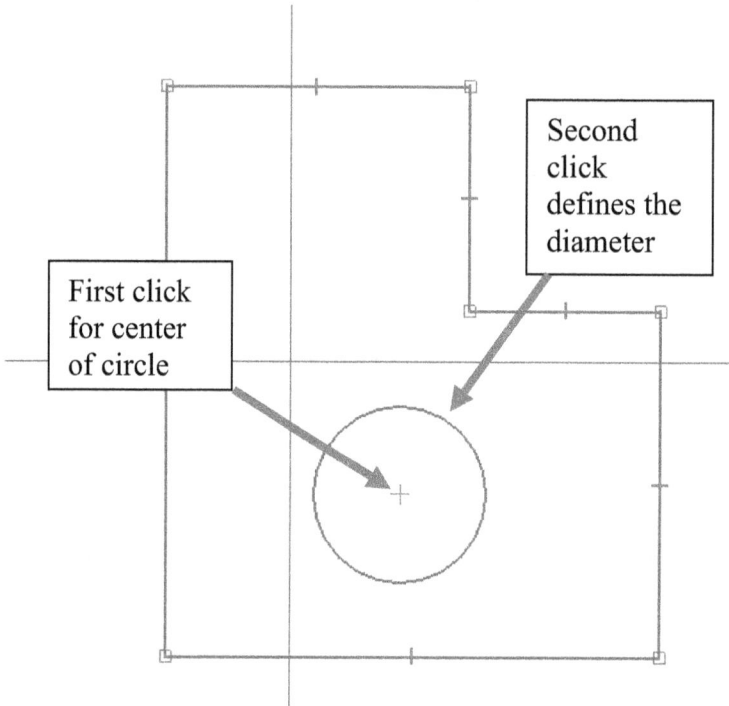

Second click defines the diameter

First click for center of circle

Using the **Smart Dimension** tool in the Dimension options box, add dimensions to the sides shown.

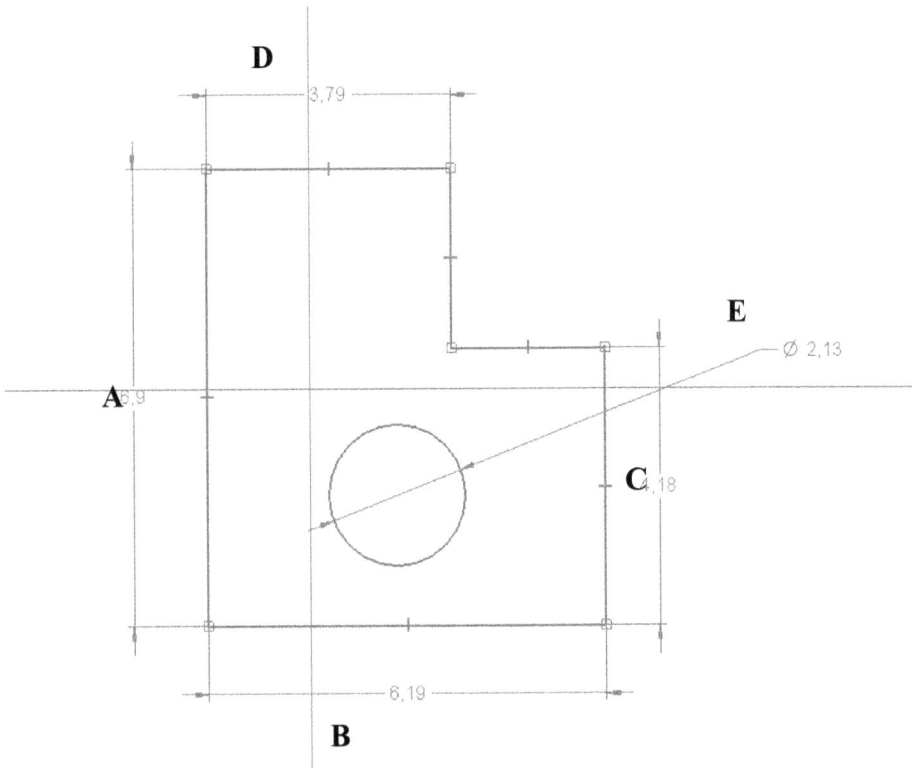

D

3.79

E

Ø 2.13

A 6.9

C 4.18

6.19

B

To change the value of the dimensions, click the select tool. Double click the dimensions, and change the value in the menu on the left just like in Exercise 1.

Solid Edge ST - Ske

Home   Inspect   Tool:

Paste   Select   Hole
Circle

Clipboard   Select

Dimension the sketch as shown in following figure using the **smart dimension** tool.

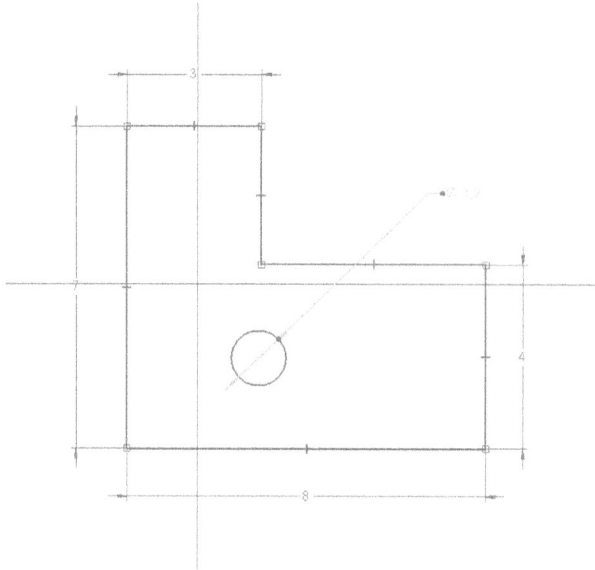

To move the circle to the correct position, use the **Smart Dimension** tool. Click on the 4in side line, and move the mouse over the edge of circle so that the center point appears. Click on the center point and the dimension will appear. Create the other dimensions shown below using the same technique between the circle and the 8in side.

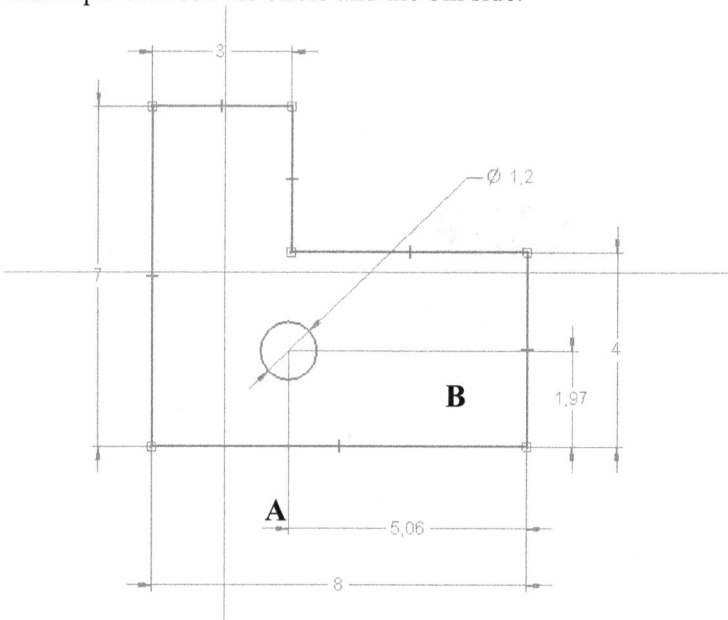

Change the values to: A= 4, B= 2

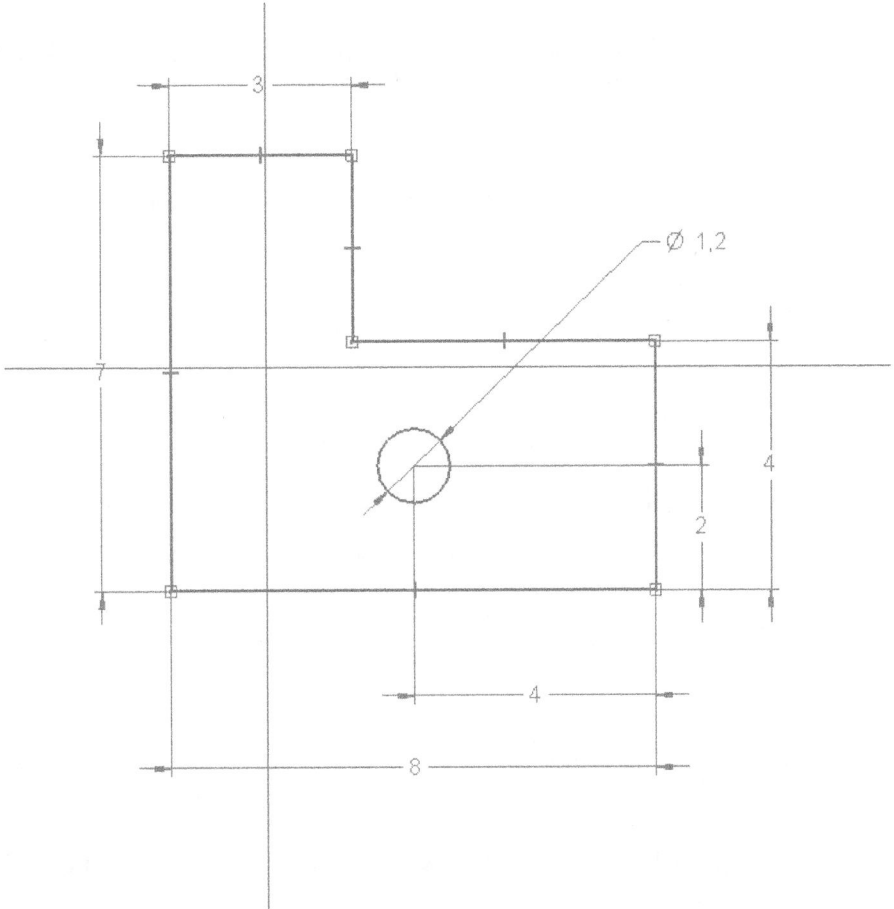

Exercise Complete

Exercise 5: Constraints - Horizontal, Vertical and How to Create Midpoints Constraints

Create a New **ANSI Part**. Create a **sketch** on the **XY plane**. Using the **Line** tool in the Draw options box create a shape that looks similar to the figure below.

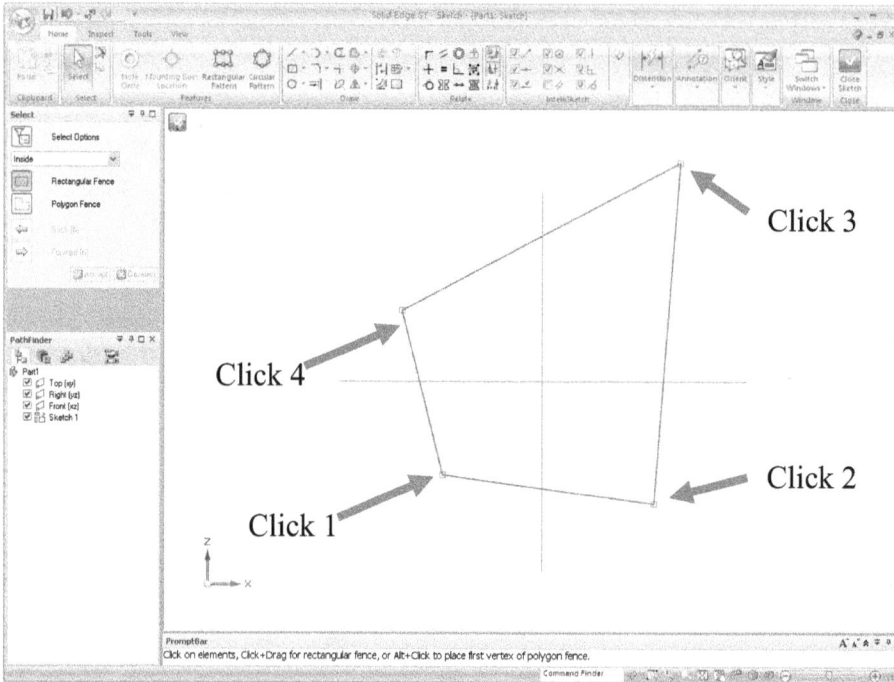

In the **Relate** options box, Click on the **Horizontal/Vertical** tool, highlighted in the following figure.

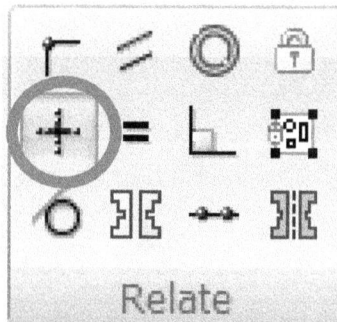

Click on all sides of the shape to make them horizontal or vertical, as demonstrated in the following figure. *Note: The software calculates the closest vector: horizontal or vertical.*

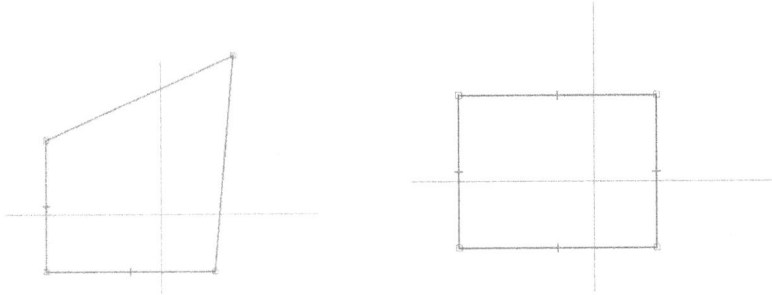

With two sides constrained   All sides constrained

To center the rectangle on the datum plane click on the **Connect** tool located in the Relate options box.

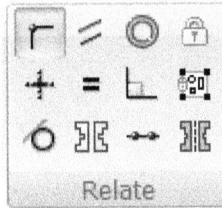

1st click the pink center mark on a side, and 2nd click on the datum plane.

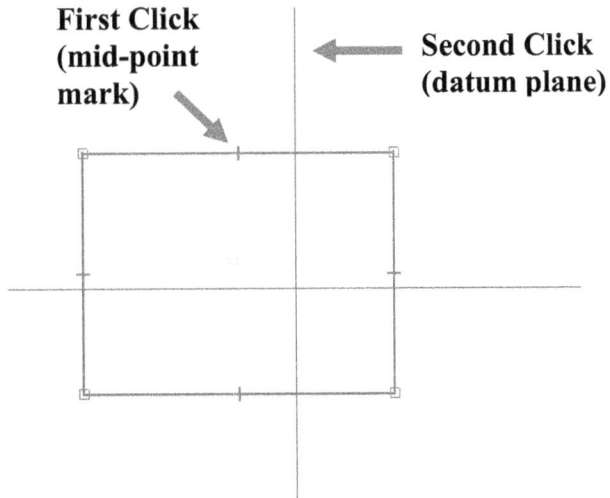

First Click (mid-point mark)   **Second Click (datum plane)**

The center mark should snap to the datum plane, and the sketch should look like the following figure.

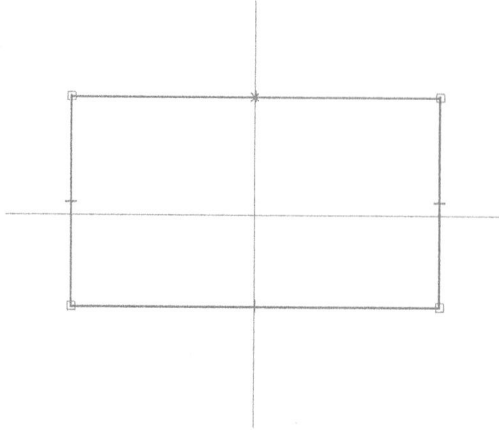

Use the same technique on the other three sides, as shown below.

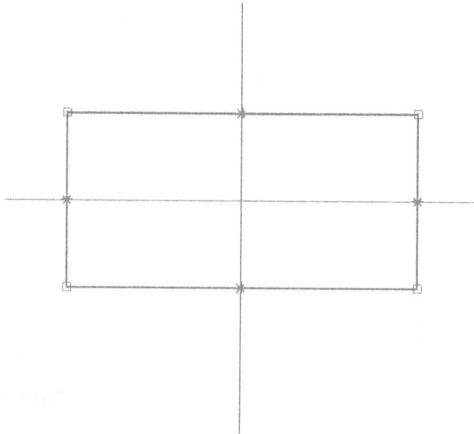

The rectangle is now centered on each datum plane.

Note: If dimensions are added to these lines the sketch will become fully constrained. The indication of a fully constrained sketch is the color. A fully constrained sketch in, an 'Out of the box' installation, SE will turn black.

Exercise Complete

Exercise 6: Constraints - Equal

Create **a new ANSI part file**. Create a sketch using the Line tool in the Draw options box that looks like the figure shown below.

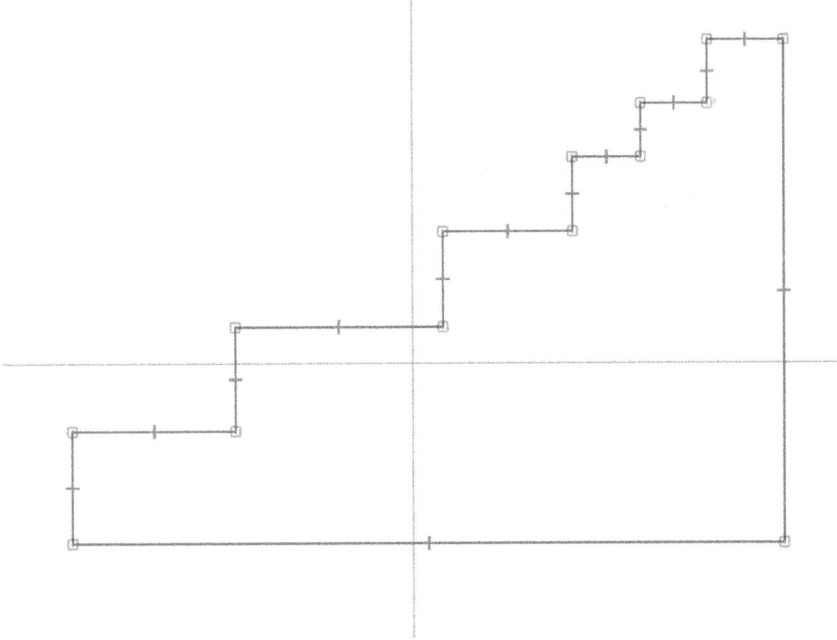

Click on the **Equal** tool in the Relate options box, shown in following figure to begin constraining the sketch.

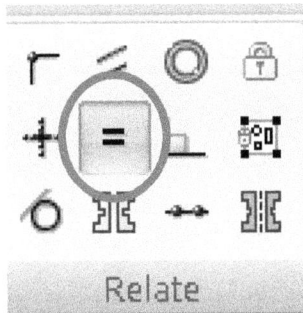

The **Equal** tool will make two lines equal in length (or two curves equal in radius); Click on the first and second lines as shown below. Following the same pattern for all the steps, they will all become equal height and length. *Note: The line clicked first will change to equal the length of the second click.*

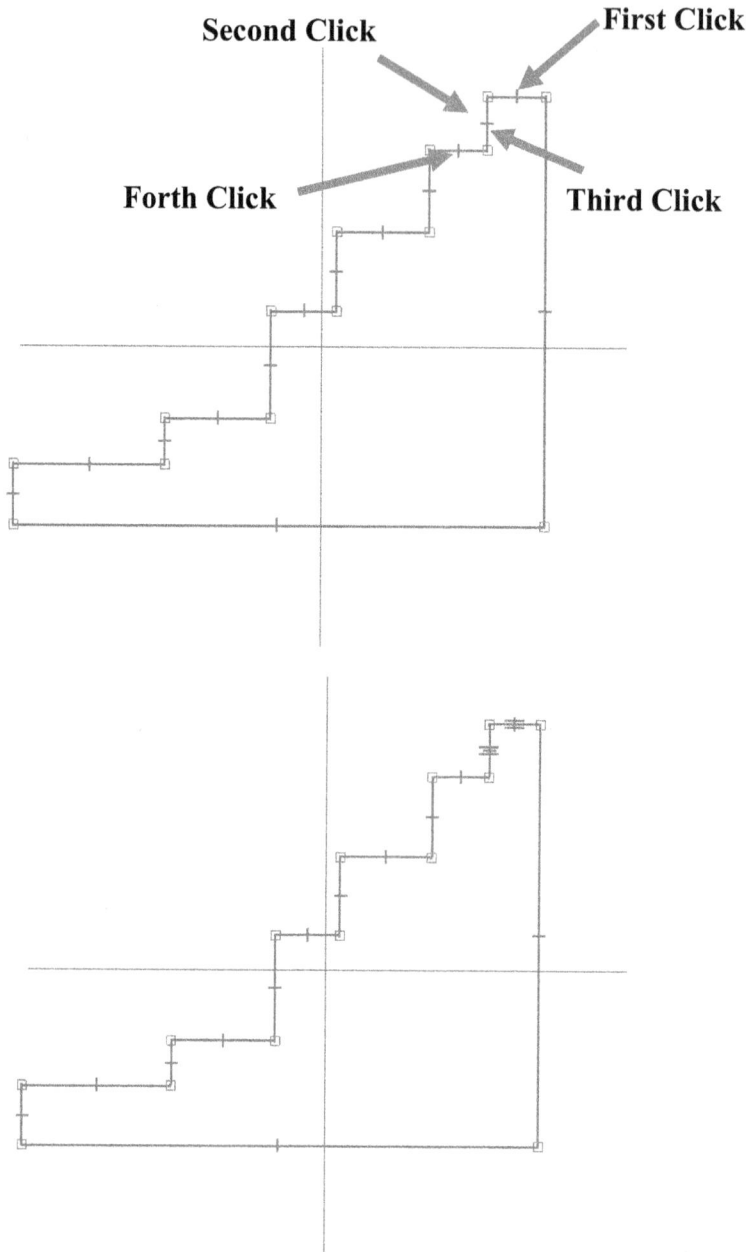

Notice: After the first 2 clicks, the lines became equal.

The Finished Sketch should look like the following figure.

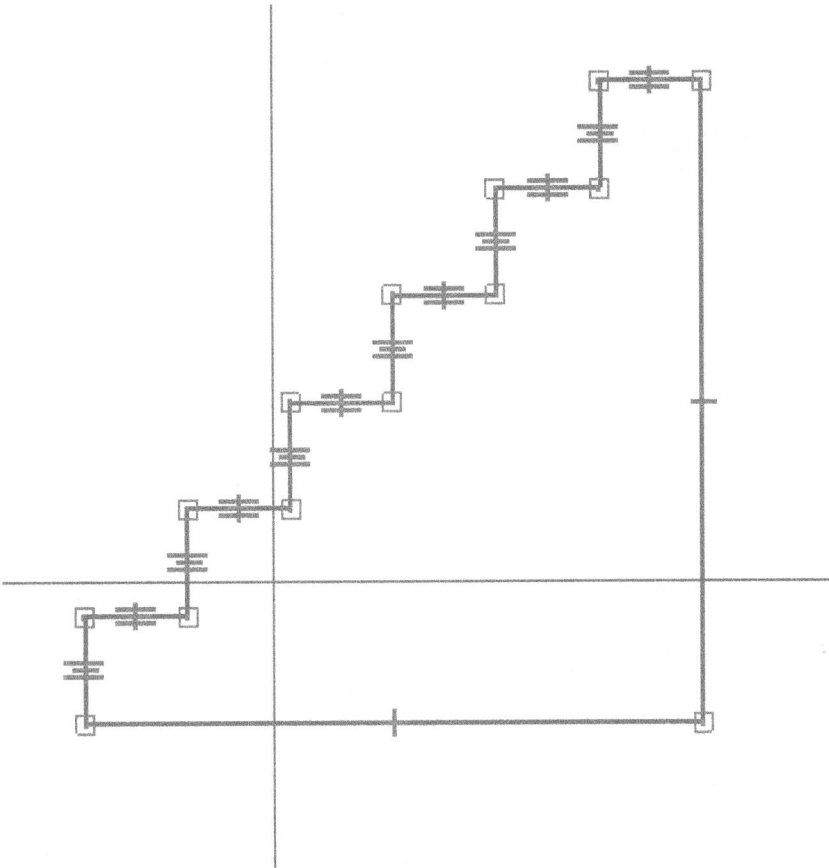

Exercise Complete

Exercise 7: Constraints - Parallel and Concentric

In a **New ANSI part file**, create a sketch using the **Line** tool and the **Arc by 3 Points** tool in the Draw options box that looks like the figure shown below.

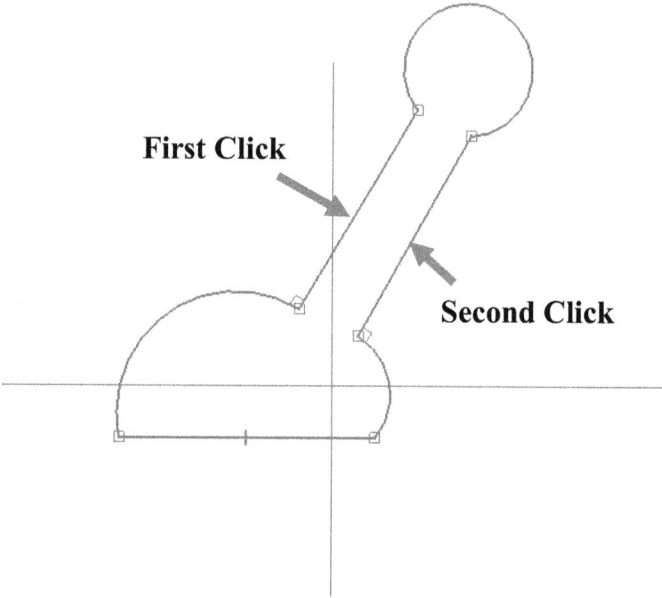

First Click

Second Click

Click on the **Parallel** tool in the Relate options box shown in the figure below. Using this tool, click on the two lines of the shaft as shown above. Once the operation is complete the parallel constraint symbols will show what sides are parallel.

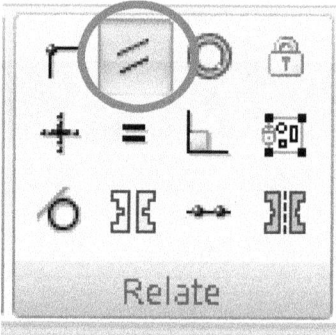

Click on the **Concentric** tool in the Relate options box..

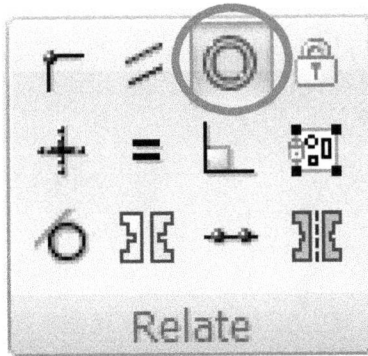

Click on the 2 lower arcs as shown below. *Note: Similar to the Equal tool we used in the last exercise, the first line selected should be the one intended to be change; this is matched to the second line selected.*

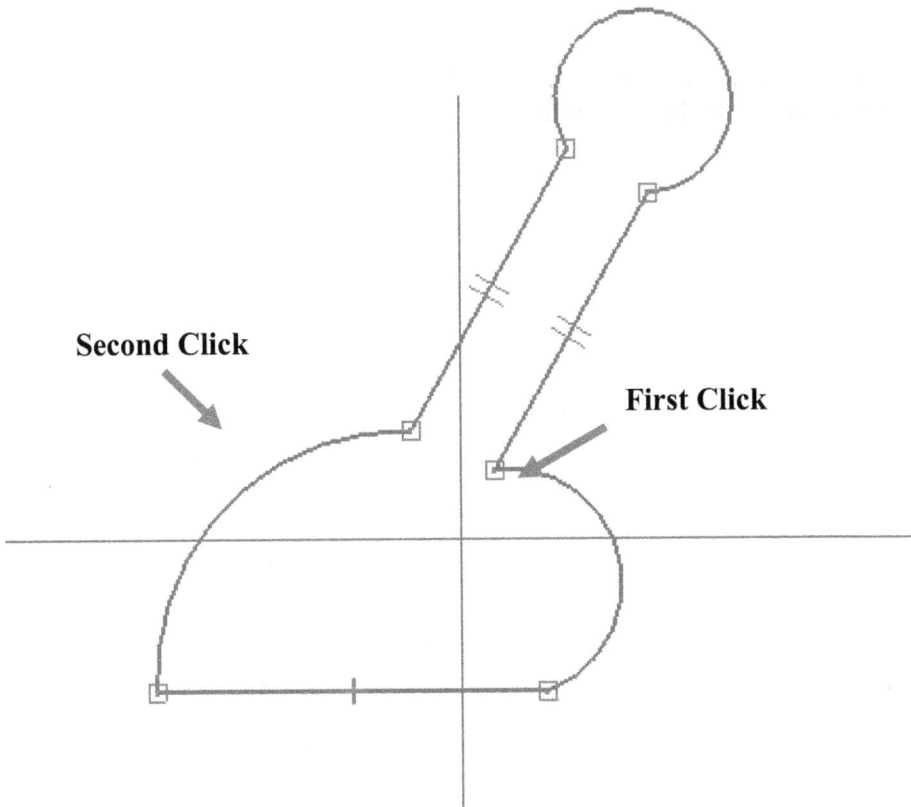

Use the **Line** tool to create a line between the centers of the lower arc and the upper arc, as shown below.

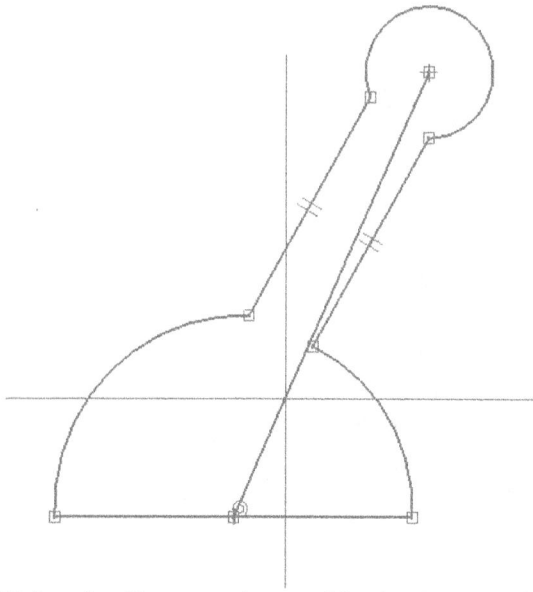

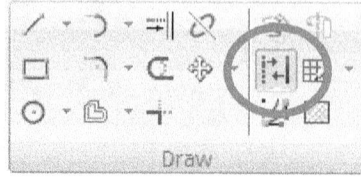

Using the **Construction** tool in the Draw options box, shown in the previous figure, click on the line just created to make it a construction line.

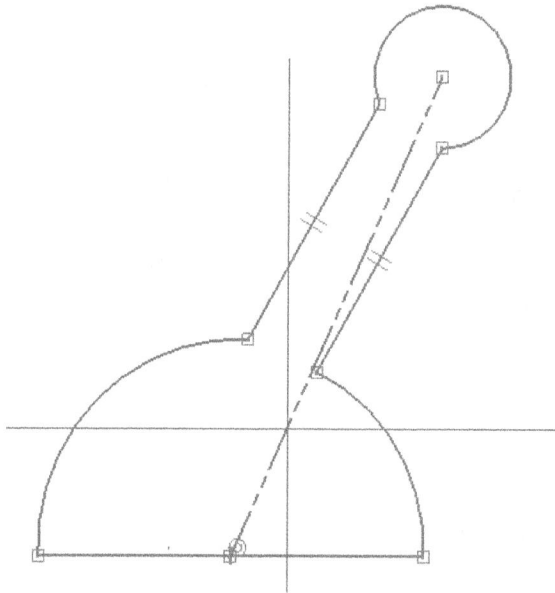

Click on the **Parallel** tool in the Relate options box to make the construction line parallel to the lines of the shaft, same as the procedure at the start of this exercise.

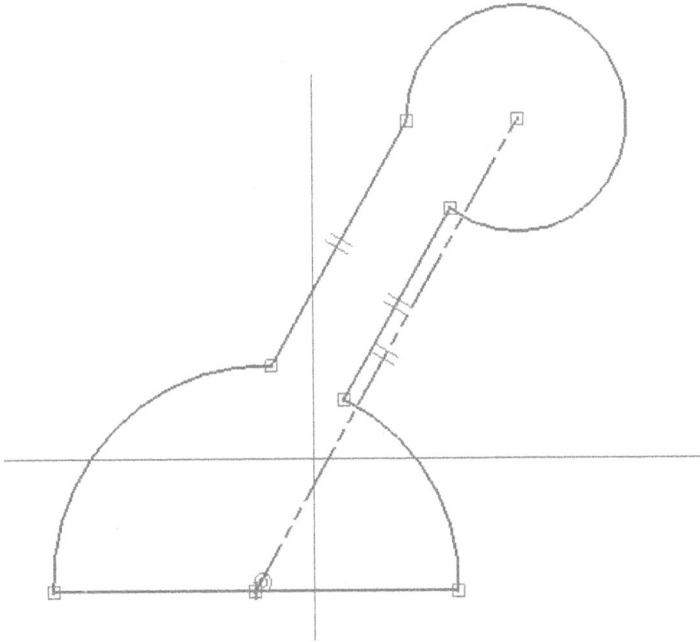

Note: if the figure goes a little crazy, don't panic. Sometimes the software does not calculate the change they way that may have predicted. It is always possible to CTRL+Z (Undo).

Click on the **Distance Between** button in the Dimension options box as shown in the following figure.

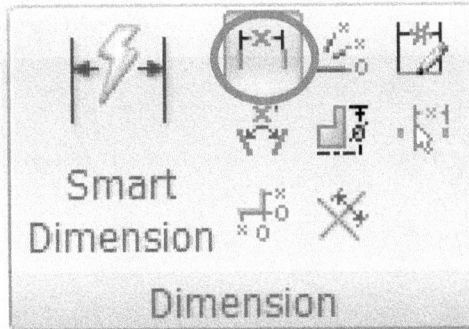

Check that the drop down menu under **'Properties'** on the left side of the screen is set to By 2 Points.

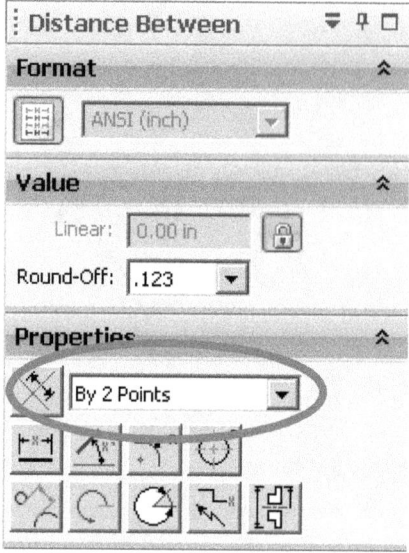

Add the dimensions as shown in the following figure.

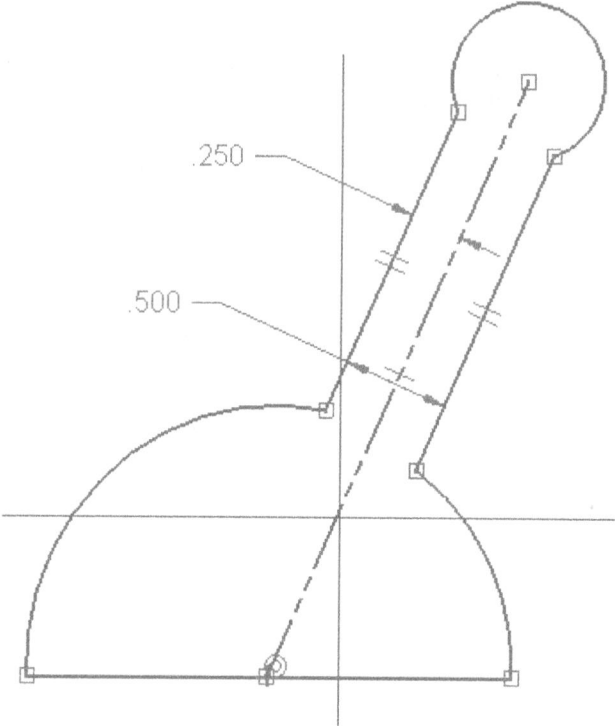

Exercise Complete

Exercise 8: Constraints – Concentric and Trim Tool

Create a **New ANSI part file**. Create a sketch using the **Circle** tool in the Draw options box that looks like the figure shown below.

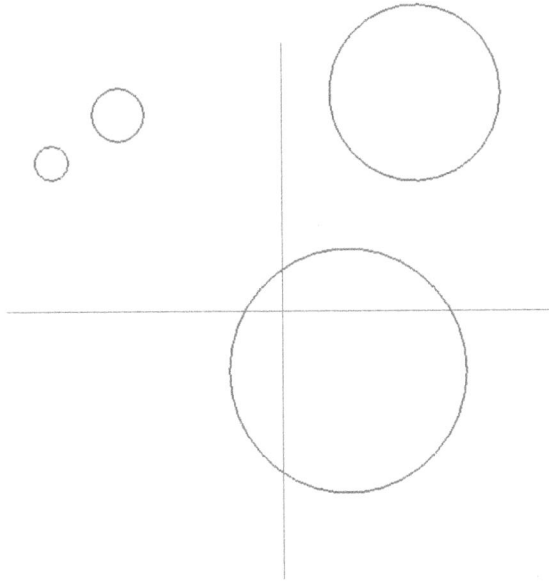

Click on the **concentric** tool in the Relate options box.

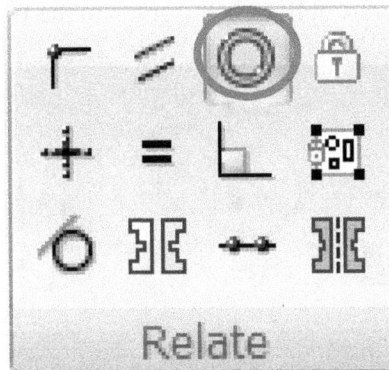

Click on two of the circles to make them concentric, repeat this process with all the circles until the sketch looks like the following figure.

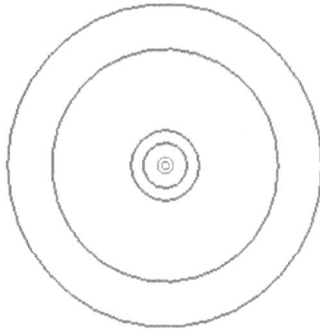

With the **Line** tool draw 2 lines as shown below.

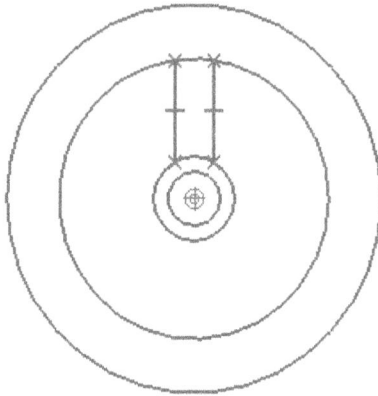

Click on the **Trim** tool in the Draw options box, shown in the figure below.

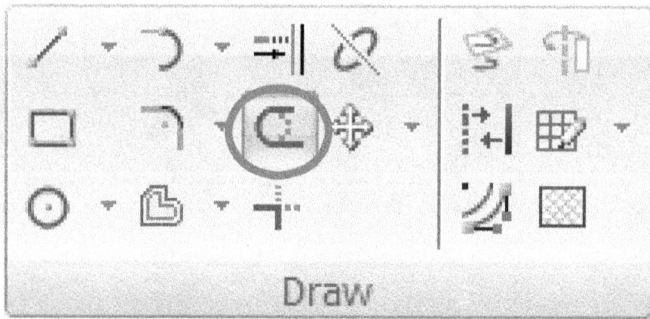

The **Trim** tool can be used 2 ways: Either click on a line segment to be erased, as shown in the figure (left). Or click and drag the cursor, like drawing with a pencil, over a segment to be erased, as shown in the figure (right).

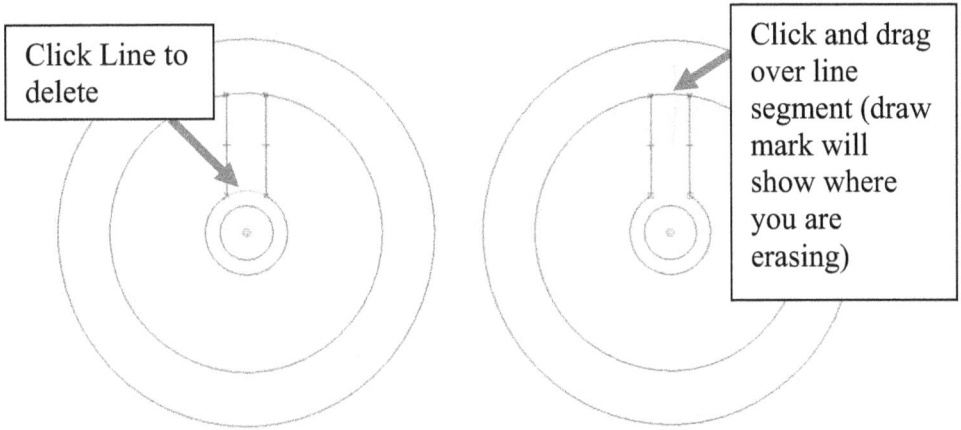

Click Line to delete

Click and drag over line segment (draw mark will show where you are erasing)

The sketch should now look like the following figure.

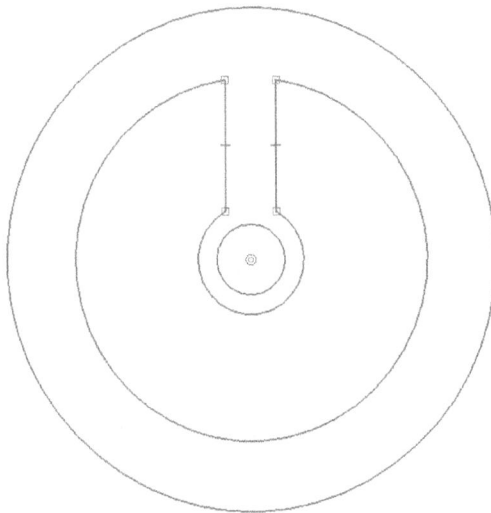

Exercise Complete

Exercise 9: Constraints - Collinear Tool

In a **New ANSI part file**, create a sketch using the **Line** tool in the Draw options box that looks like the figure shown below.

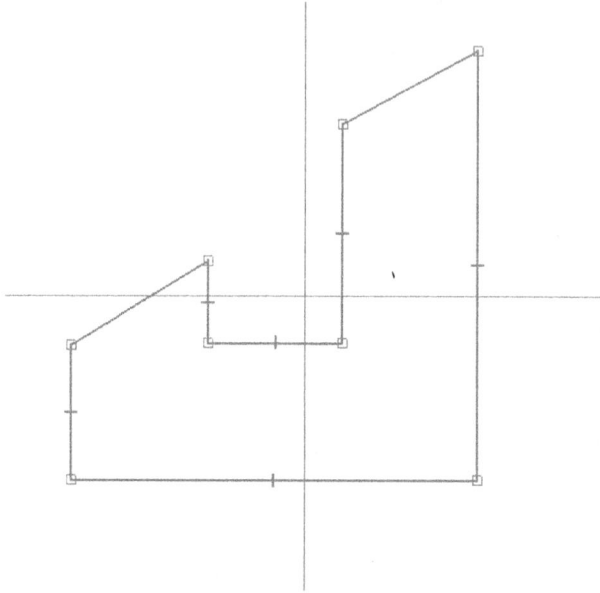

Click on the **collinear** tool in the Relate options box.

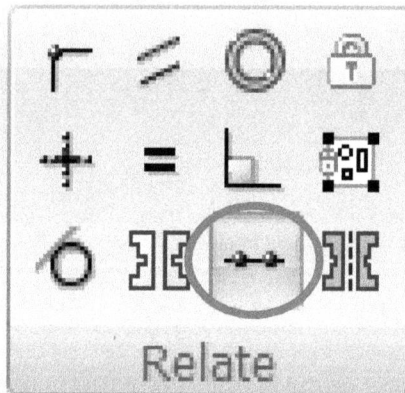

Click on the lines shown in the following figure. The sketch should now look like the following figure.

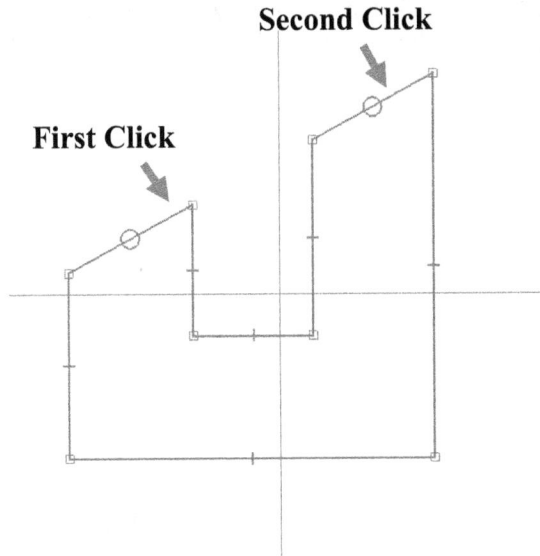

Second Click

First Click

Note: The first line clicked will be the line that is changed.

Exercise Complete

Exercise 10: Constraints -Tangent

In a **New ANSI part file** create a sketch using the **Line** tool and the **Arc by 3 Points** Tool in the Draw options box that looks like the figure shown below.

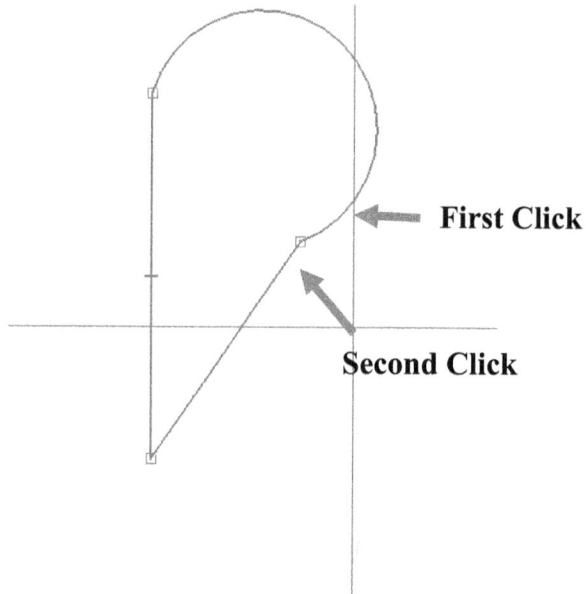

Click on the Tangent tool in the Relate options box, as shown below. Create a tangency between the arc and the lower line as shown above. *Note: It works best when you click the line close to the end points nearest each other.*

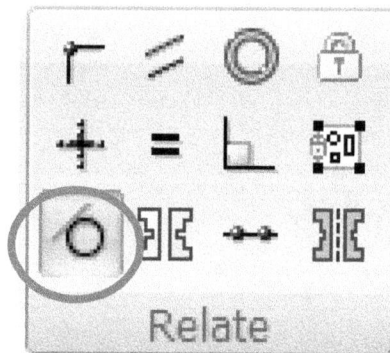

The sketch should now look like the following figure.

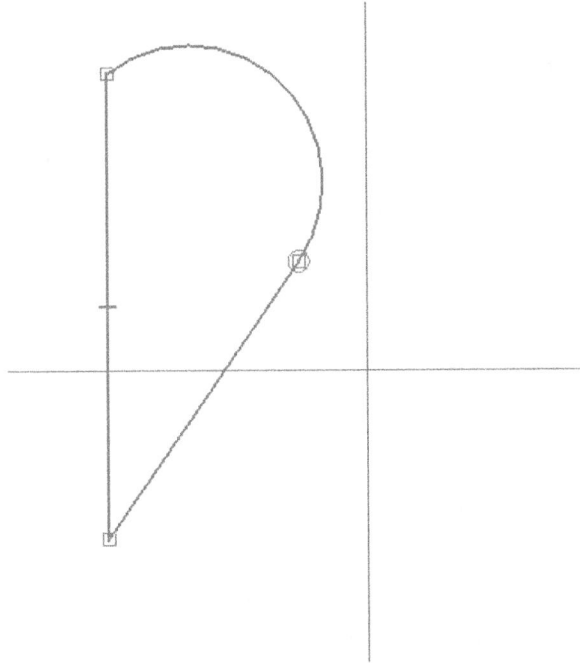

Click the little arrow next to **Move/Mirror** tool in the Draw options box, as shown in the previous and following figure.

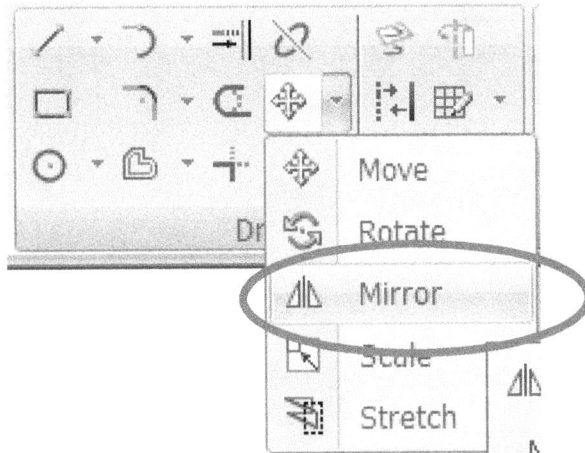

Click and drag the cursor to create a box around the part as shown below.

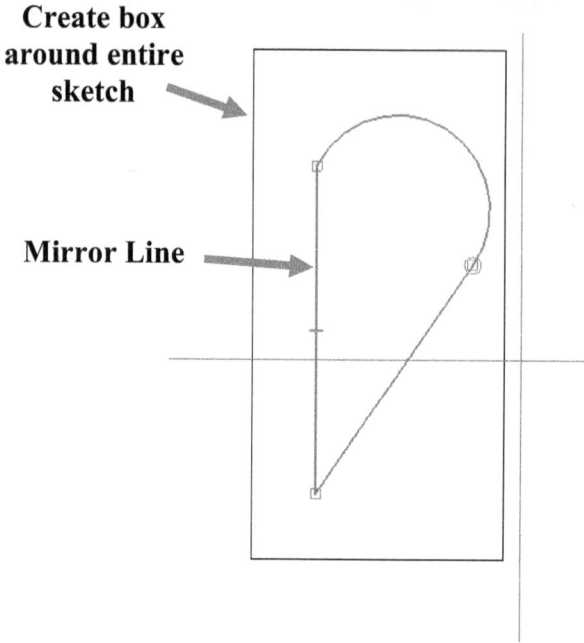

**Create box
around entire
sketch**

Mirror Line

To complete, click on the mirror line as shown above. Don't you just love this tool?

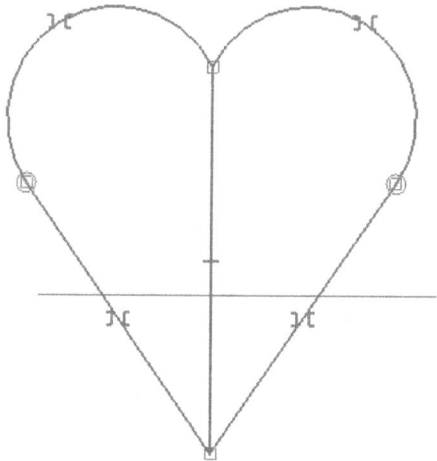

Exercise Complete

Exercise 11: Constraints - Connect and Angle

In a **New ANSI part file**, create a sketch using the **Line** tool and the Arc by 3 Points tool in the Draw options box that looks like the figure shown below.

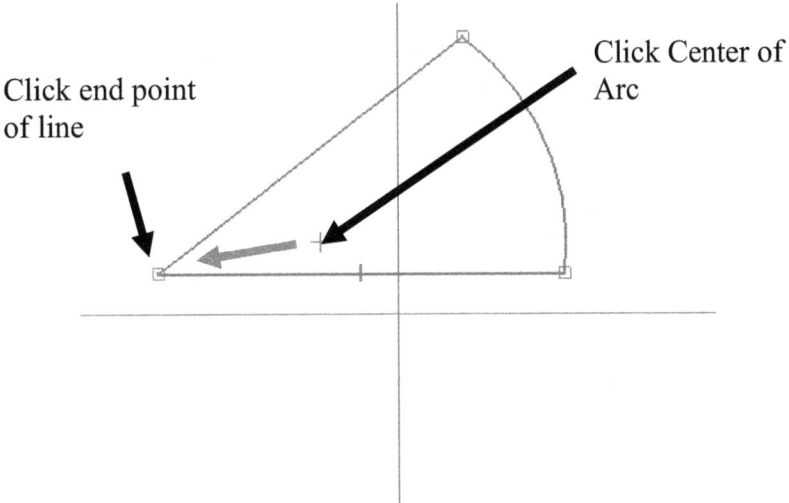

Click end point
of line

Click Center of
Arc

To move the center mark of the arc to the point shown above, select the **Connect** tool in the Relate options box.

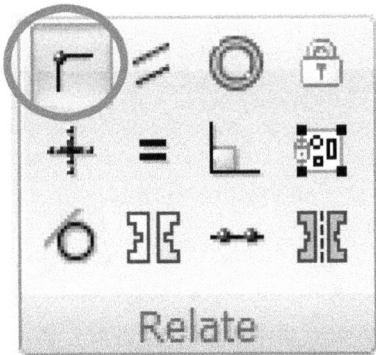

Relate

Hover the cursor over the edge of the arc to make the center mark appear. Click the center point and then the point connecting the 2 lines, as shown on the previous page. The sketch should now look like the following figure.

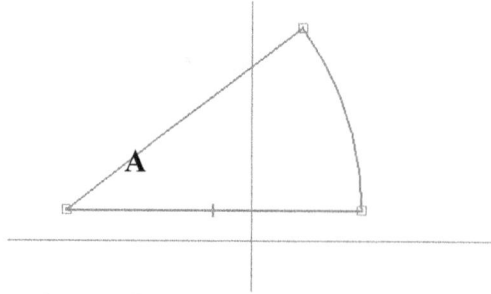

To change the angle between the two lines (A) click on the **Angle Between** tool in the Dimension options box and click the sides as shown below.

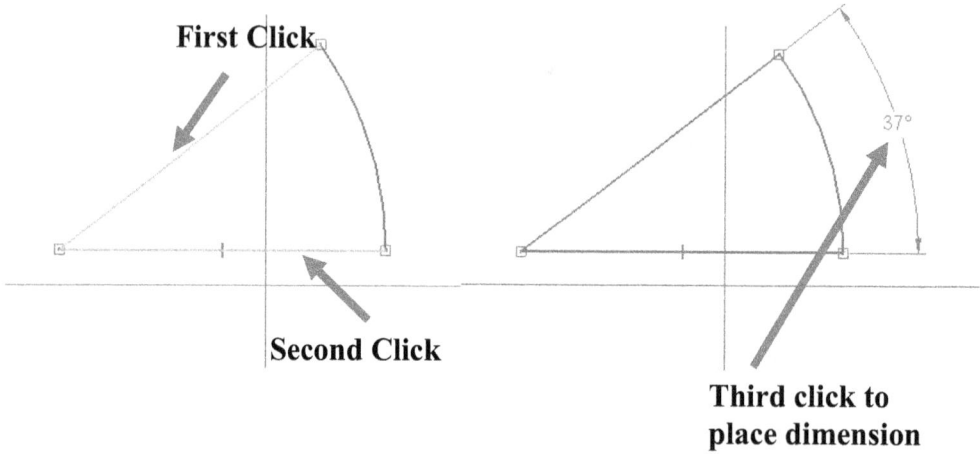

Change the angle to 20°

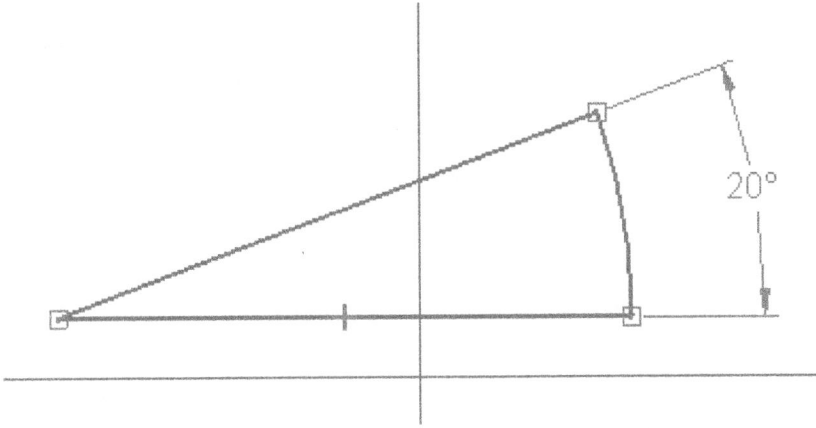

Exercise Complete

Exercise 12: Creating Relationships in a Sketch

In a **New ANSI part file** create the following sketch using the **Line** tool and add the dimensions shown below

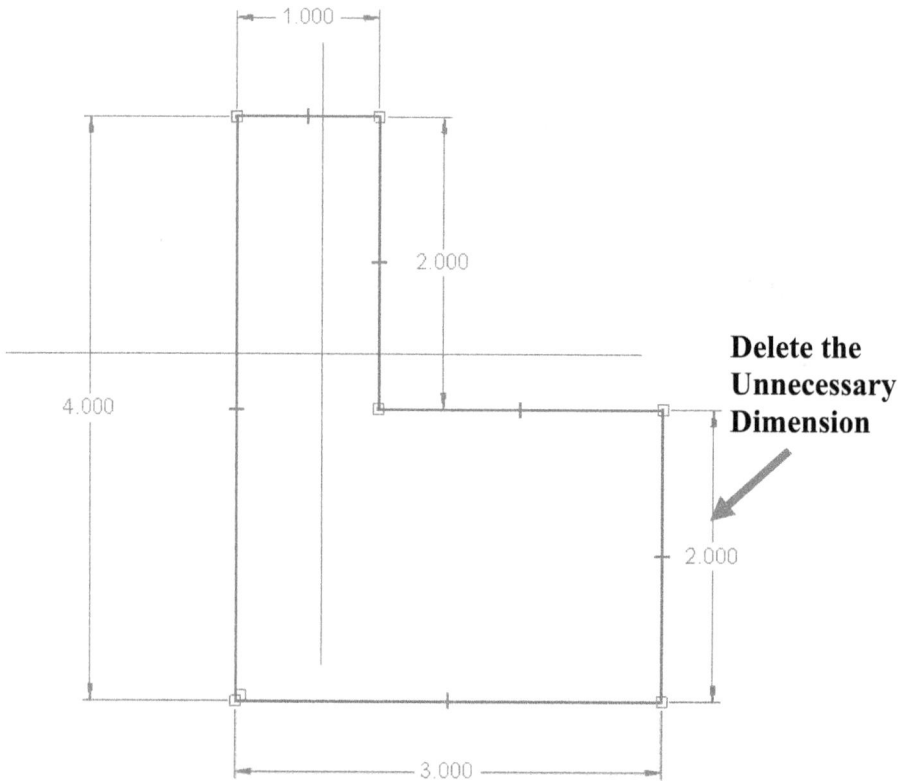

Delete the Unnecessary Dimension

Note: When a sketch is over dimensioned any redundant dimensions will appear in blue.

Delete the unnecessary dimension by selecting it and either right click/delete or pressing the delete button on the keyboard.

In the following section we will learn how to create relationships between dimensions. Double-click on the 4in dimension and a menu labeled **Edit Formula** will appear in the dock on the left.

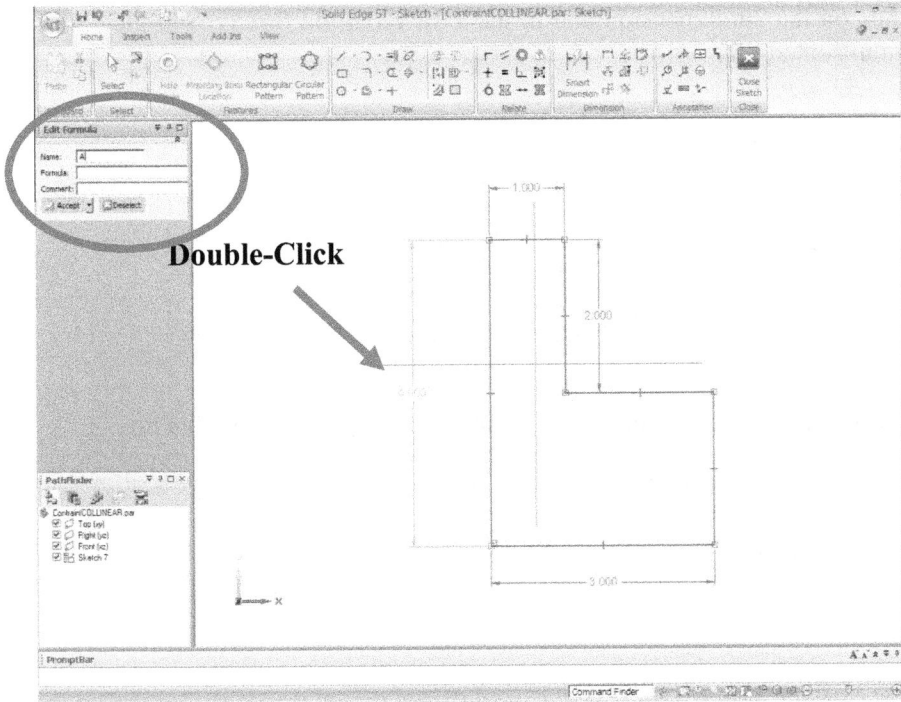

Double-Click

Change the name to **A** and hit **Enter** *(the Enter is important!!)*

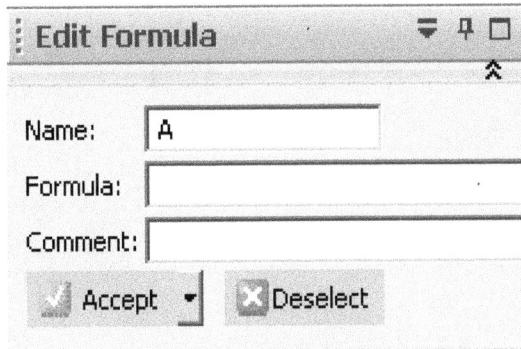

Right-click on the screen to exit the **Edit Formula box**.

Double-click on the 2in dimension and in the **Edit Formula box** type in the Name as **B** and under Formula enter **A/2**.

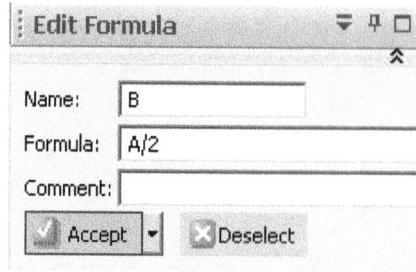

Edit Formula

Name: B
Formula: A/2
Comment:
Accept Deselect

Then select **Accept** and Right-click on the screen to exit the Edit Formula box.

Congratulations, a relationship between sides A & B has been created.

Test the relationship by changing the value of side A to 5.00.

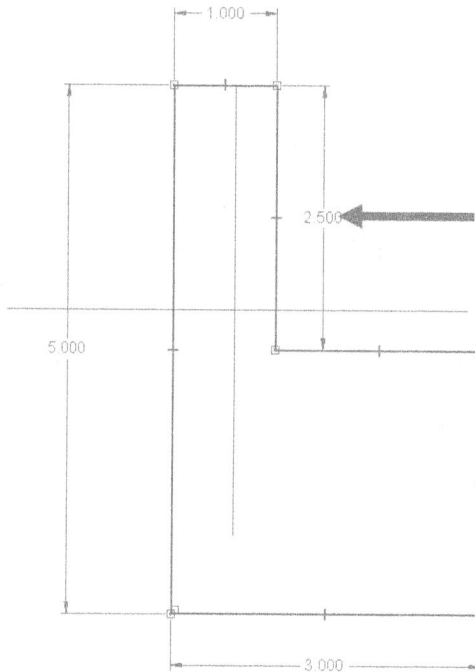

Notice: when you create a relationship, the dimension turns blue. Remember: If the arrows turn blue as well it means the sketch is over-constrained

The value of side B should change to 2.50, as shown above.

Exercise Complete

Exercise 13: Extruding

Click the Extrude button in the Solids options box, in a **New ANSI part** file, and go straight to creating an extruded part. The **extrude** command is located in the Solids menu under the **Home** tab as shown below.

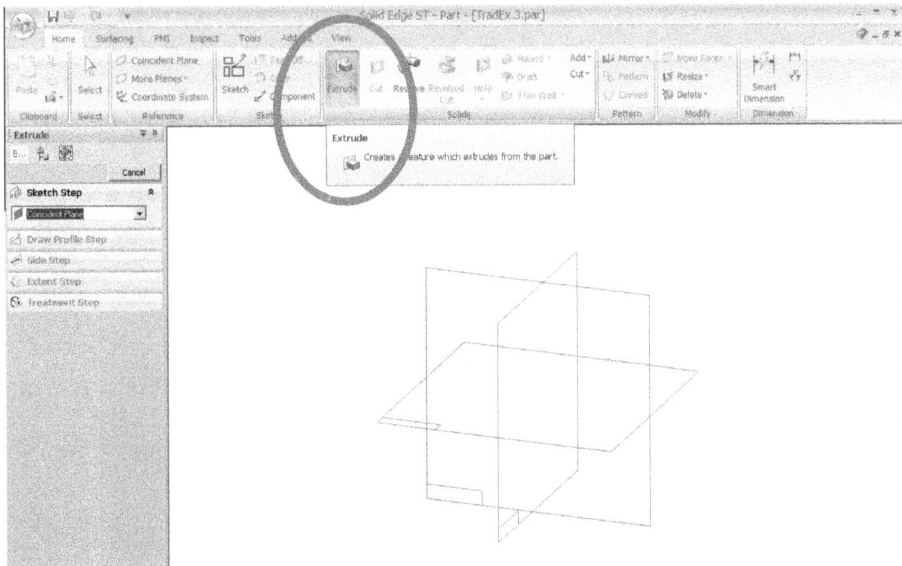

The first step is to select a sketch plane. Pick the **XY plane**, as shown in the following figure, to begin sketching.

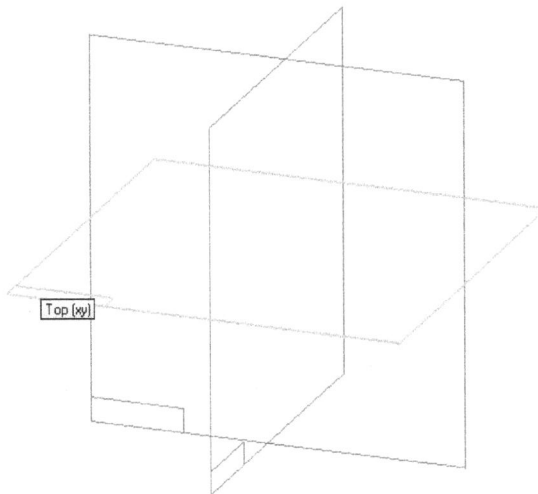

Using the **Line** tool in the Draw options box, draw a shape similar to the one shown in the following figure.

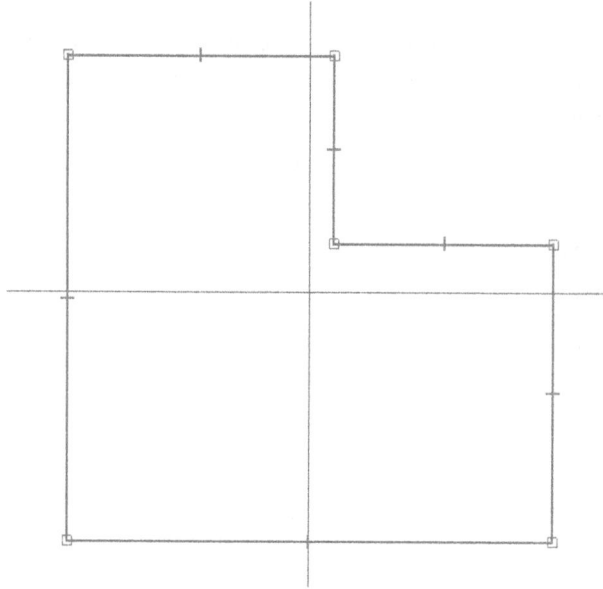

Then dimension the shape with the **Smart Dimension** tool in the Dimension options box with the values shown in the figure below.

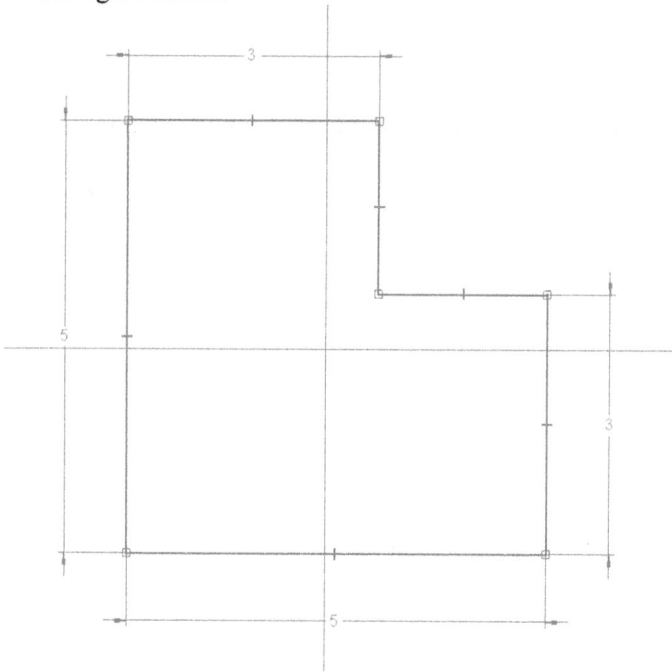

Once satisfied with the sketch, click the **Close Sketch** button in the Close options box.

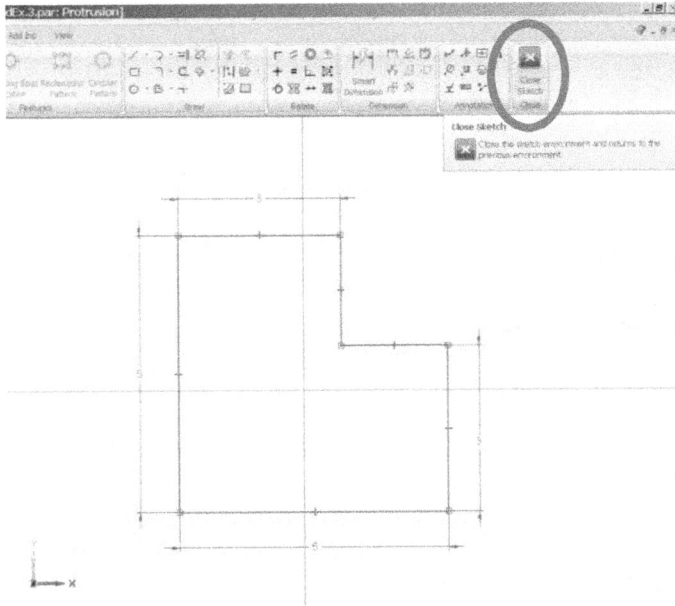

The screen should now look like the following figure.

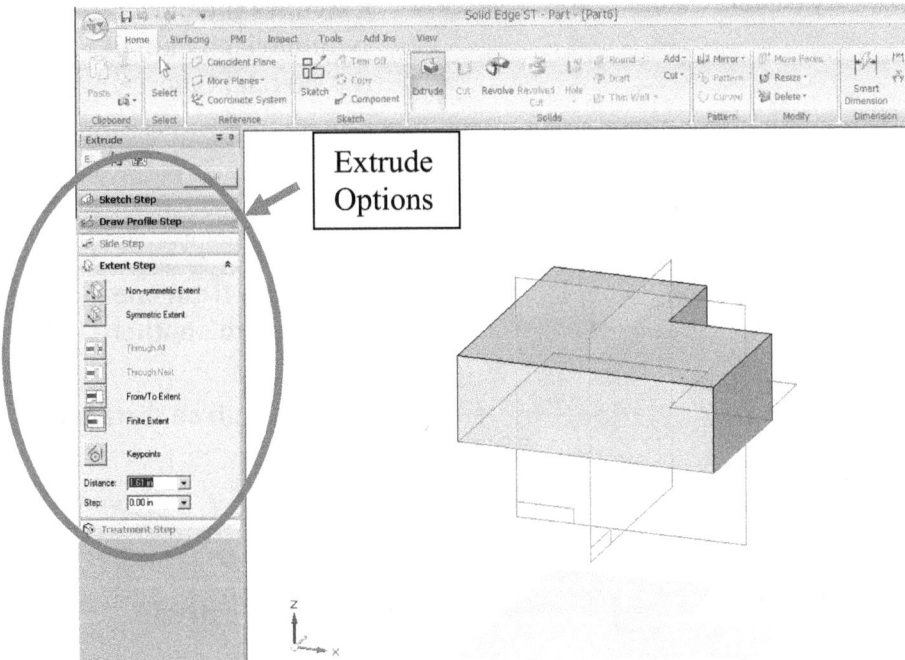

Extrude
Options

Notice the **Extent** options in the menu on the left.

Extrude		
E...		

Finish

🗐 **Sketch Step**

✏️ **Draw Profile Step**

🗐 Side Step

📎 **Extent Step** ⌃

| | Non-symmetric Extent | ← | Extrudes the part about the sketch plane non-symmetrically |

| | Symmetric Extent | ← | Extrudes the part symmetrically about the sketch plane |

| | Through All |
| | Through Next |

| | From/To Extent | ← | Extrudes from one distance to another |

| | Finite Extent | ← | Extrudes a given distance |

| | Keypoints |

Distance: [1.61 in ▼] ← **Distance to extrude**

Step: [0.00 in ▼]

Enter **1** into the **Distance** drop-down menu and hit enter. The part should snap to a depth of 1in. Then specify which direction to extrude. Accomplish this by clicking above the sketch plane. *Notice: if the cursor is positioned below the sketch plane the extrude changes direction as shown in the following figure.*

Click here to extrude above the sketch plane

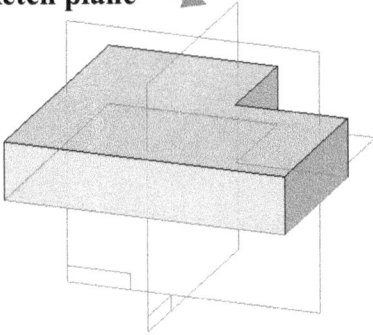

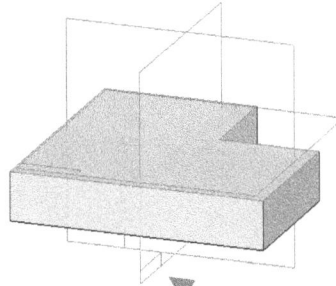

Click here to extrude below the sketch plane

The screen should look like the figure shown below.

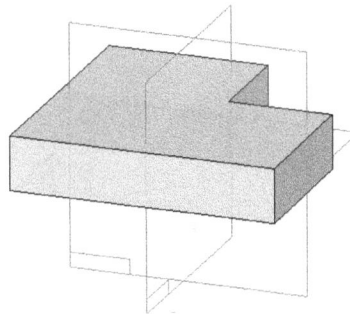

Click the Finish button to complete the **Extrude,** as circled above.

Exercise Complete

Exercise 14: More Extruding

Start a **new ANSI part file**. Click the **Extrude** button in the Solids options box, and select a plane to sketch on.

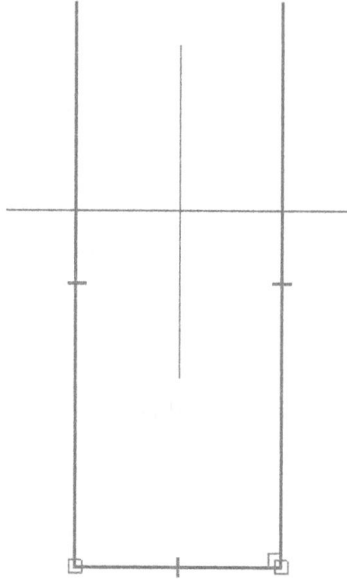

Using the **Line** tool in the Draw options box, draw a shape similar to the one shown above.

Click on the **Tangent Arc** tool in the Draw options box.

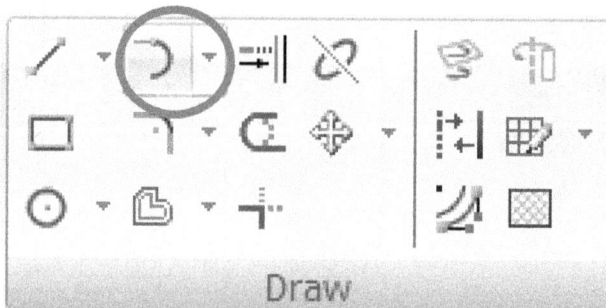

Click on the starting point, to begin the arc.

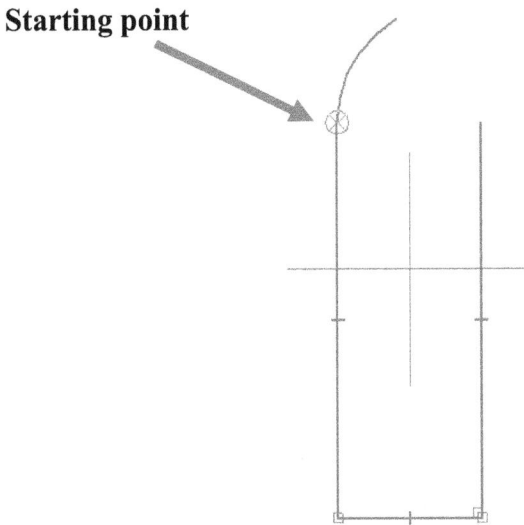

Starting point

Now here's where it gets tricky. Drag the cursor up through the top quarter of the circle as shown below. *Note: Pretend to actually draw it with a pencil, following the desired shape, it will help create the correct arc.*

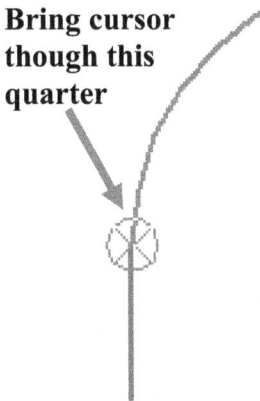

Bring cursor though this quarter

NOT through the following quarters:

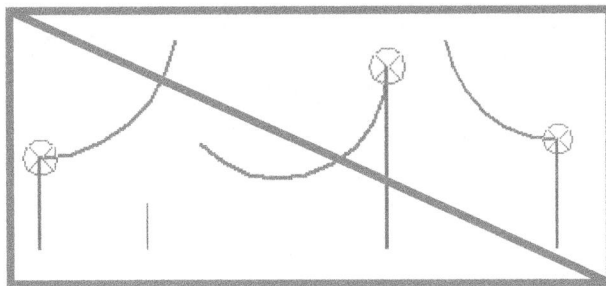

Once the tangency is achieved, click to define the end point as shown in the following figure.

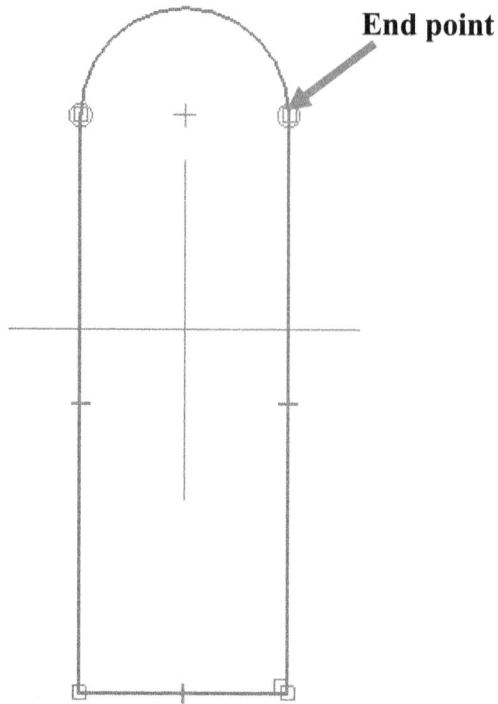

It is also possible to create an arc using the **Arc by 3 Points** tool found by clicking the little arrow next to the **Tangent Arc** tool in the Draw options box.

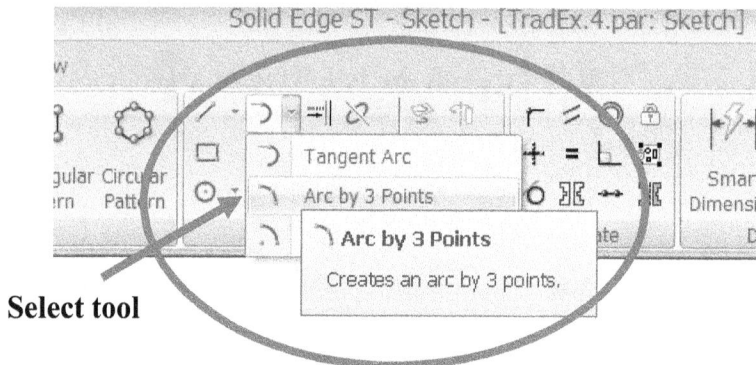

To use this tool first select the start point of the arc as shown below on the left. Then select the end point of the arc. Finally select a location between those two points to define the radius as shown below on the right.

First Click on end point of line

Second Click on end point of other line

Third Click to specify radius

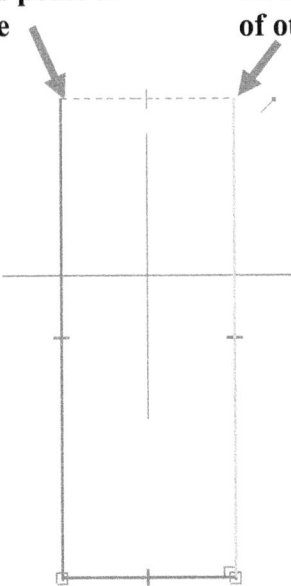

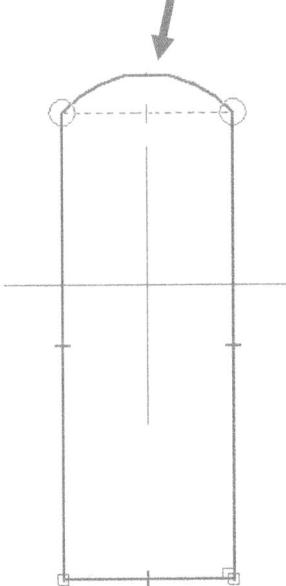

To make the arc tangent to the other lines, click on the **Tangent** tool in the Relate options box.

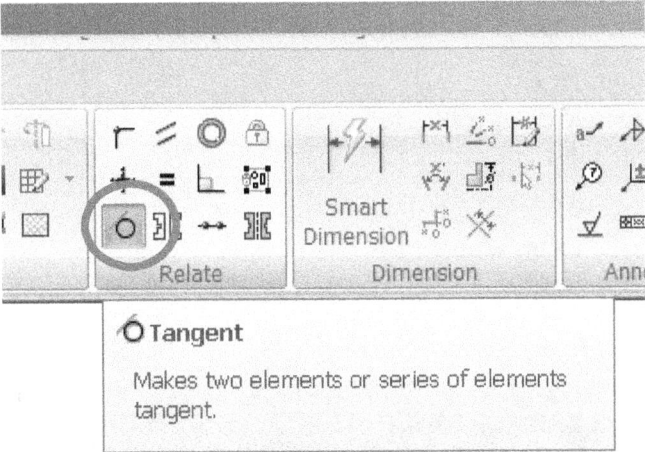

Click toward the end of the arc that is going to be tangent then click towards the end of the line that the arc will be tangent too.

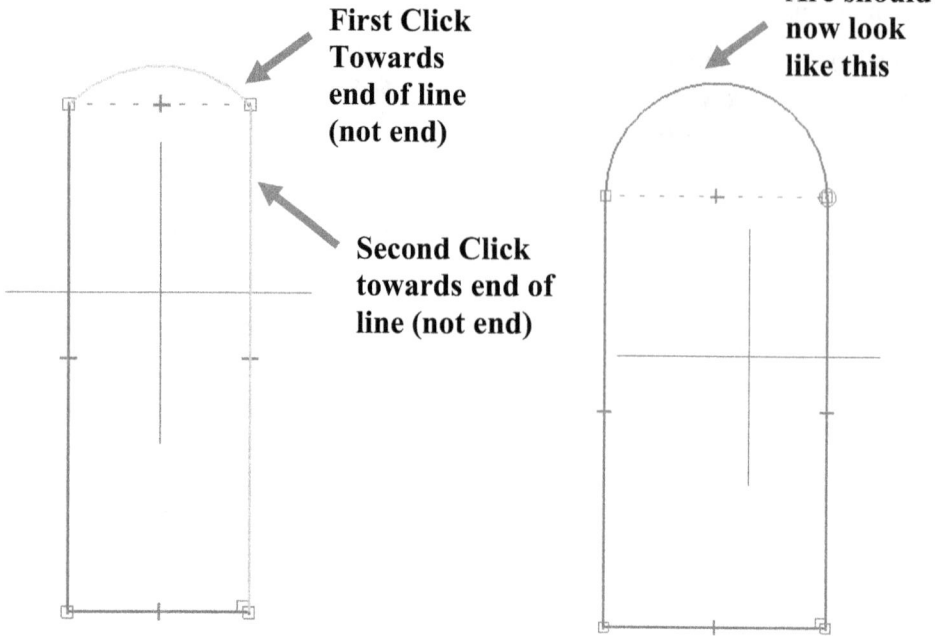

**First Click
Towards
end of line
(not end)**

**Arc should
now look
like this**

**Second Click
towards end of
line (not end)**

Repeat this operation for the other end of the arc. Using the **Smart Dimension** tool in the Dimension options box, add dimensions to the base of the sketch as shown below.

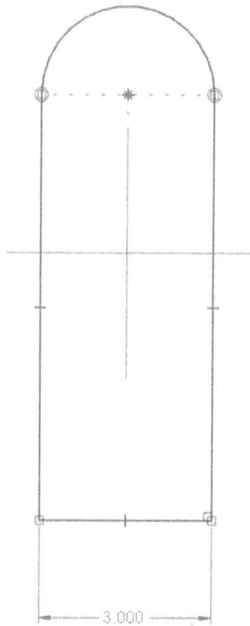

3.000

To create the overall height dimension, use the **Distance Between** tool in the Dimension options box. *Note: Using this tool is an easy way to pick up the tangency of an arc rather than the center of a circle or an arc.*

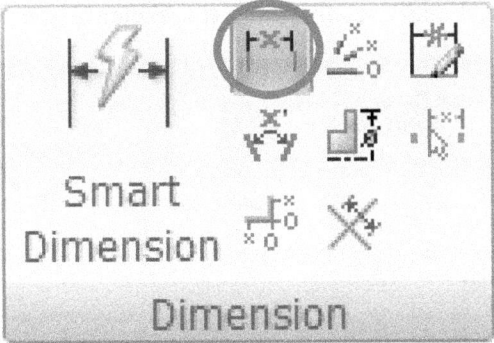

Click on the bottom line of the sketch, and the arc to create the dimension shown below.

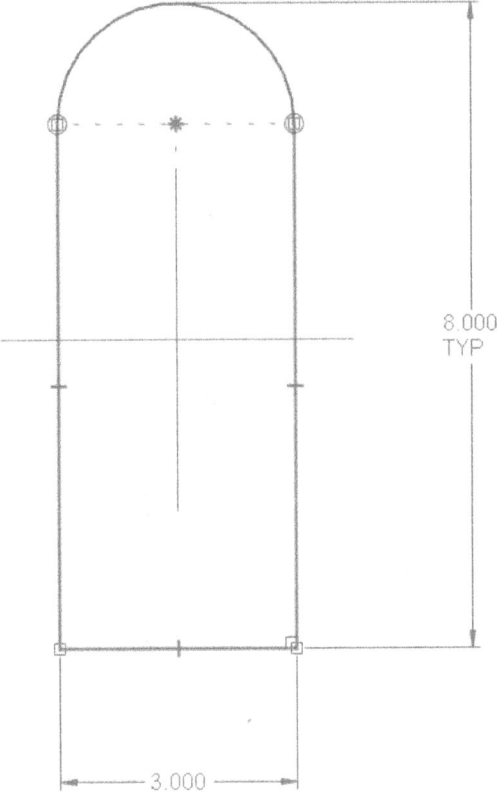

8.000
TYP

3.000

Now that the sketch is dimensioned, close the sketch with the **Close Sketch** button, as shown in the previous exercise.

The next step in extruding is to type 0.5 into the distance box under the **Extent Step** in the left column, as shown below. Finally, click above the sketch plane to extrude the part in the tangent plane.

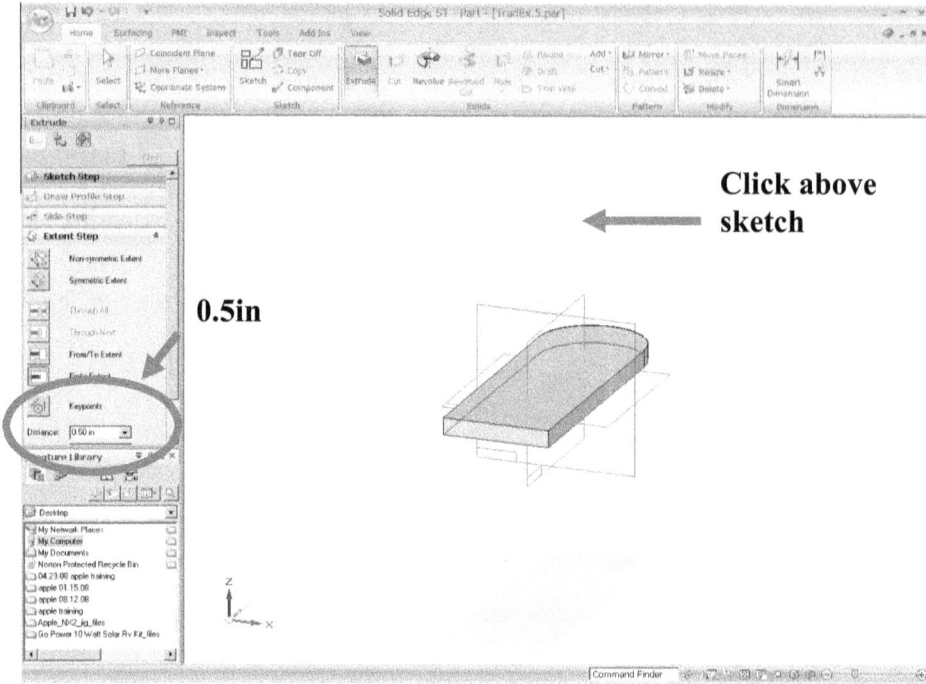

Your model should now look like the following figure.

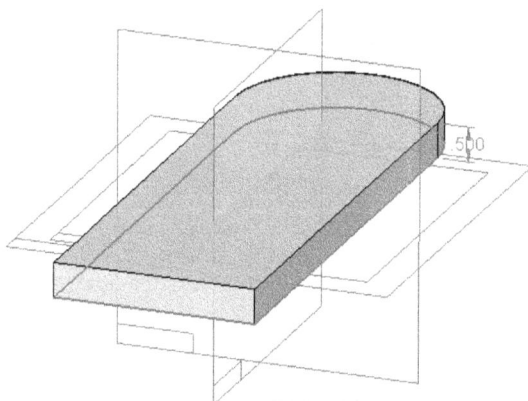

Exercise complete

Exercise 15: Creating Holes

Create a New **ANSI part file**. Click the **Extrude** button in the Solids options box and select a plane to sketch on. Create a sketch using the Line tool in the Draw options box that looks like the figure shown below. Dimensions are not important in this exercise.
Click on the **Symmetric** tool in the Relate options box.

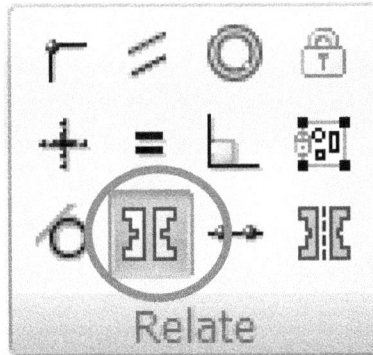

First select the axis that the lines will be symmetric about. Then select the line to change and finally select the edge to mimic. The mouse clicks are shown graphically in the following figure.

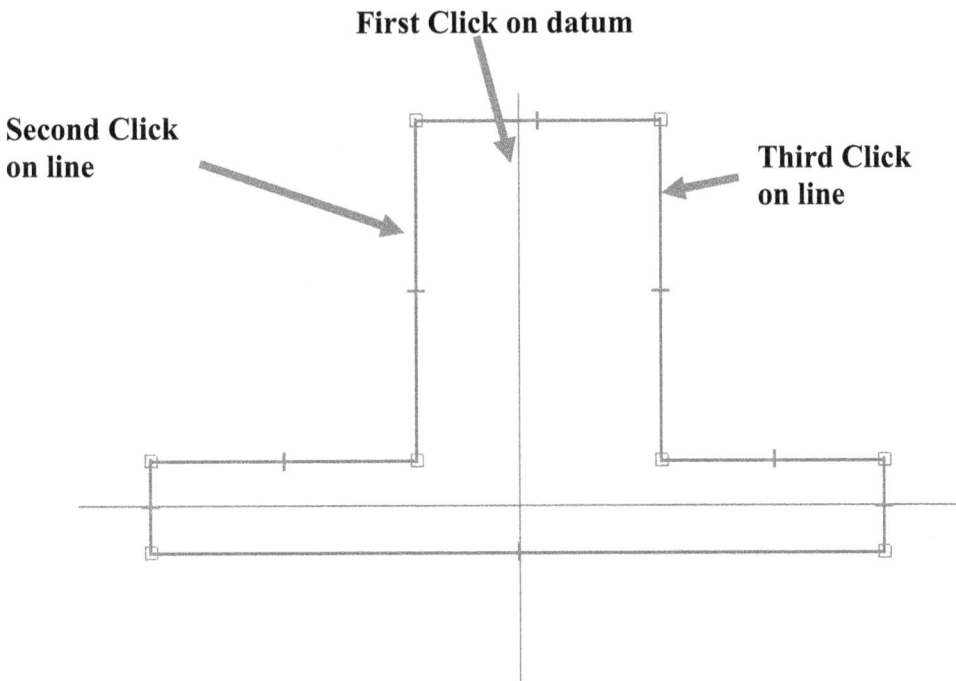

Repeat this process for the lines shown above, it is not necessary to select the axis again, unless you have quit out of the symmetric tool.

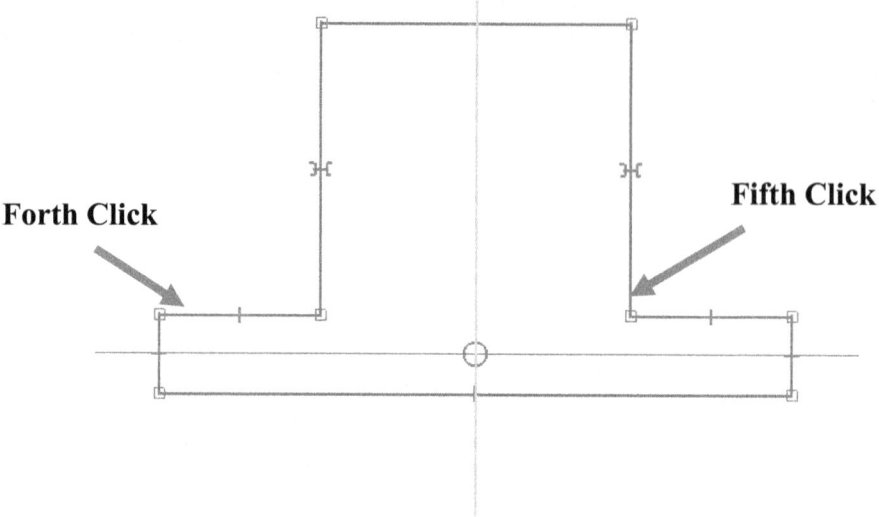

Forth Click

Fifth Click

Draw a circle on the vertical datum plane as shown below.

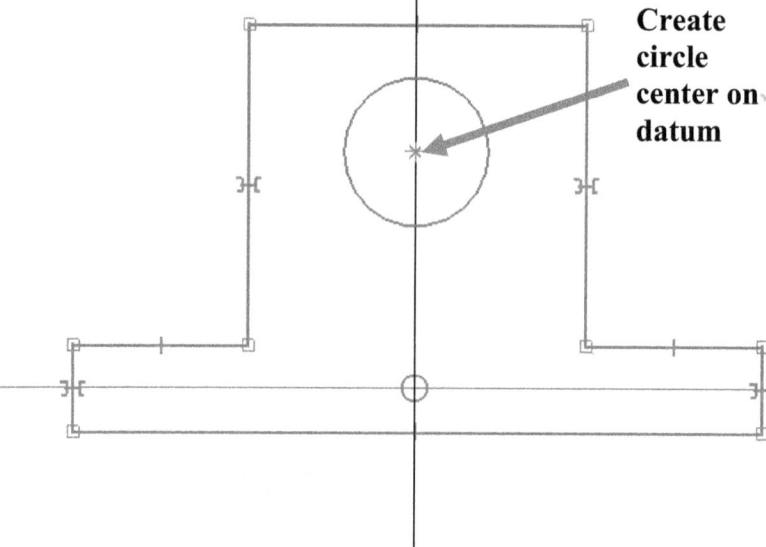

Create circle center on datum

Close the sketch by clicking the **Close Sketch** button.

Click the **Symmetric extent** button and set the distance to 50.

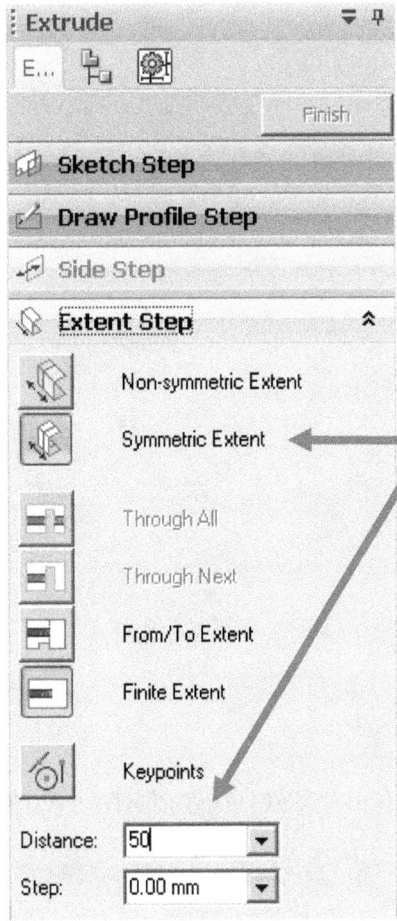

Click the
**Symmetric
Extent**
button, and
set the
distance to **50**

Click **Finish** to finish the extrude feature, as shown in the following figure.

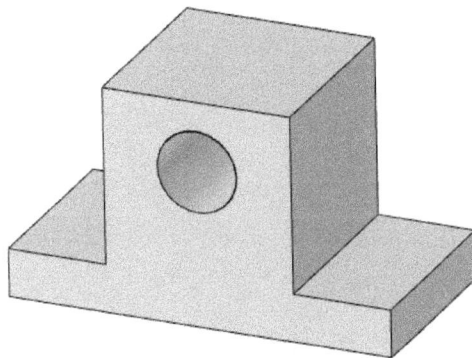

Click the **Hole** button in the Solids options box and select the face shown below.

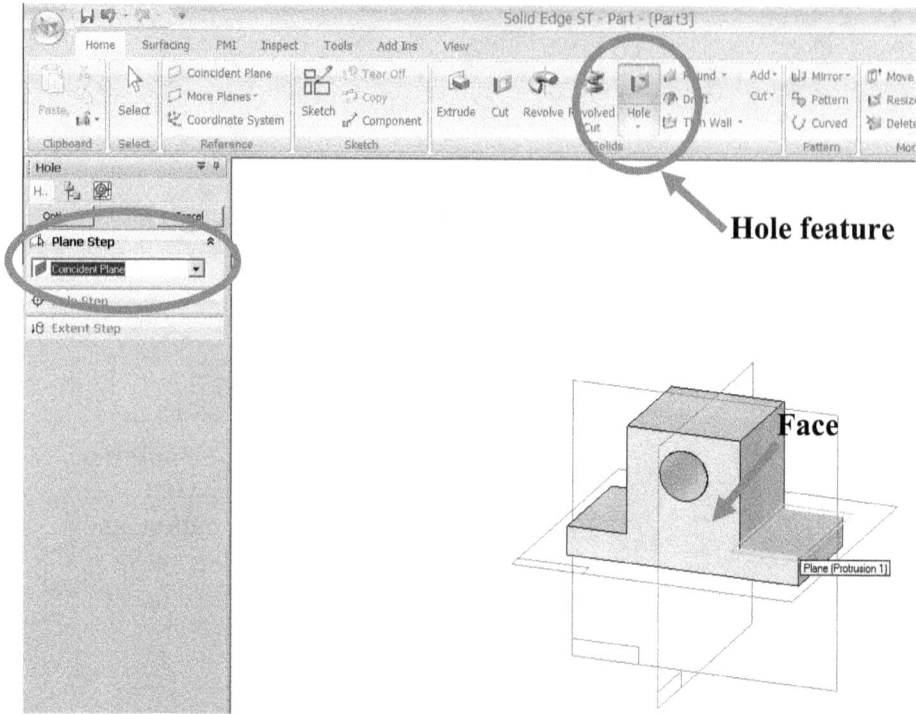

Note: use the Fit button at the lower right corner of screen to view entire part

The screen should now look like the following figure.

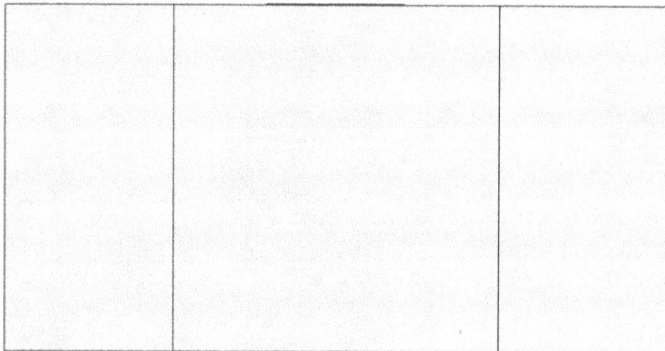

Place 2 holes on the horizontal datum plane as shown below, and click **Close Sketch**

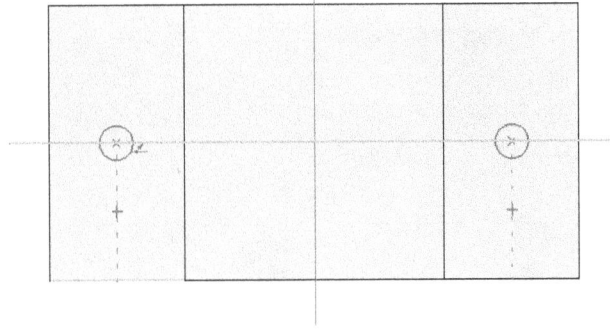

To Change the size of the holes, click **Options** at the top of the menu on the left hand side of the screen, change the Diameter to 8.

Click **Through All** in the Extent Step.

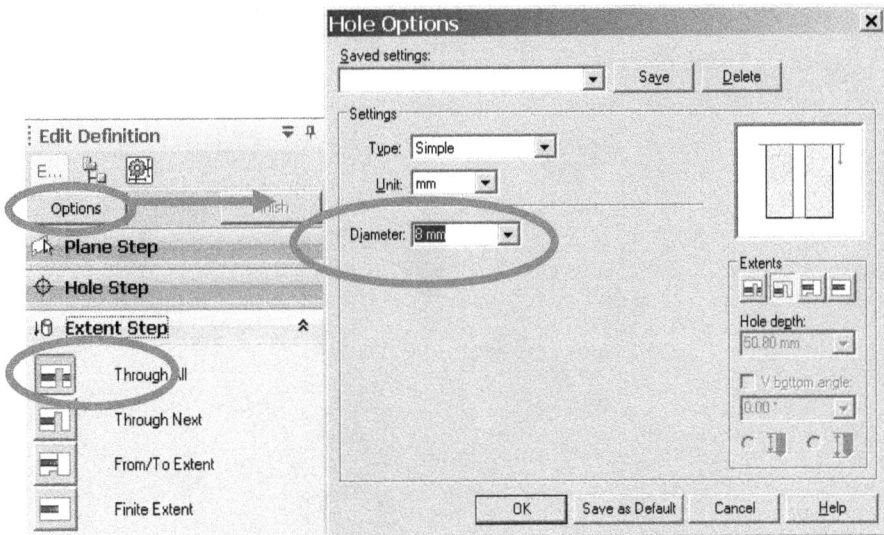

Note: It is also possible to change the type of hole from this menu. (For this exercise we will use a simple hole).

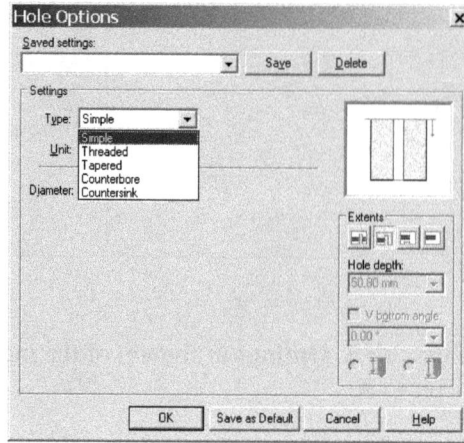

The last step is to specify the direction for the hole to be created.

Note: In this case only one direction will work so if the cursor is held above the sketch plane an error sign (!) will show.

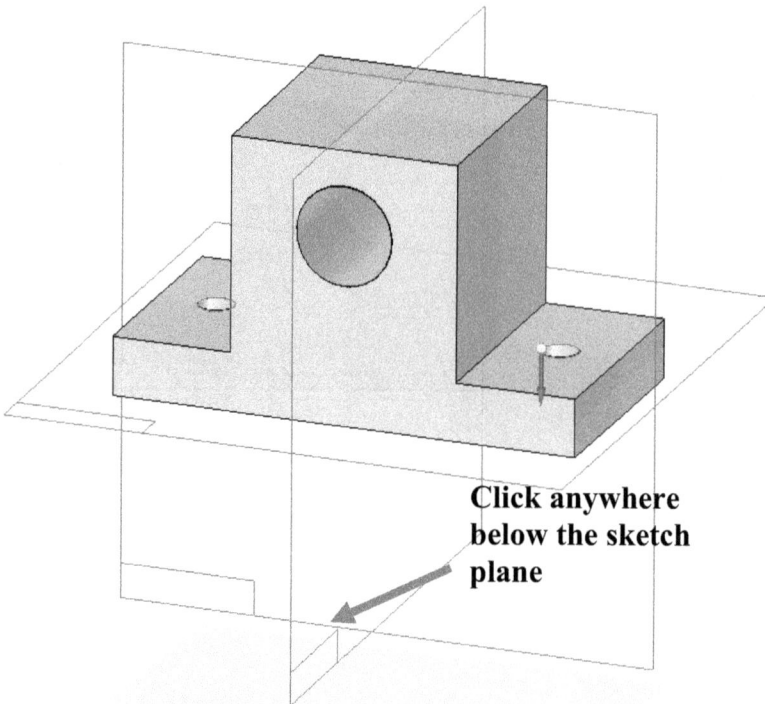

Click anywhere below the sketch plane

Click below the sketch plane to specify the direction as shown in the previous figure and click **Finish**.

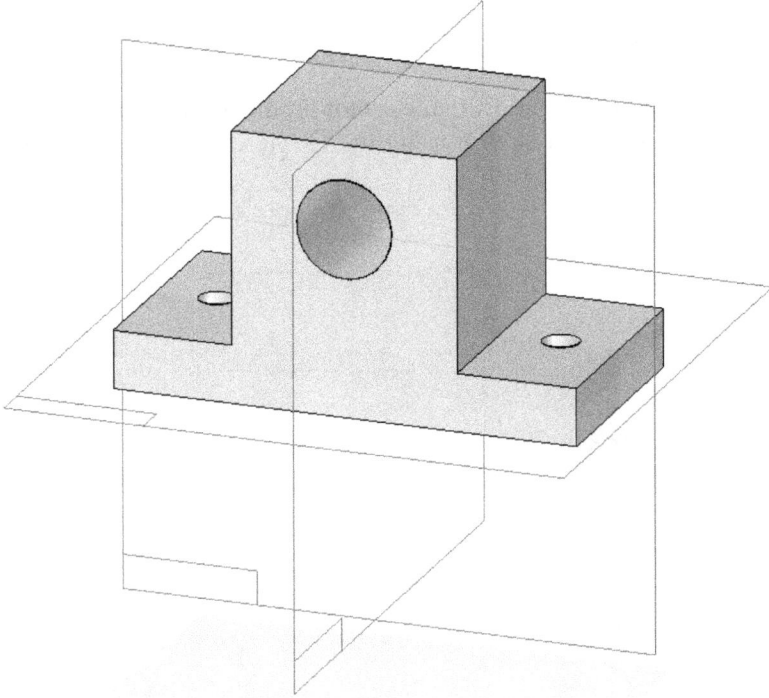

Exercise Complete

Exercise 16: More Holes

Create a **new ANSI part**. Click the **Extrude** button in the Solids options box and select the XY plane to sketch on. Create a **sketch** using the **Circle** tool in the Draw options box that looks like the figure shown below.

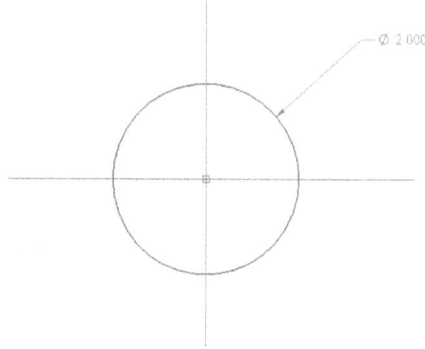

Close the Sketch and **Extrude** symmetrically about the sketch plane a distance of 3.00, as shown in the following figure.

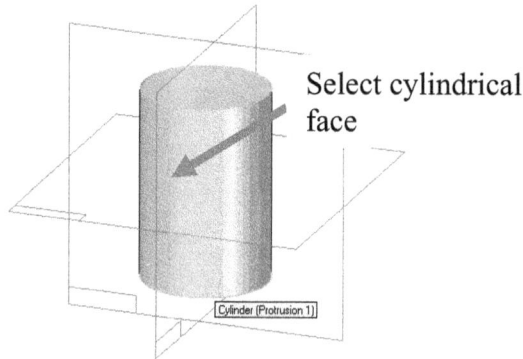

Select cylindrical face

Select **Tangent** from the More Planes dropdown menu in the Reference options box, and click the cylindrical face of the cylinder, as shown above.

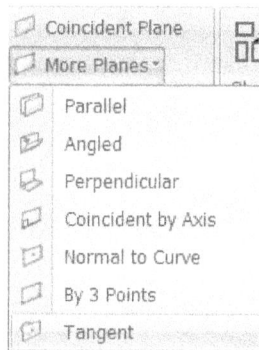

Click on the figure to begin the process, input **90°** into the **Angle** box, then click on the figure to set the tangent plane.

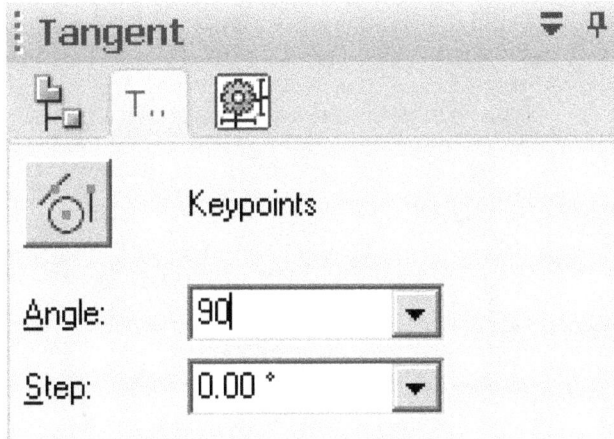

The plane added should look like the following figure.

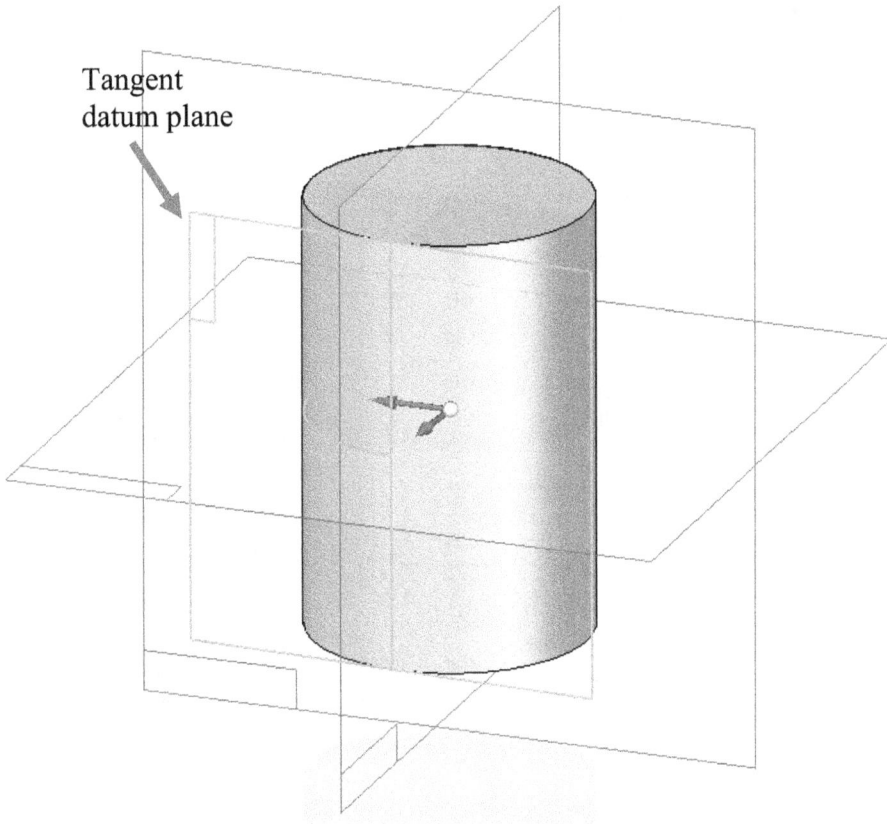

Tangent
datum plane

Click on the **Hole** button in the Solids options box and choose the tangent plane created to sketch on.

Place a hole in the center of the cylinder as shown below then close the sketch.

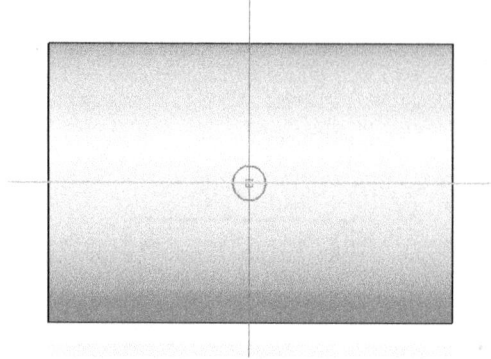

Click the **Hole** Options button in the top of the left menu and input the following details.

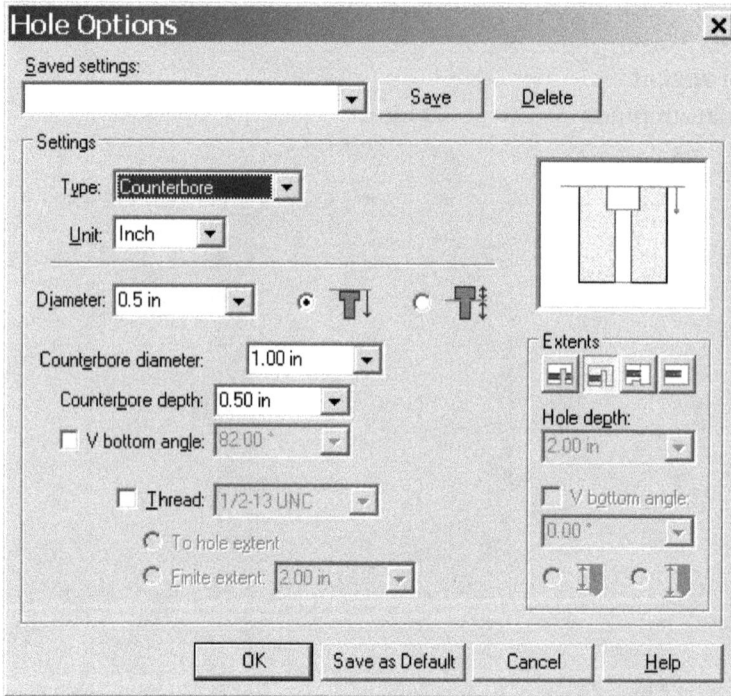

Select **Through All** as the extent step and click to define the cut direction to create the hole shown in the following figure.

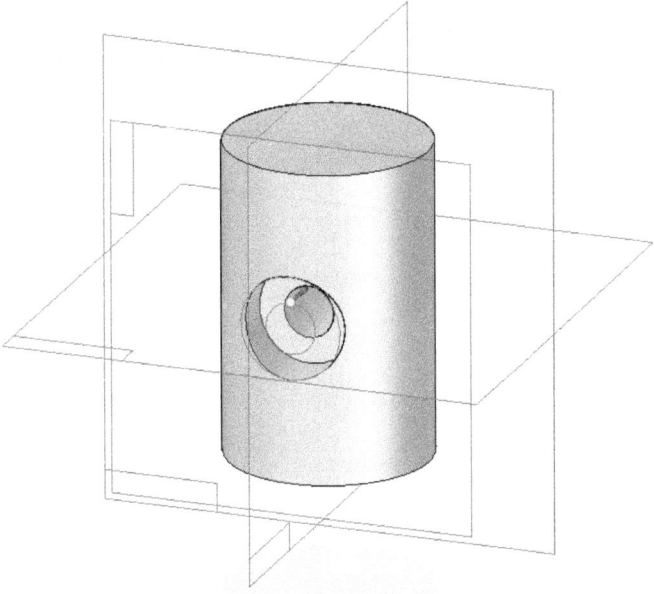

Click **Finish** to complete the Hole feature.

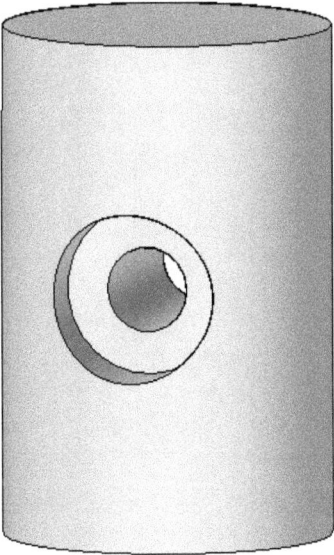

Exercise Complete

Exercise 17: Cutting

Create a **new ANSI part**. Click the **Extrude** button in the Solids options box and select a plane to sketch on. Create a sketch that looks like the figure shown below. Feel free to be creative with the dimensions.

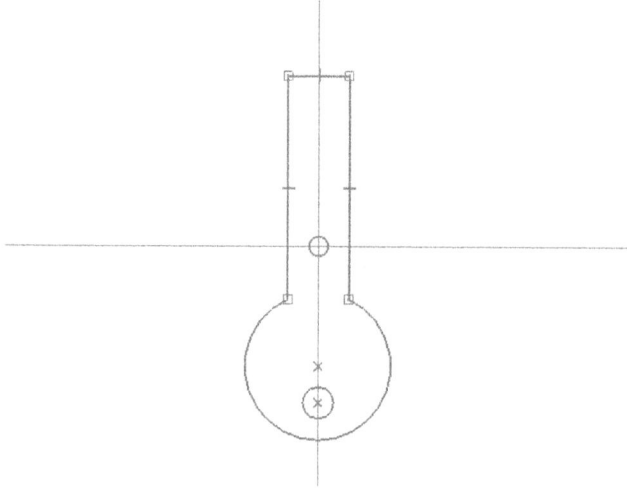

Exit the sketch by clicking **Close Sketch** and extrude the figure approximately the distance shown below.

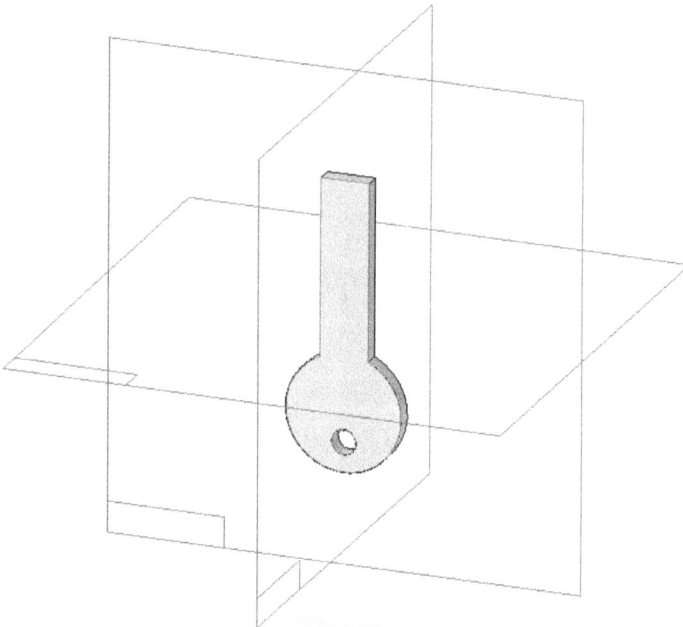

Click the button in the Sketch options box and select the surface shown below to sketch on.

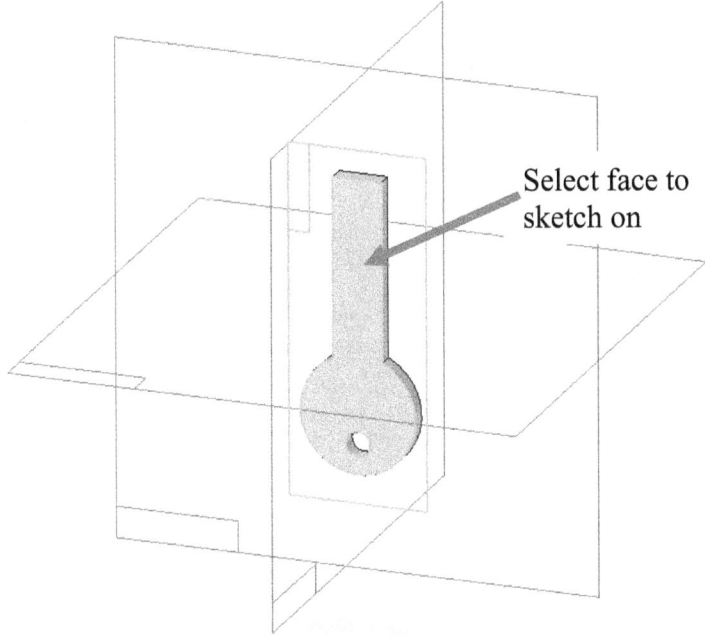

Select face to sketch on

Create a sketch that looks similar to the following figure. Be creative.

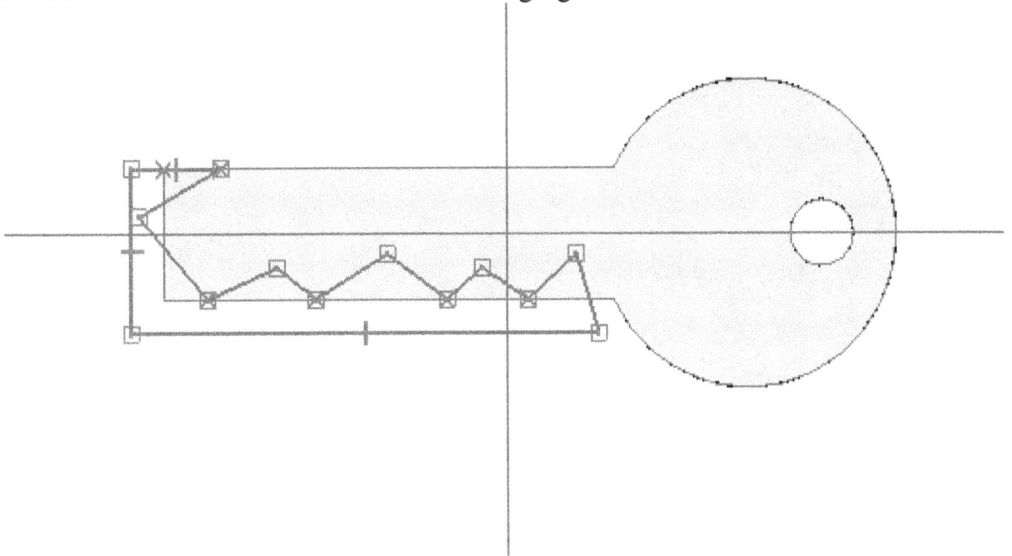

Click Close Sketch.

Click the **Cut** button in the Solids options box

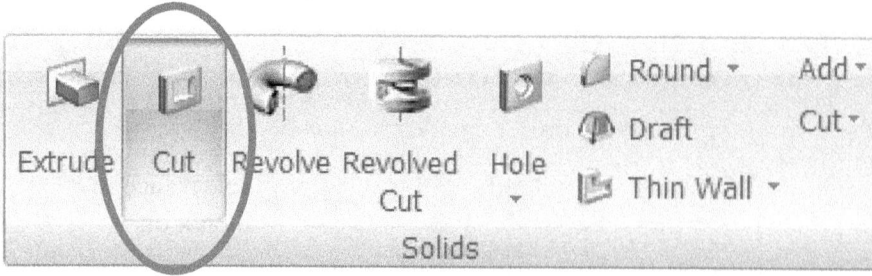

In the **Sketch Step** in the menu on the left choose **Select from Sketch** from the drop down menu and select the sketch as shown in the following figure.

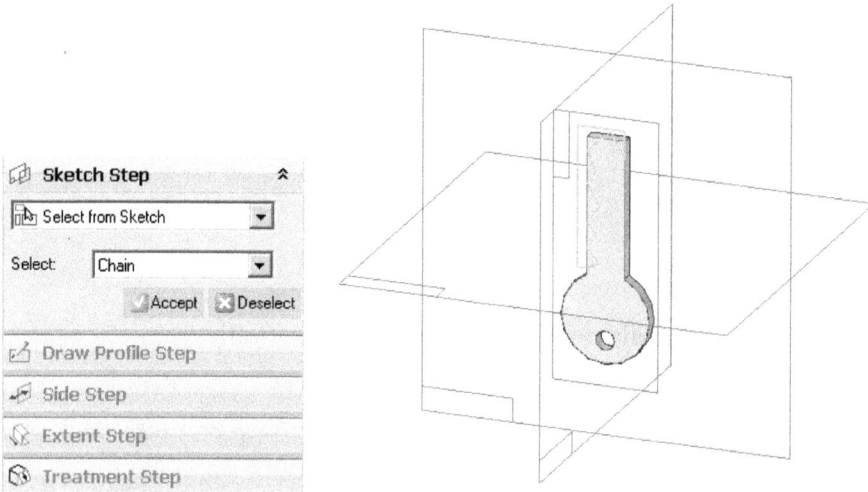

Click **Accept**.

In the **Extent** Step choose **Through All** and click as shown below to specify the direction, the preview will show the cut.

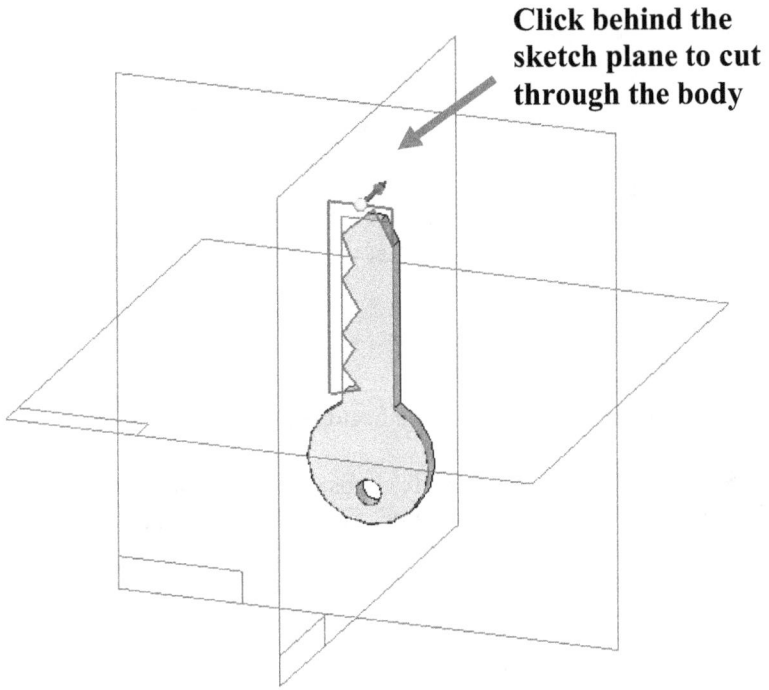

Click behind the sketch plane to cut through the body

Click the part should now look like the following figure.

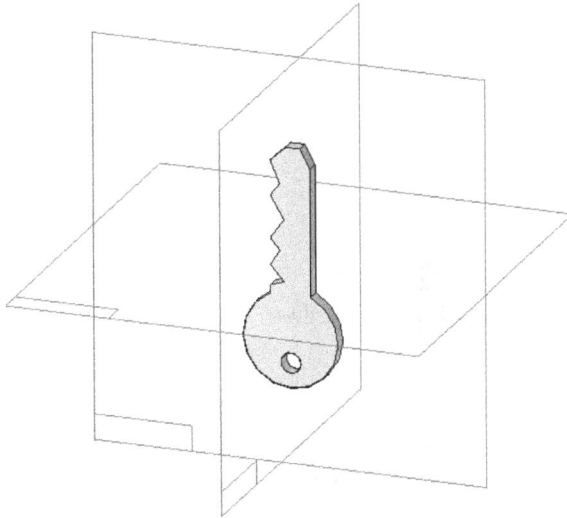

Now let's create another **cut**. Create a sketch on the front surface that looks like the figure below.

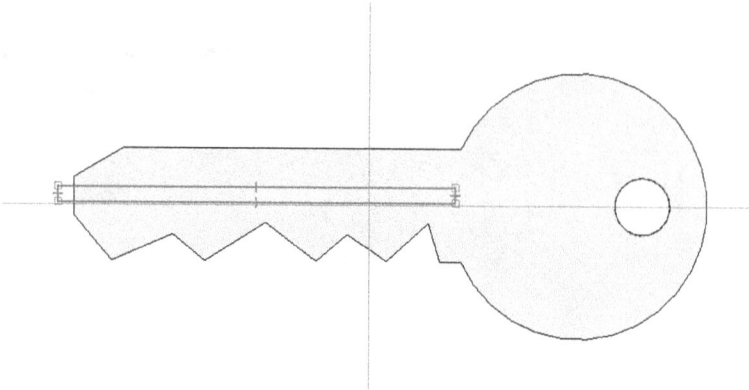

Close Sketch and click the **Cut** button in the Solids options box.

In the **Sketch Step** in the menu on the left choose **Select from Sketch** from the drop down menu and select the sketch profile you created. Select **Finite Extent** and enter a Distance 50% through your body.

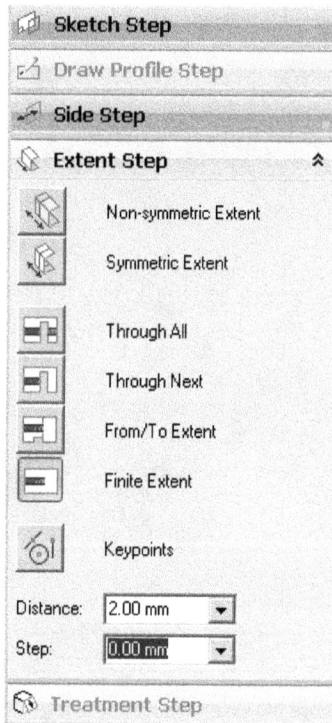

Specify the direction by clicking behind the part.

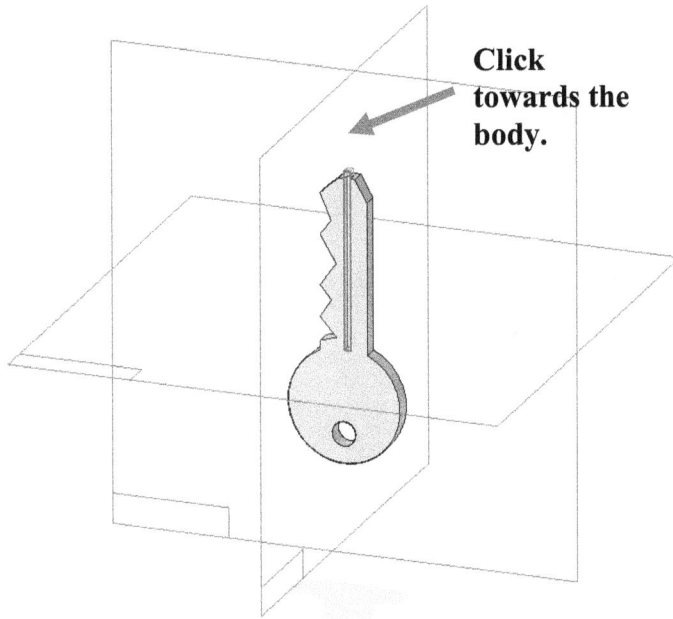

Click towards the body.

Click Finish.

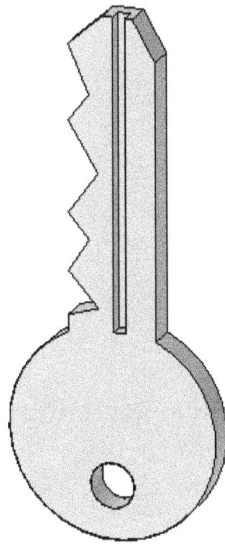

Exercise Complete

Exercise 18: Revolving

Create a **new ANSI part**. Click the **Revolve** button in the Solids options box and select the **XZ plane** to sketch on.

Create a sketch using the **Line** tool in the Draw options box that looks like the figure shown below. Add the dimensions shown.

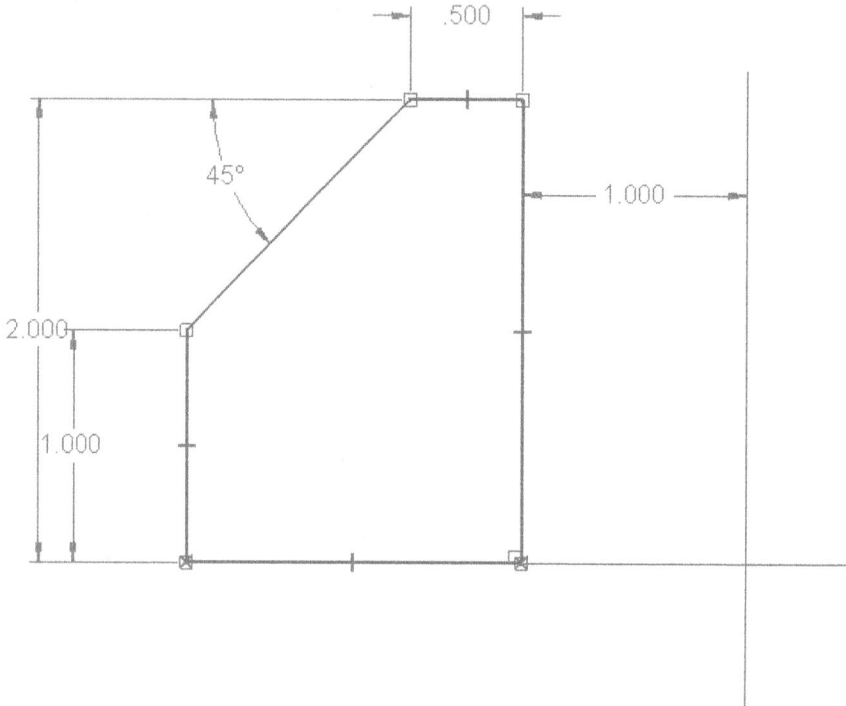

Before exiting the sketch, click the **Axis of Revolution** tool in the Draw options box.

Click on the vertical datum shown in the following figure.

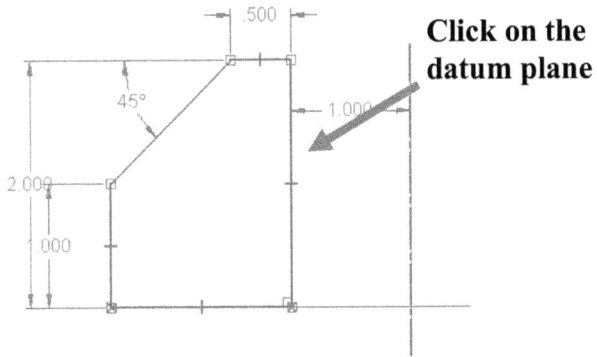

Click on the datum plane

Exit the sketch using the **Close Sketch** button, to view the following screen. *Notice: Moving the cursor around the graphics display will change the degree of revolution.*

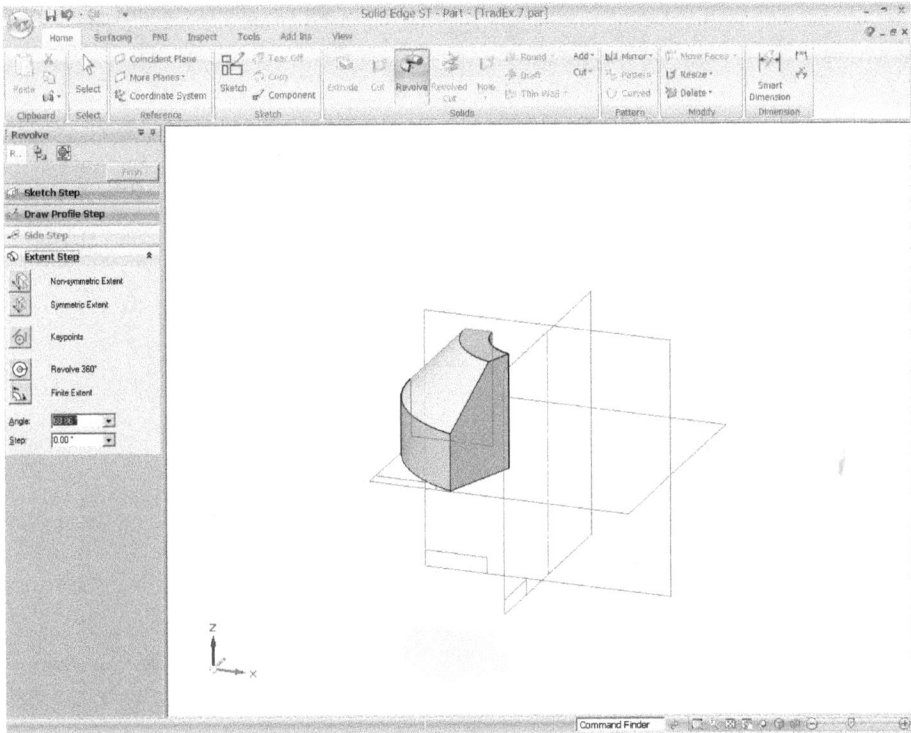

Select **Revolve 360⁰** and click Finish to exit the completed revolve feature.

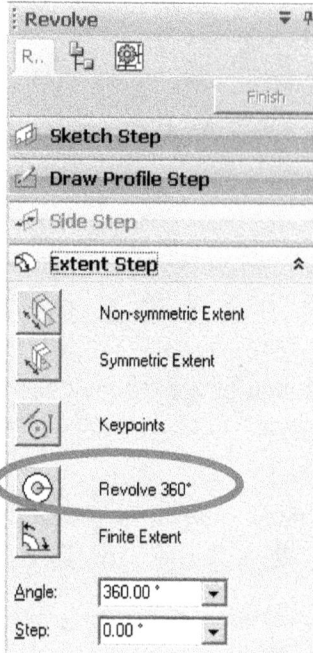

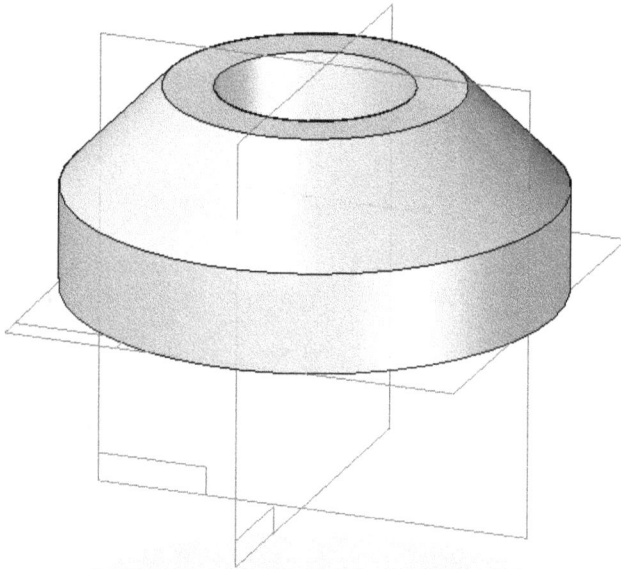

Exercise Complete

Exercise 19: Revolved Cut with Angled Datum Planes

Create a **new ANSI part**. Click the **Extrude** button in the Solids options box and select a plane to sketch on.

Create a sketch using the **Rectangle** tool in the Draw options box that looks like the figure shown below. Add the midpoint constraints using the **Connect** tool as shown in a previous exercise. Exit the sketch and extrude 2in.

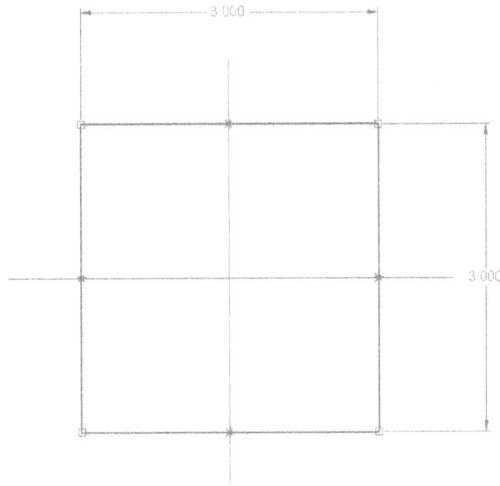

Select **Angled from the More Planes** dropdown menu in the Reference options box.

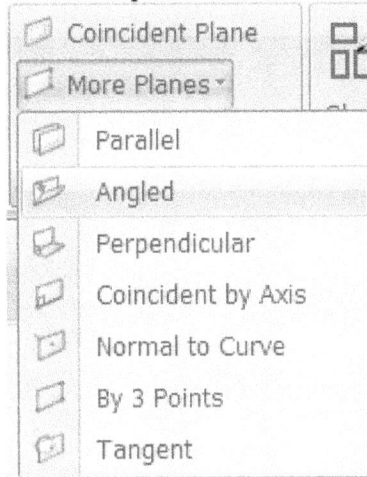

Click the following datum planes, in the order shown below. Moving the cursor around the graphics display will change the angle of the plane.

First Click

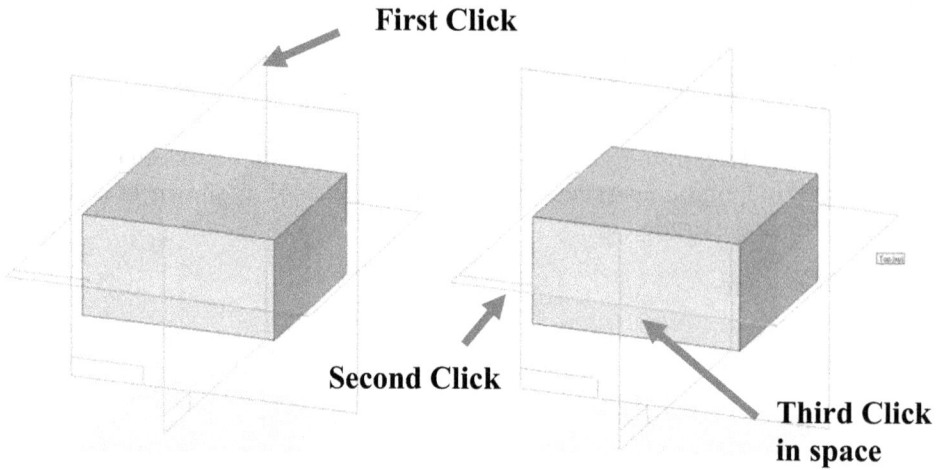

Second Click

Third Click in space

Enter 20° into the angle box in the left menu.

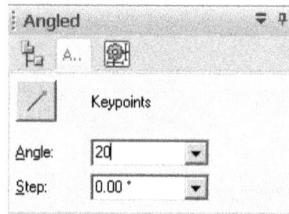

Click on the right side of the graphics display (in space). This will select the direction of the plane as shown in the following figure.

New plane

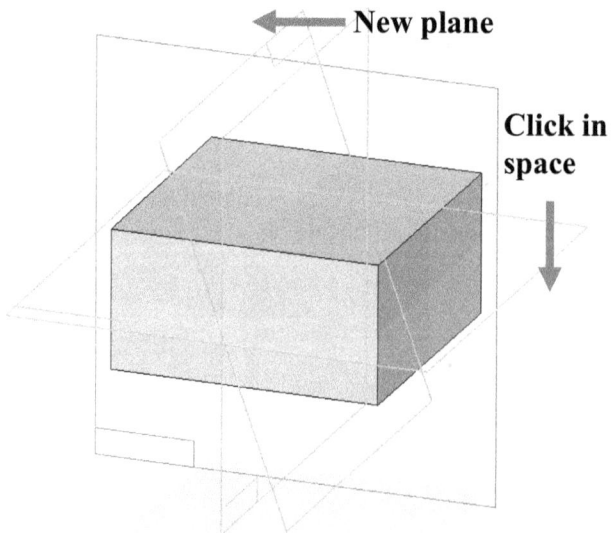

Click in space

Create another **angled plane** by selecting the planes shown below.

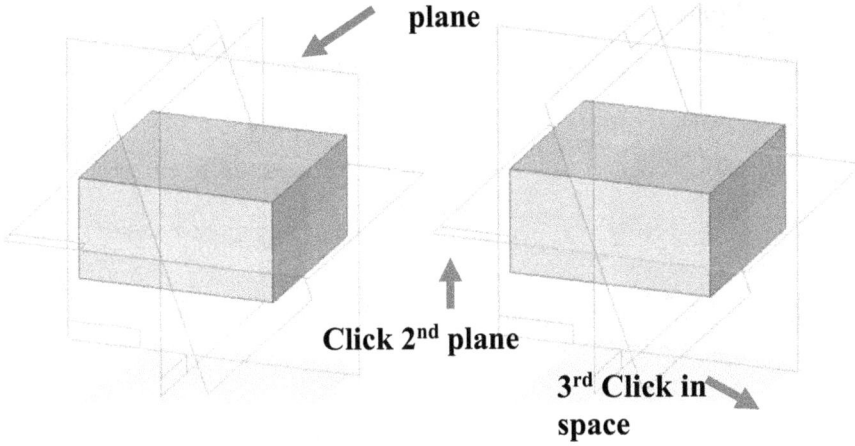

Click 1st plane

Click 2nd plane

3rd Click in space

Enter 10° into the angle box in the left menu, and click in front of the part to select the direction to produce the datum shown in the following figure.

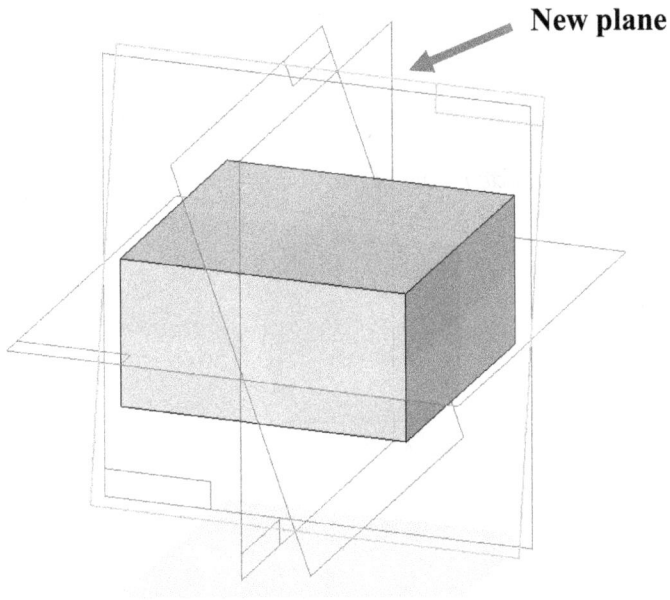

New plane

Right click on the vertical **YZ plane** and select Hide as shown below.

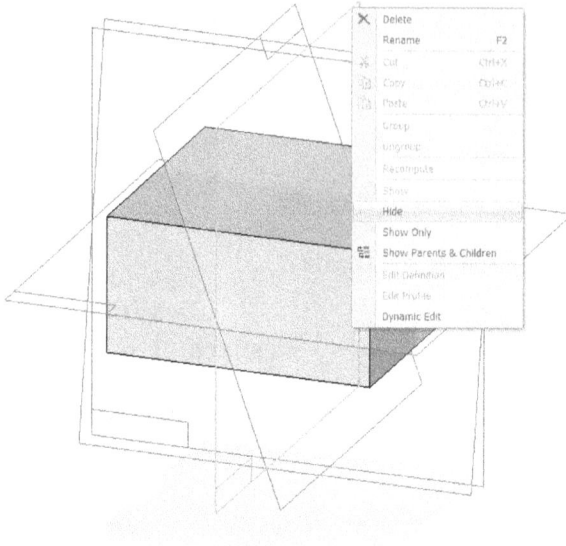

Hide the XZ and XY planes using the same method.

Note: The other planes are hidden from clarity.

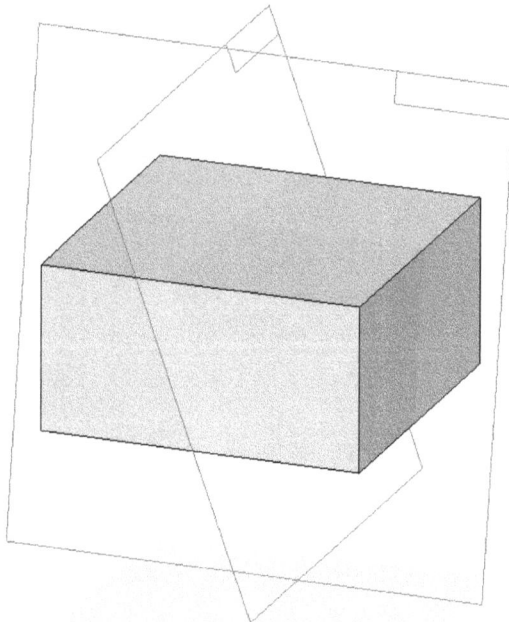

In the **Surfacing** tab click the **Intersection** tool in the **Curves** options box.

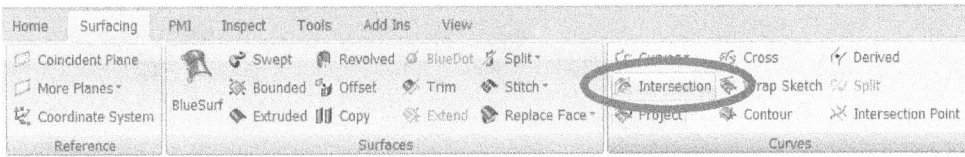

In the Select Set 1 step in the left menu, click on one of the angled datum planes you created.

Click on the other plane in the Select Set 2 step.

Click **Finish** to create the intersection line.

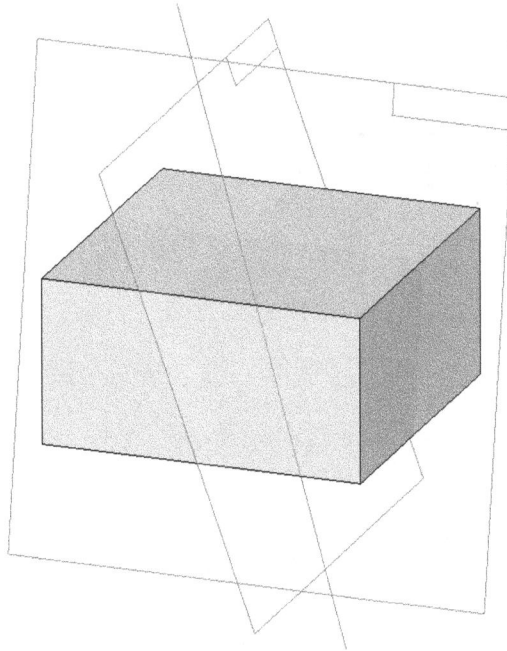

Go back to the **Home** tab and create a sketch on either one of the planes.

Create a **sketch** that looks like the figure below.

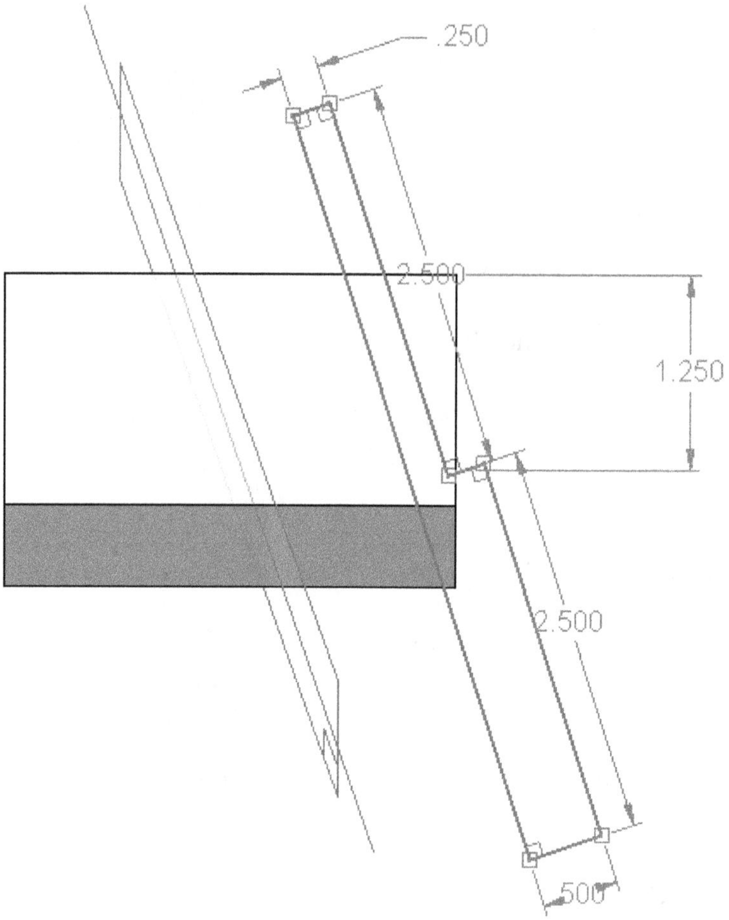

Select the **Collinear** tool in the Relate options box.

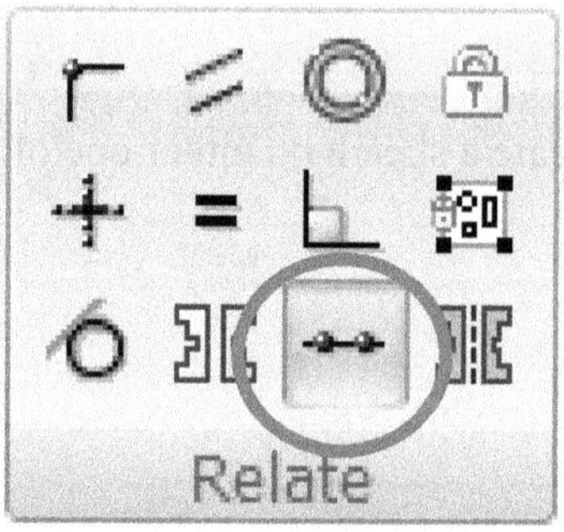

Click on the longest line of the sketch and the intersection line as shown below.

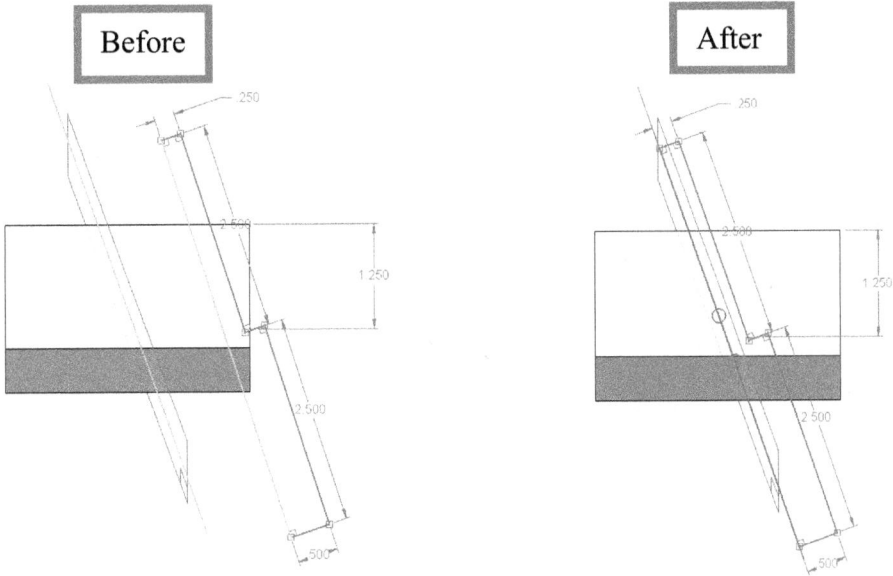

Before

After

Close Sketch. Click **Revolved Cut** in the Solids options box.

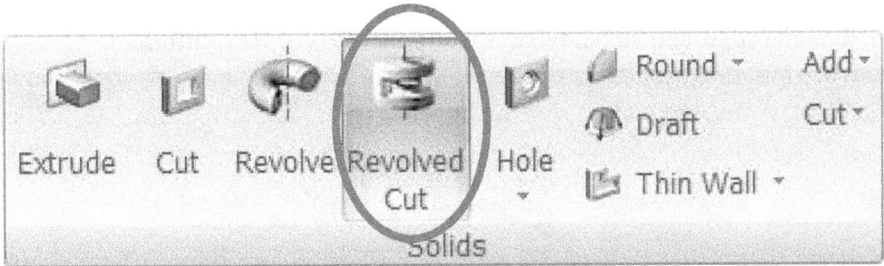

Extrude Cut Revolve Revolved Cut Hole Round Draft Thin Wall Add Cut

Solids

In the **Sketch Step** make sure **Select from Sketch** is selected in the dropdown menu and click on the sketch.

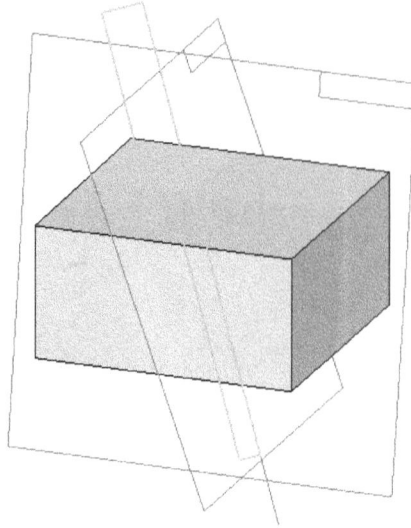

Click Accept.

Now click the long edge, shown below, to specify the **Axis of Revolution**.

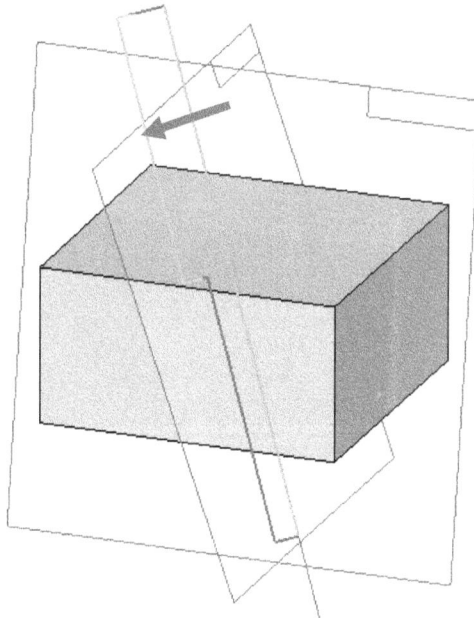

In the **Extent Step** click the **Revolve 360°** button then the Finish button at the top of the left menu.

To view the model without all the construction lines etc. right click on the screen and select **Hide All / Sketches**, as shown below. **Hide** the **Reference Planes** and **Curves** with the same method.

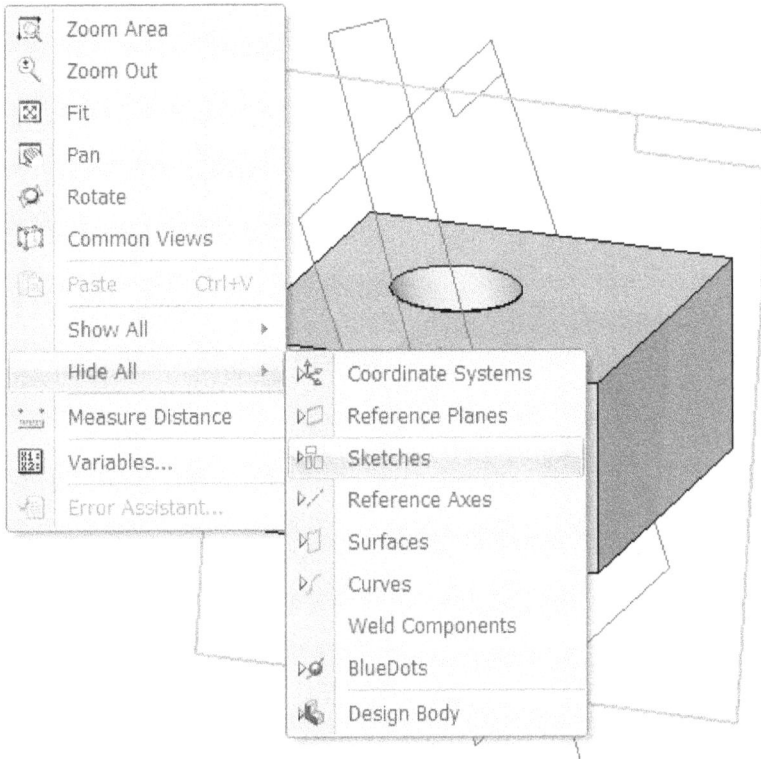

Select the **Visible and Hidden Edge** button from the menu in the bottom right corner of the screen, as shown in the following figure.

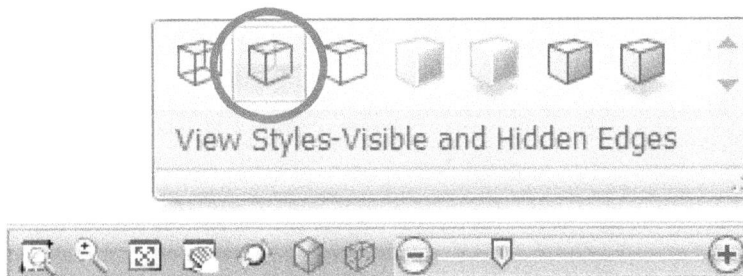

The graphics display will now show the wireframe view of the model.

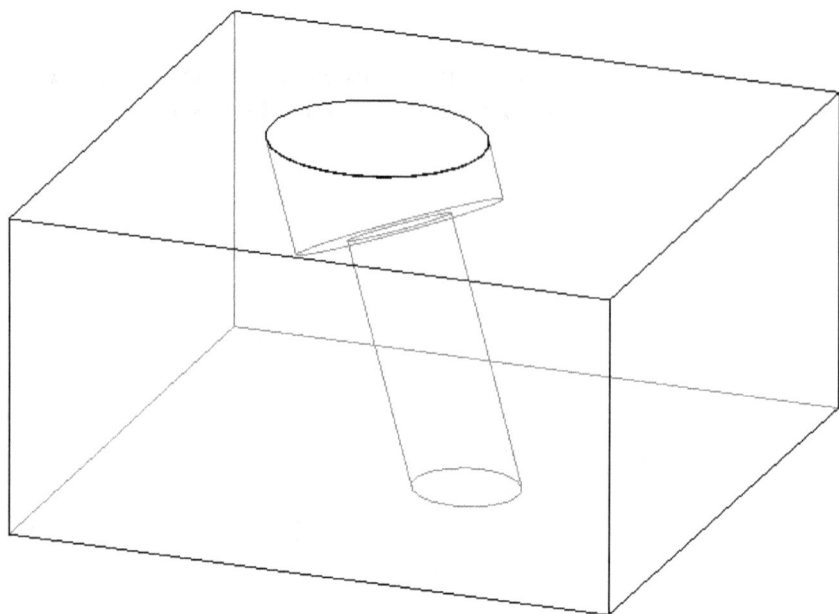

Exercise Complete

Exercise 20: Mirror

Start a **new ANSI part.** Click the **Extrude** button and select a plane to sketch on.

Create a sketch that looks like the figure shown below.

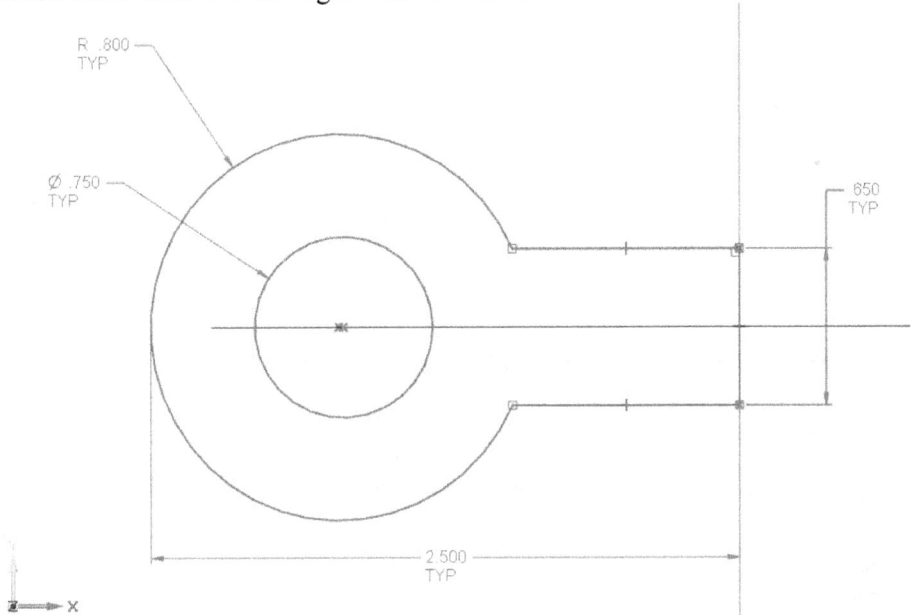

Close the sketch.

Extrude above the sketch plane a distance of 0.5 and click Finish to exit the extrude feature.

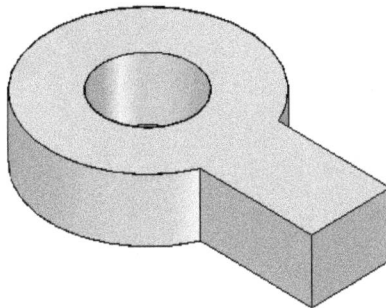

Note: The planes are hidden for clarity purposes only.

Click the little arrow next to the **Mirror** button in the **Pattern** options box and select **Mirror Copy Part**.

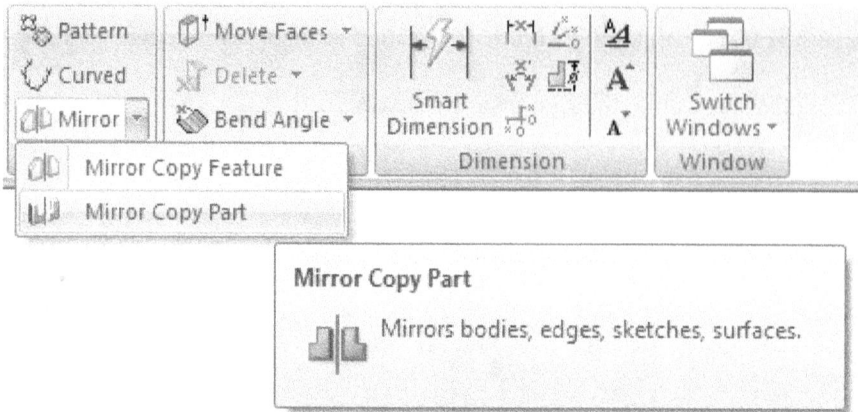

The menu on the left will instruct you to select a body, so click on the part.

The screen should look like the following figure.

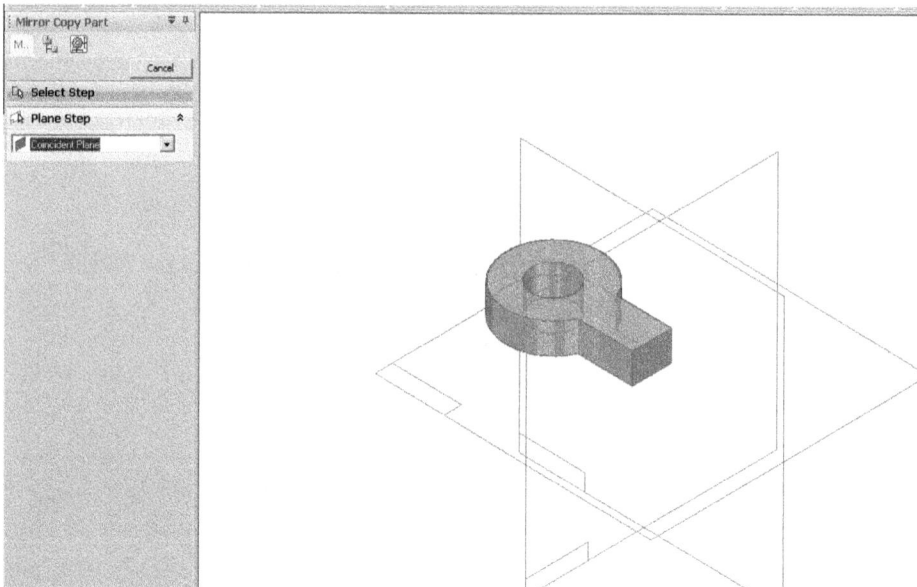

In the **Plane Step** select the plane to mirror about, as shown in the following figure.

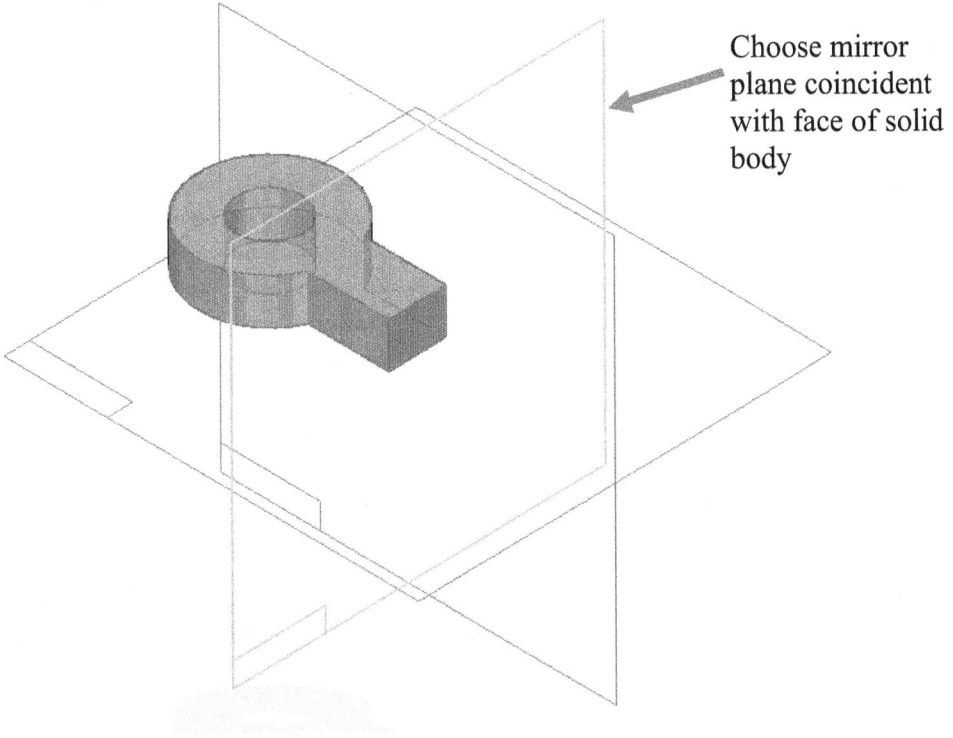

Choose mirror
plane coincident
with face of solid
body

Click the **Finish** Button to exit.
The part should look like the figure below.

Click the **Extrude** button in the Solids options box and select the top surface to sketch on.

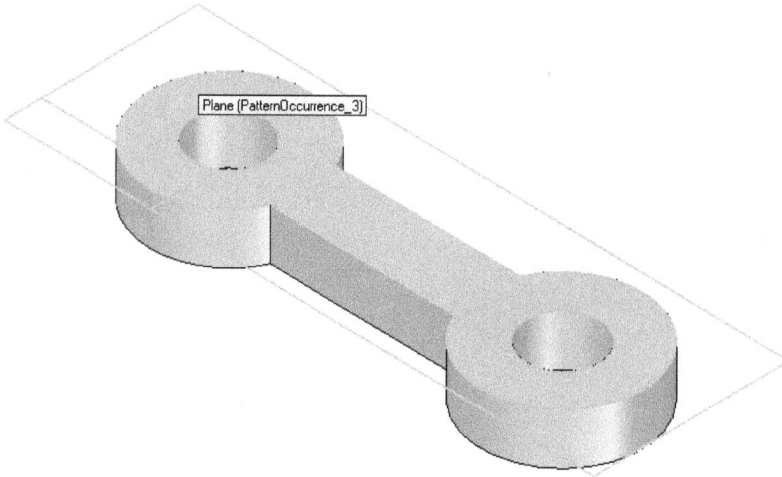

Using the **Circle** tool, sketch the 4 circles shown below. *Note: You can easily achieve concentric circles by clicking the center of the existing circles to define the center of the new one. And click the edges of the existing circles to define the radius.*

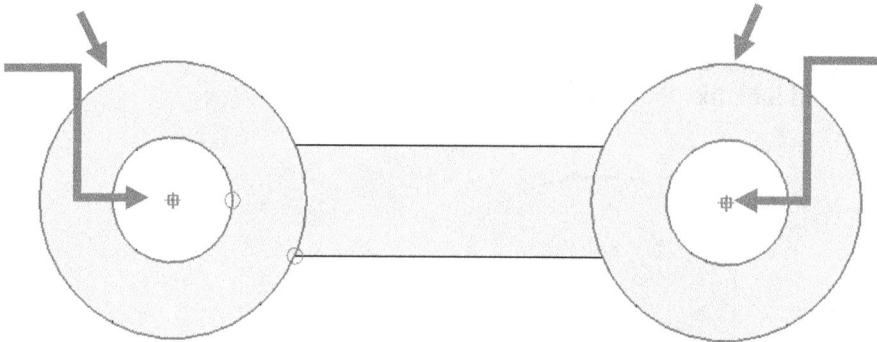

Remember: to make the center marks of the circles appear run the cursor over the edge of the circle.

Exit sketch by clicking **Close Sketch.**

Extrude above the sketch plane a distance of 0.5 and click **Finish.**

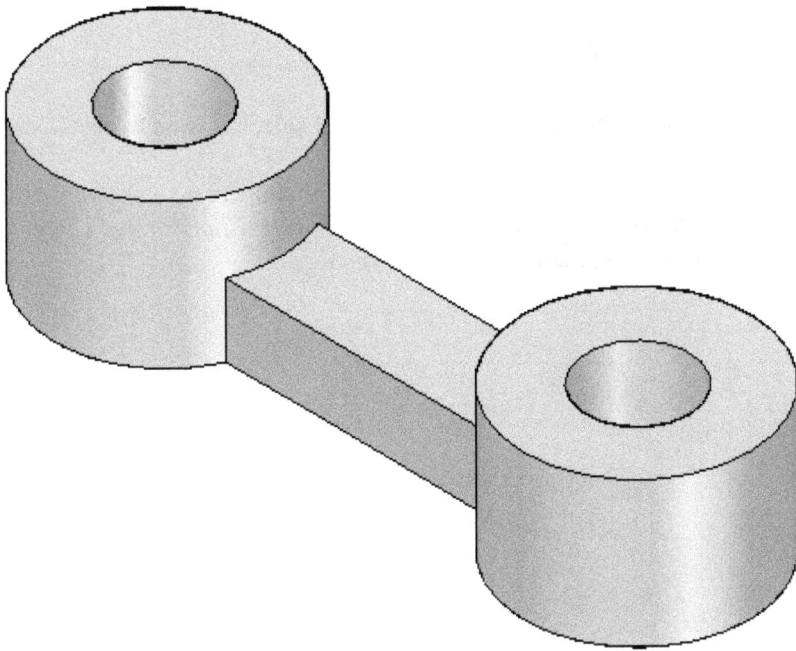

Exercise Complete

Exercise 21: Draft, Round and Chamfer

Click the **Extrude** button in the Solids options box and select a plane to sketch on.

Using the **Ellipse** tool located in the **Circle** dropdown menu, create a sketch that looks like the figure shown below. *Note: Ellipse by 3 points works well. Click for the center then define 2 points of the major radius then the radius of the minor.*

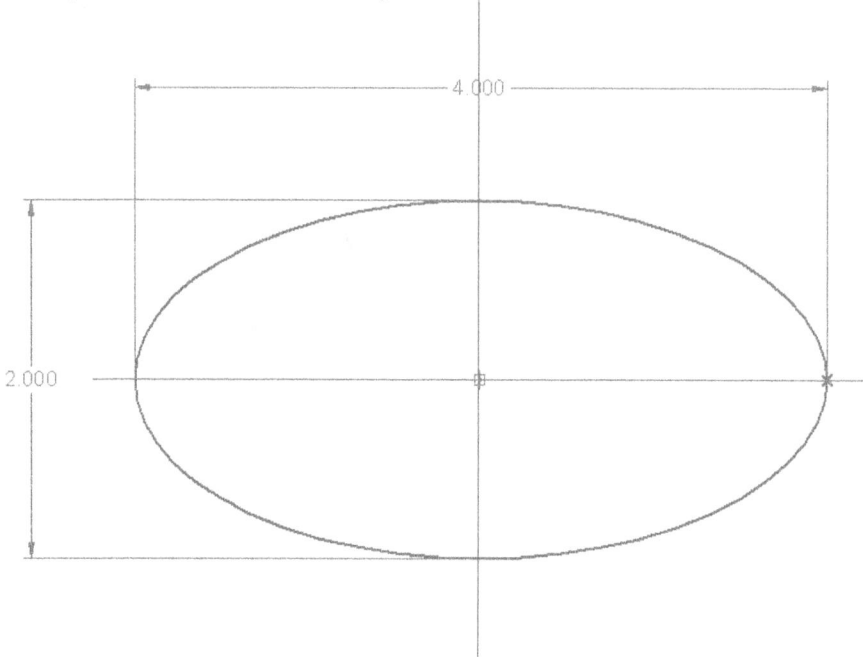

Close Sketch.

Extrude above the sketch plane 1in.

Click on the **Draft** button in the Solids options box, shown in the following figure.

In the **Draft** Plane Step select the bottom face of the part as shown in the following figure.

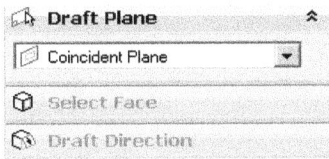

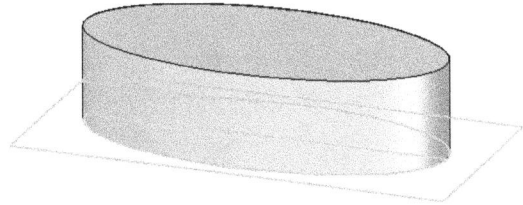

In the **Select Face** Step select the sides of the extruded ellipse.

Enter 15.00° into the Draft Angle box and click Accept then Next at the top of the menu.

To specify the **Draft** Direction, in the Draft Direction Step, click on the screen so the top arrow is pointing towards the part.

Click here ⟶

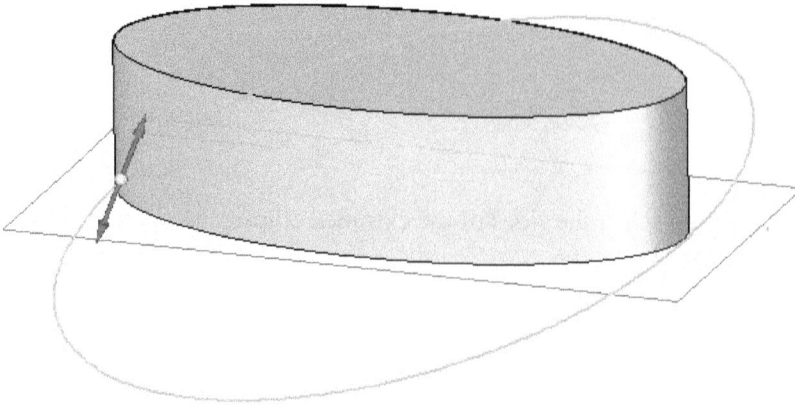

Click Finish.

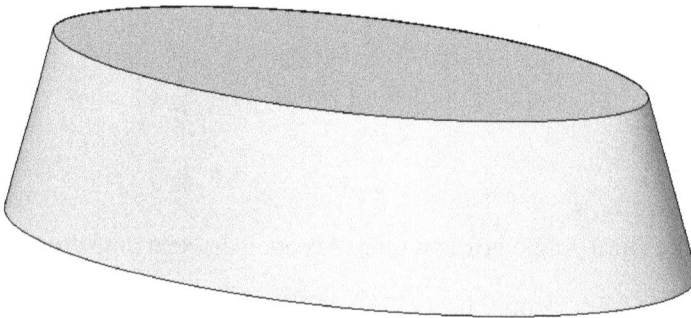

Click on the **Round** button in the Solids options box.

In the **Select Step** click on the top edge of the part and enter 0.25 in into the Radius box.

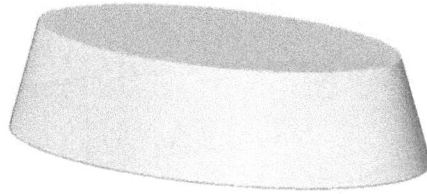

Click **Preview** then **Finish**. The part should look like the following figure.

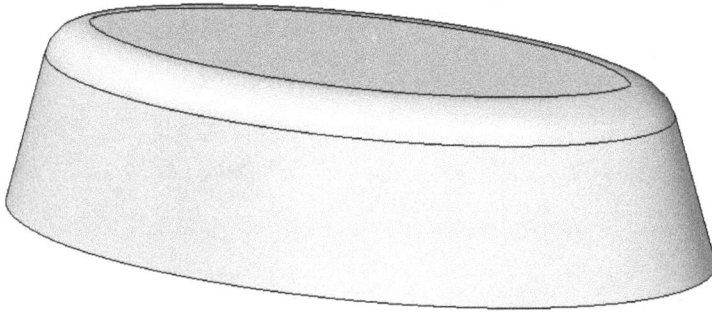

Click on the **Chamfer** button located in the Round dropdown menu.

In the Select Edge Step select the bottom edge of the part, and enter 0.50in into the Setback box.

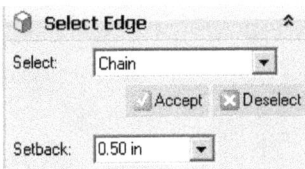

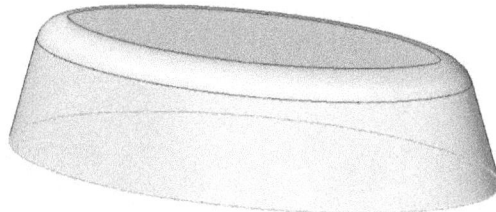

Click **Accept** then **Finish**. The part should look like the following figure.

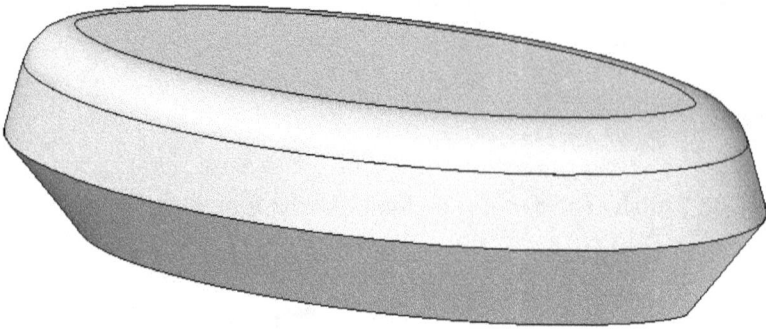

Exercise Complete

Exercise 22: Sweep Along a Path

To create a sweep, like most solid features, it is possible to either sketch before the sweep, or sketch within the sweep tool. In this exercise we will sketch before the sweep.

Choose the **XZ Plane** to sketch on and draw a square as shown below.

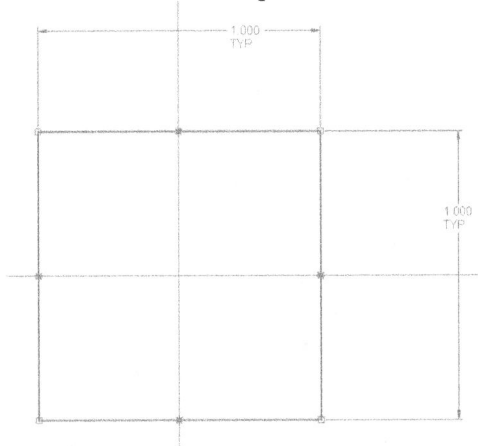

Close Sketch and create a new sketch on the **ZY Plane** as shown below.

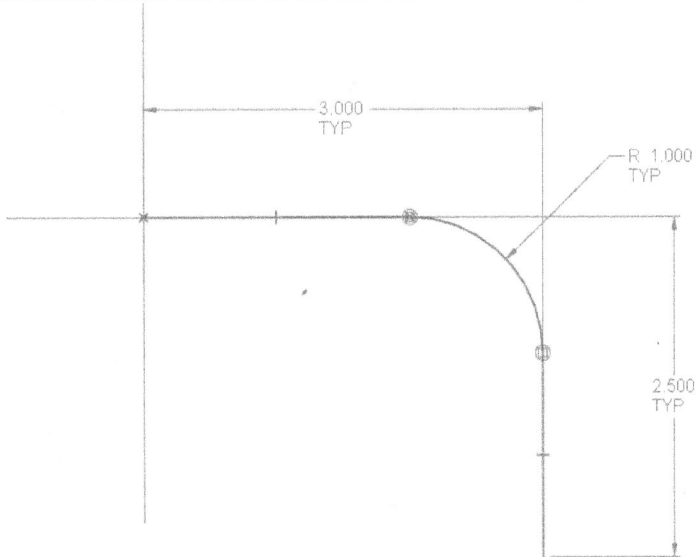

Note: Sweep works better if the guided curve (sketch #2) starts at the plane of the cross section (sketch #1).

Close Sketch.

Click the little arrow next to the sweep symbol in the Solids options box and select **Sweep.**

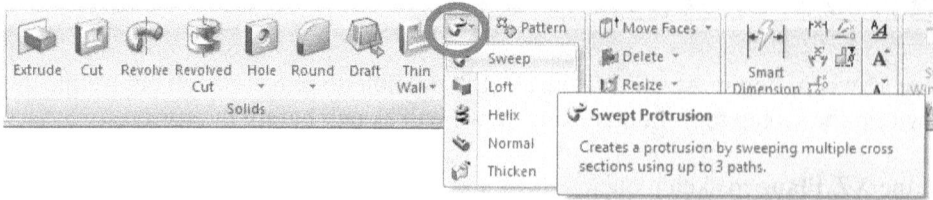

The following figure should appear.

Make sure **Single path** and **cross section** are chosen and hit **OK.**

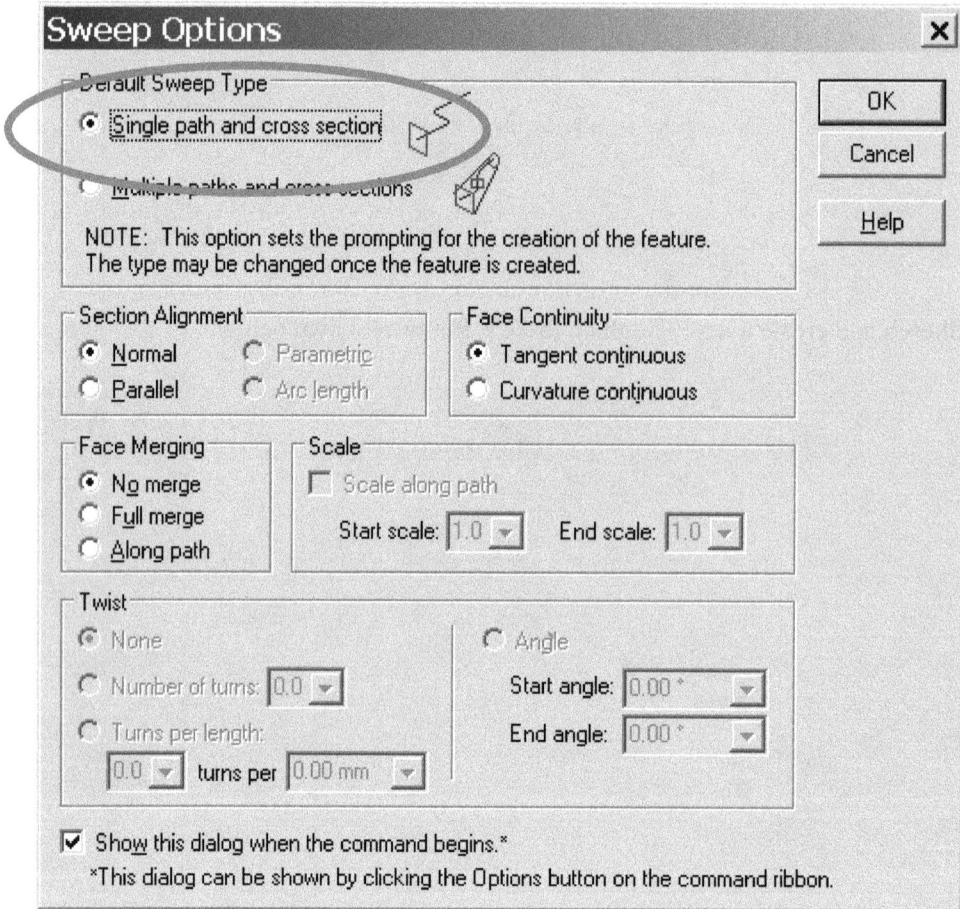

In the **Path Step** menu choose **Sketch from Sketch/Part Edges** in the drop down menu and **'Chain'** as the selection intent in the following drop down menu.

Now click the second sketch as the path.

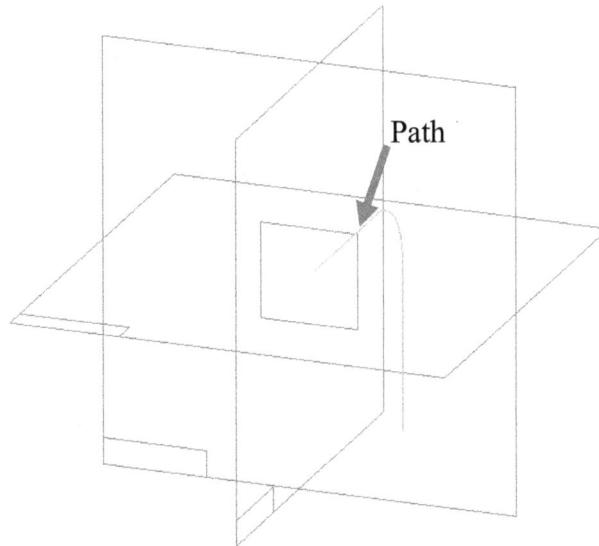

Click Accept

In the **Cross Section** Step, make sure that **Sketch from 'Sketch/Part Edges'** and the **'Chain'** option are chosen in the drop down menu.

Click on the square.

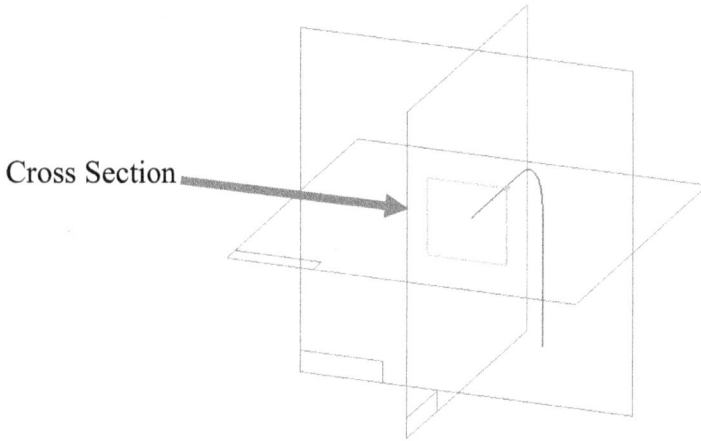

Cross Section

Click **Finish** at the top of the menu and the part will look like the figure shown below.

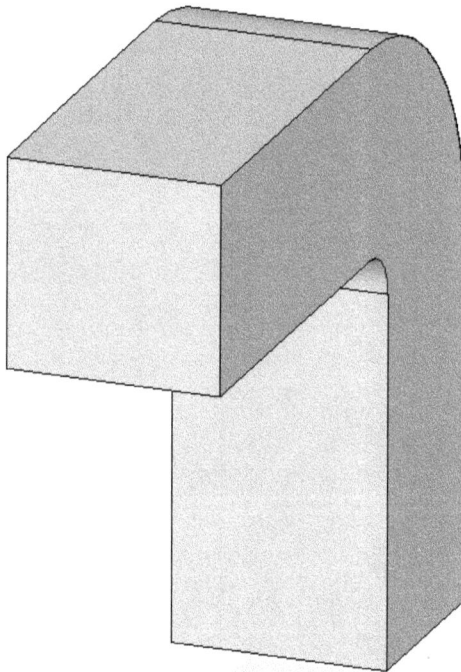

Exercise Complete

Exercise 23: Thin Wall

Click the **Extrude** button in the Solids options box and select the **XY plane**.

Create a **sketch** that looks like the figure shown below.

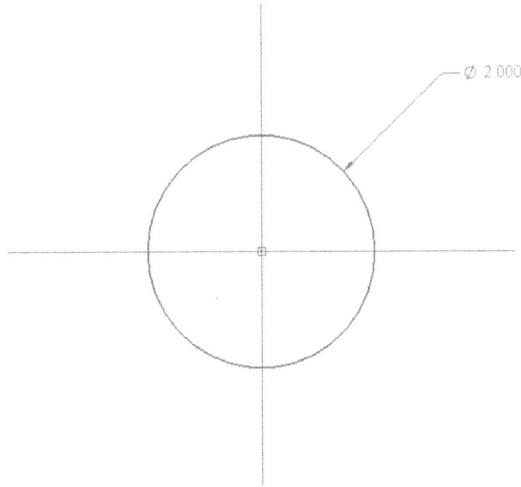

Extrude a distance of 4in as shown below.

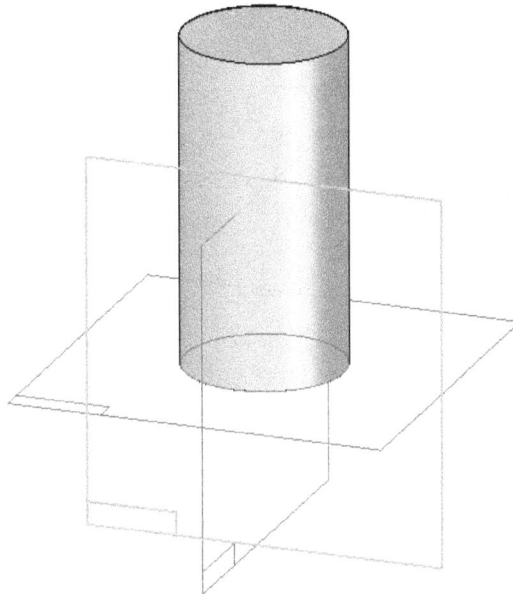

Create a **sketch** on the **XZ Plane** as shown in the following figure.

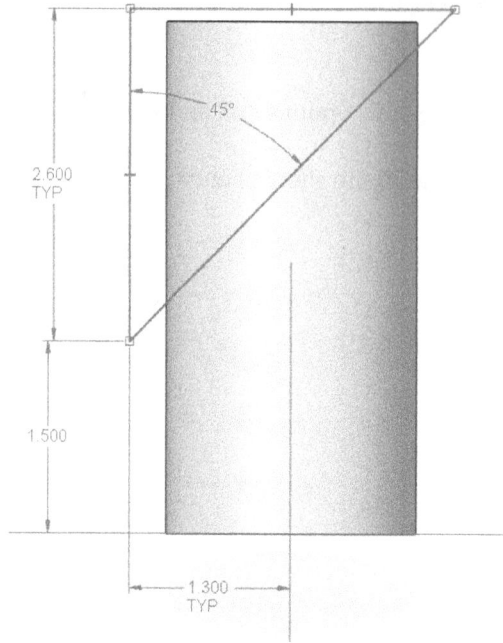

Click the **Cut** button in the Solids options box and make sure the **'Select from Sketch'** option is chosen in the drop down menu. Click **Accept.**

Select '**Through all**' in the Extend Step as shown below.

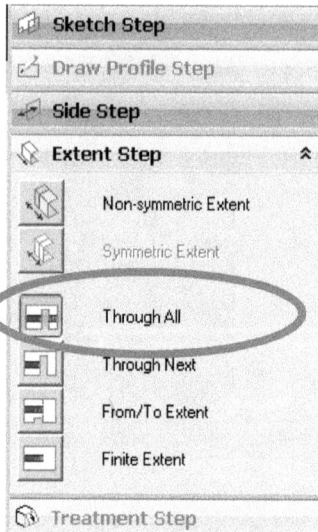

Hold the cursor on the X-axis and click when the double sided arrow is shown, illustrated in the following figure.

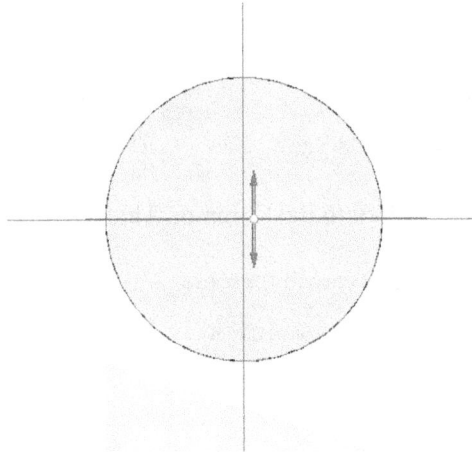

Click **Finish** and reset the view by hitting Ctrl+J

To rotate the part so we can see the diagonal face, click the **Rotate** tool in the bottom right corner of the screen.

An XYZ coordinate system should appear in the center of the screen. Click and drag the Z axis to the right until the screen looks like the following figure.

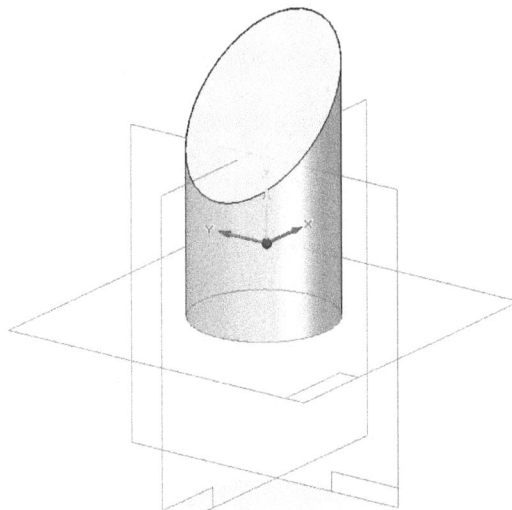

Right Click to exit the **Rotate** command and select **Thin Wall** form the Solids options box.

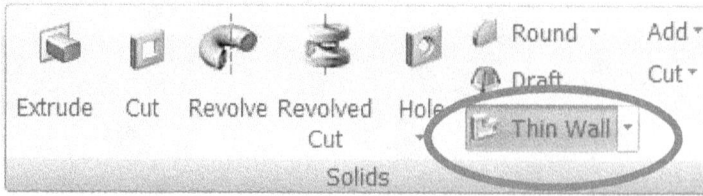

Select **Offset Inside** and type 0.125 into the Common Thickness box.

Hit **Enter** and select the **Open Face ('Chain')** as shown below.

Select Face

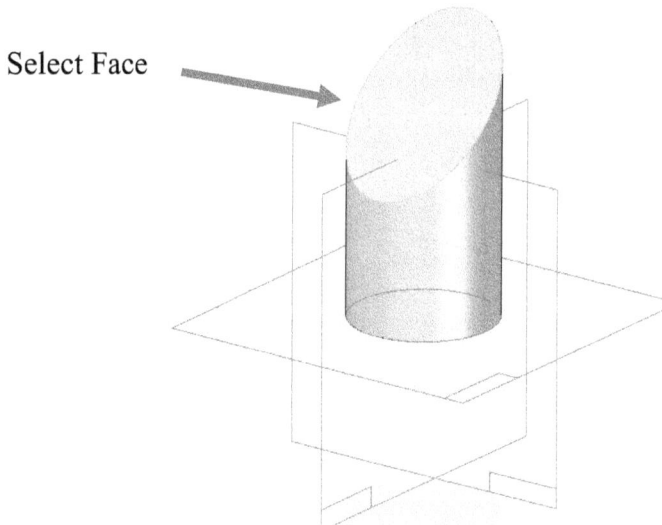

Click **Accept**, then **Finish.**

The finished part should look like the following model.

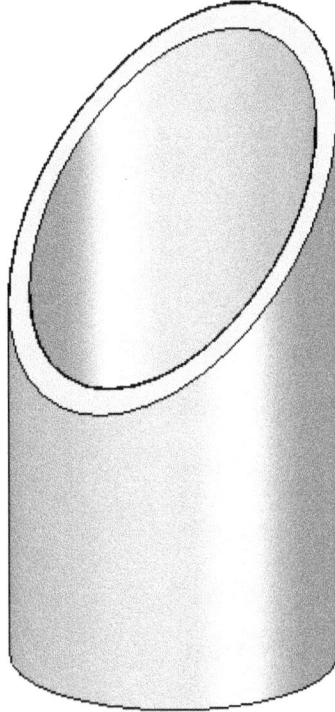

Exercise Complete

Exercise 24: Mounting Bosses

Create the following sketch and extrude by 0.5in. Then create a **Thin Wall** thickness of 0.125in as shown below.

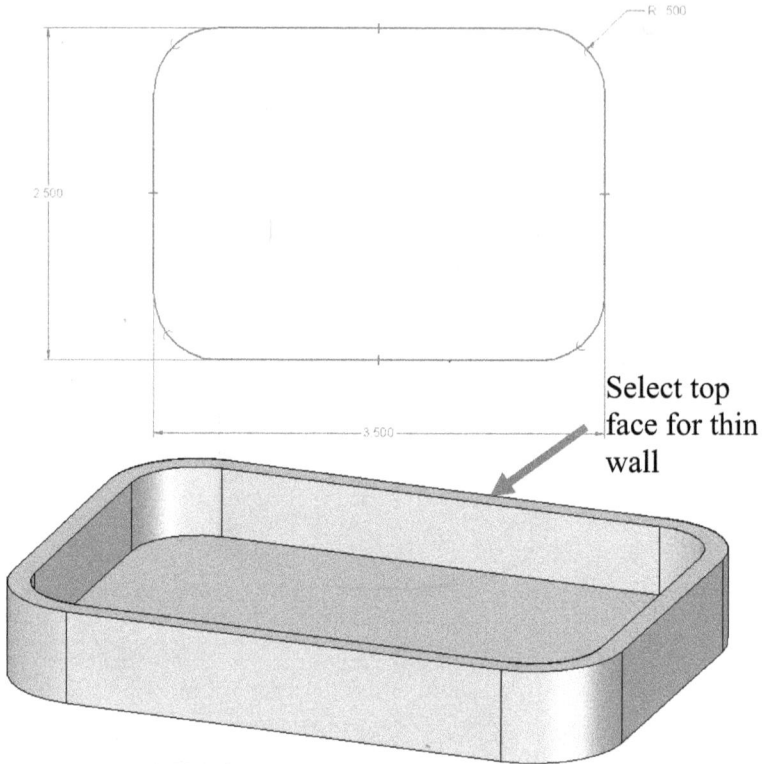

Select top face for thin wall

Click on the **Mounting Boss** tool under the **Thin Wall** dropdown menu shown below.

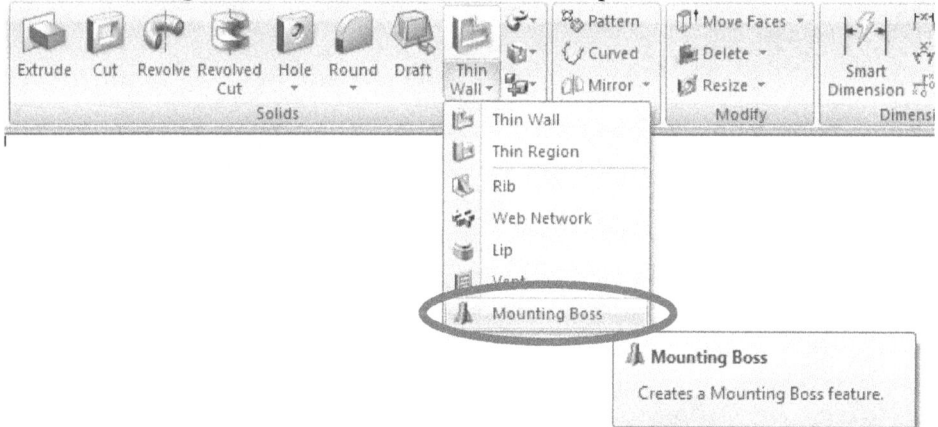

Click on the **Options** button in the upper left corner of the left menu.

Enter the values shown in the diagram below, then click **OK**.

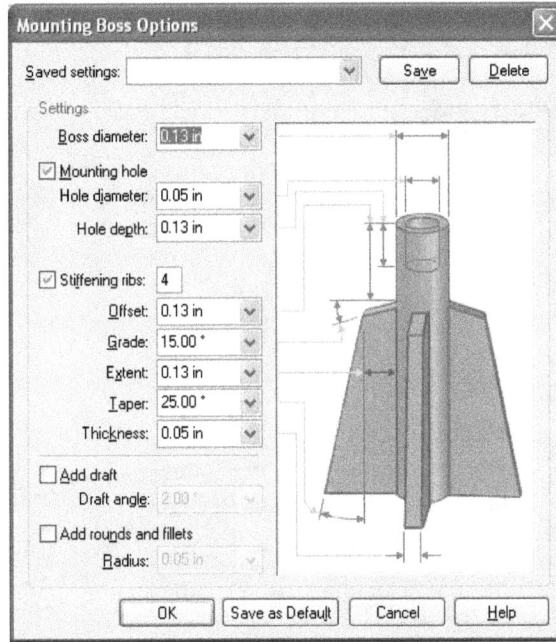

In the **Plane Step** make sure **Coincident Plane** is selected, and click on the top of the thin wall plane as shown below. Make sure the mouse hovers over the middle of the wall, otherwise the correct plane will not be selected.

Hover the cursor over one of the corners to make the center mark appear. Click on the center mark to place the boss. Repeat this for the other corners. **Close Sketch.**

To complete the bosses, in the Extent Step, click below the part to specify the direction.

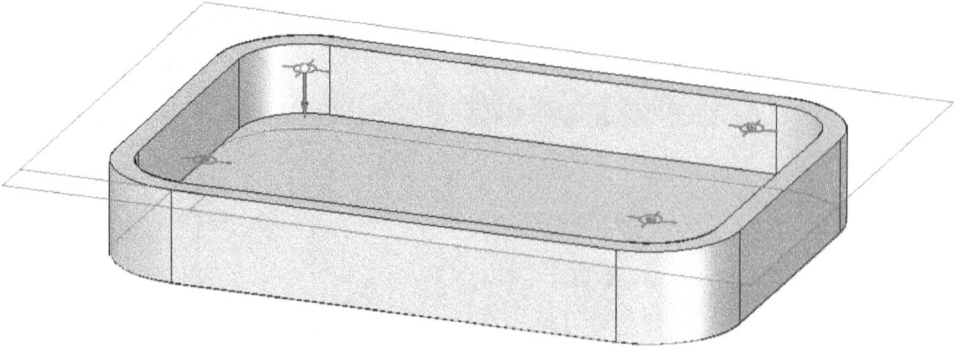

Click Finish.

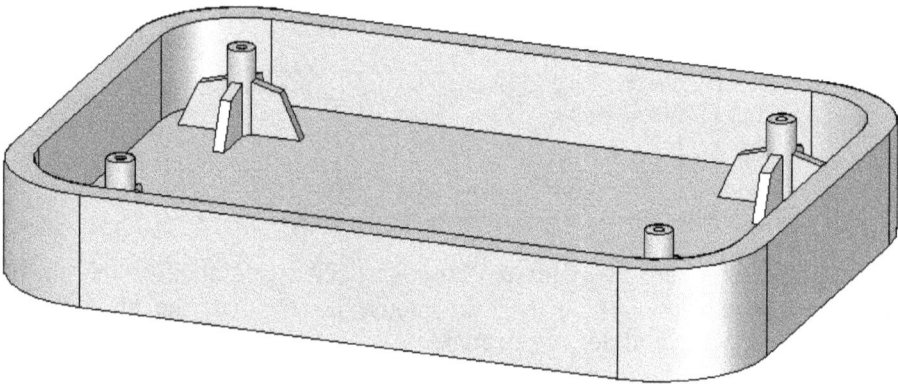

Exercise Complete

Exercise 25: Loft and Web Networks

In a ANSI part file, create a **sketch** on the **XY** Plane using the **Circle** tool in the **Draw** options box that looks like the figure shown below.

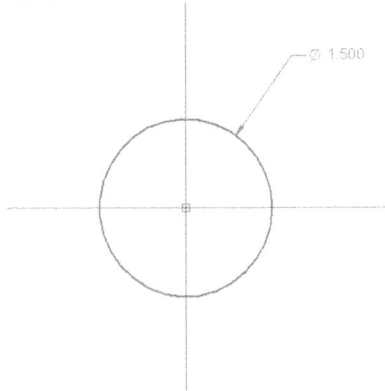

Close Sketch and create another sketch on a **plane** Parallel to the XY Plane offset by 1.0in.
Hint: Use the parallel datum creator.
Draw a concentric circle that has a 3.0in diameter, shown in the following figure.

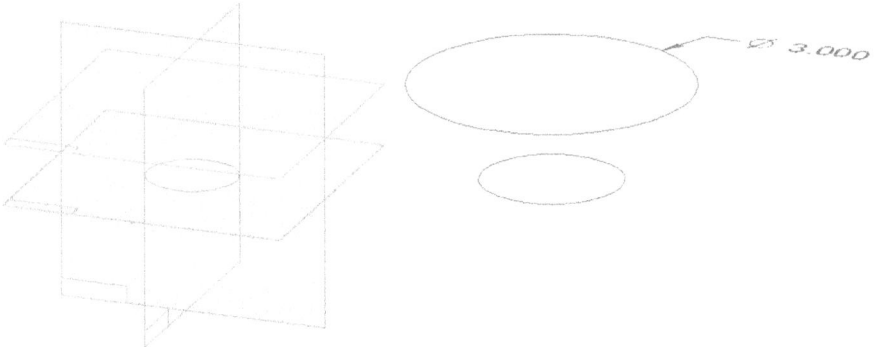

Close Sketch and create another sketch on a **plane** Parallel to the XY Plane offset by 2.0in
Draw a concentric circle that has a 4.0in diameter, shown in the following figure.

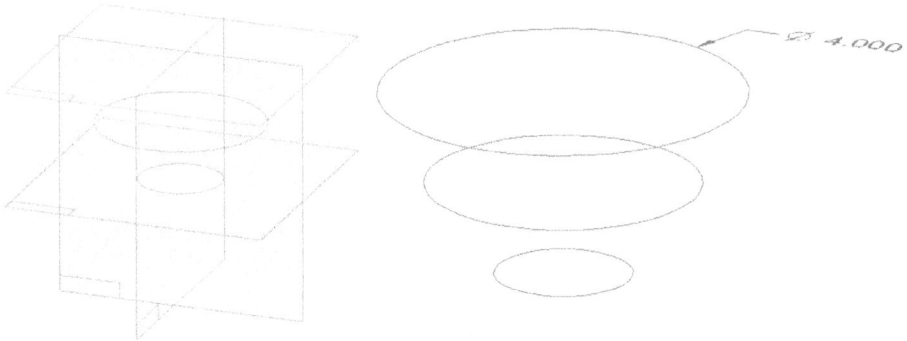

Close Sketch.

Click on the **Loft** tool under the 'Add' dropdown menu in the Solids options box.

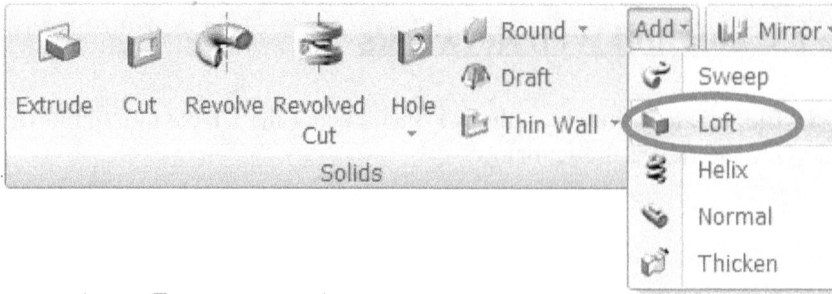

In the **Cross Section** Step ('Chain') click on all three circles.

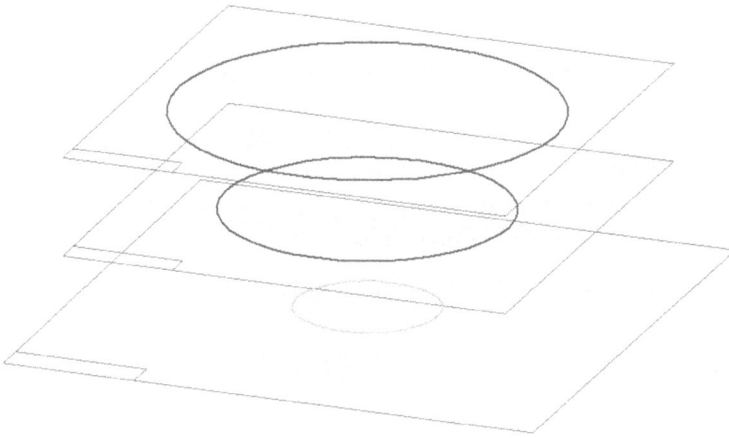

Click **Preview** then **Finish.**

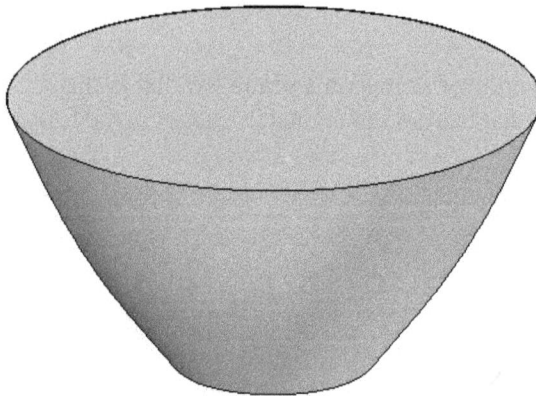

Click on the **Thin Wall** tool.

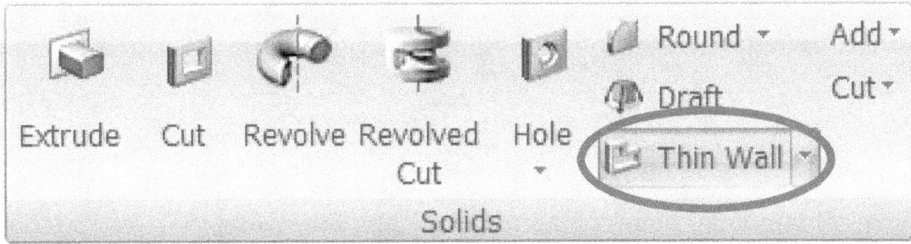

Enter a **Common Thickness** of 0.12 in and make sure Offset Inside is selected hit then **Enter** on the keyboard to accept.

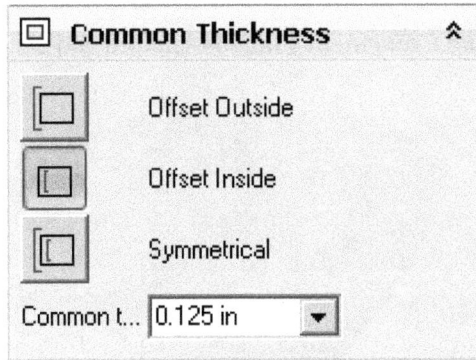

In the **Open Faces** ('Chain') Step click on the top face and click **Accept.**

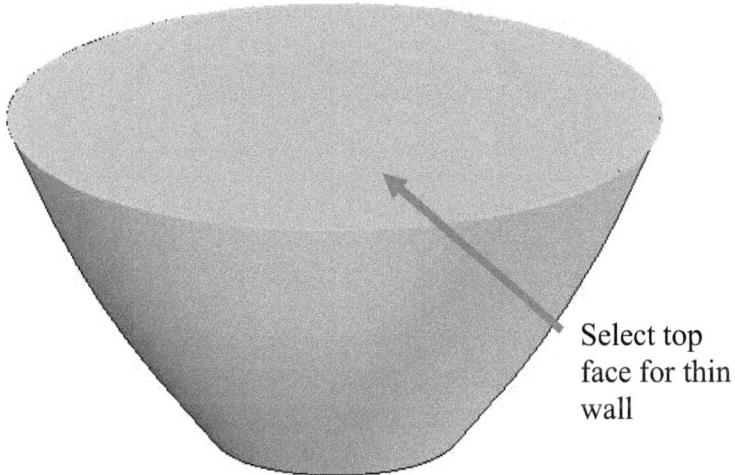

Select top face for thin wall

Click **Preview** then **Finish.**

Click on the **Web Network** tool under the **Thin Wall** dropdown menu in the Solids options box, as shown below.

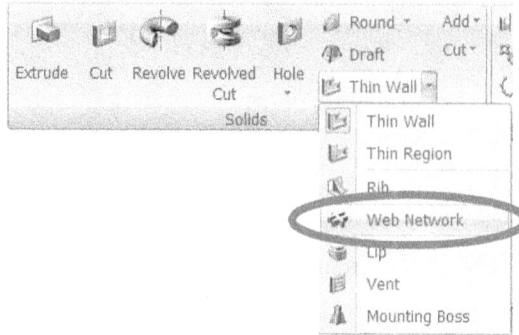

In the **Sketch Step** make sure **Coincident Plane** is selected and click on the top plane.

Create a **sketch** that looks like the figure below. Make the lines evenly spaced.

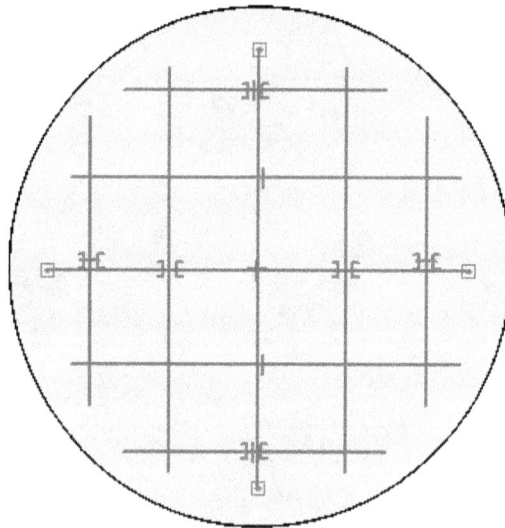

In the **Direction Step** select '**Extend Profile**' and '**Finite Depth**'. Enter a thickness of 0.25 in and a depth of 0.125 in.

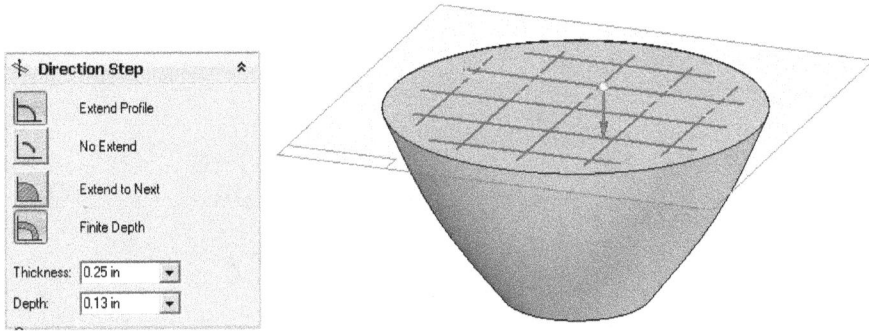

Click below the part to specify the direction.

Click **Finish.**

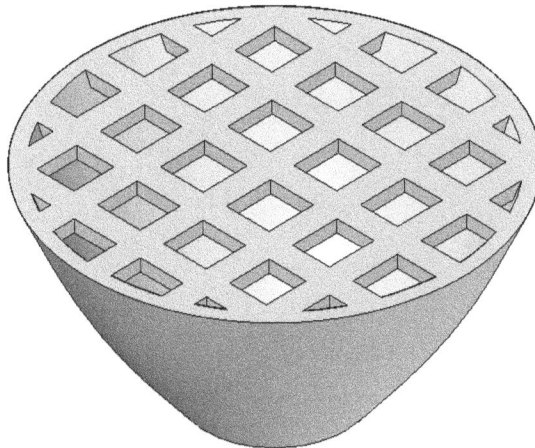

Exercise Complete

Exercise 26: Vent tool

In a new ANSI part file create an **extruded circle** that is 3.5in in diameter and extruded 0.5 inches like the figure shown below.

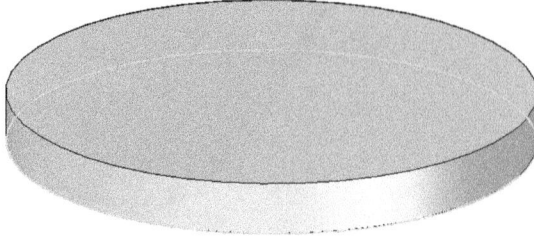

Use the **Thin Wall** tool. In the **Open Faces** Step, select the bottom face of the cylinder click **Finish.** In the Common Thickness Step, enter 0.125 in.

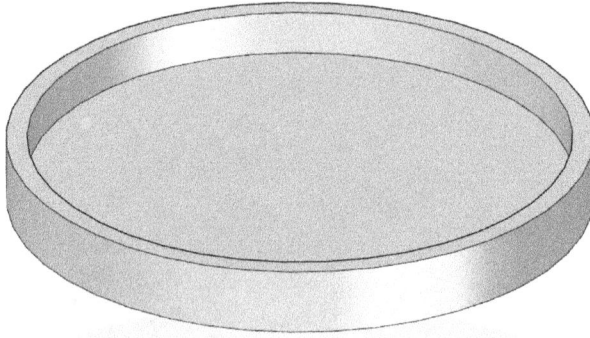

Using the **Round** tool in the Solids options box create a 0.25in round on the top edge of the circle as shown below.

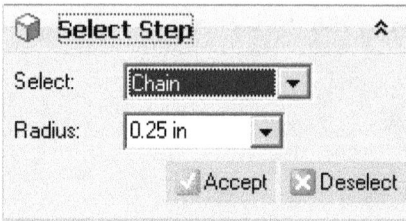

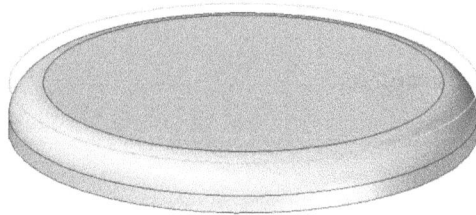

Create the following sketch on the top of the cylinder. The circle has to match the inside edge of edge of the round. Then create a series of equally spaced lines from the center.

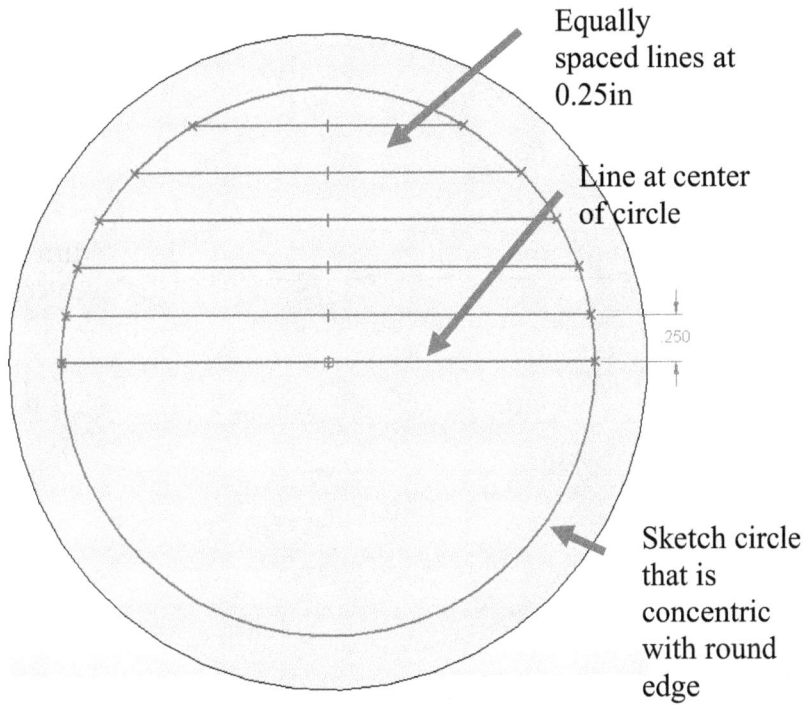

Equally spaced lines at 0.25in

Line at center of circle

.250

Sketch circle that is concentric with round edge

Click on the **Mirror** tool in the Draw options box.

Draw

Mirror the lines shown below across the center line as shown below. You can achieve by dragging a box over all the lines that you with to mirror and selecting the centerline to mirror about.

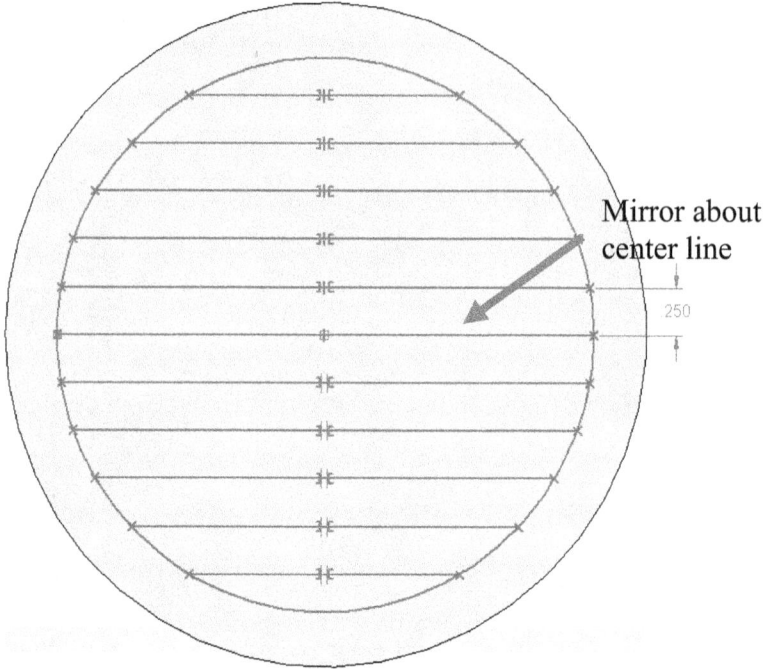

Mirror about center line

250

Note: select the multiple features by holding shift, or click and drag a box around all the features that you wish to mirror.

Click on the **Rotate** tool under the Move dropdown menu.

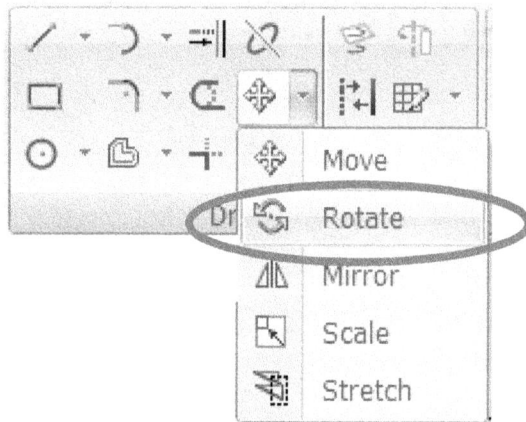

Highlight the entire sketch, click in the center of the circle and then click the perimeter of the circle in order to select the information required.

In the menu on the left, enter **45** into the **angle box** and hit **enter,** then click on the upper right side of the screen to place the rotation as shown below.

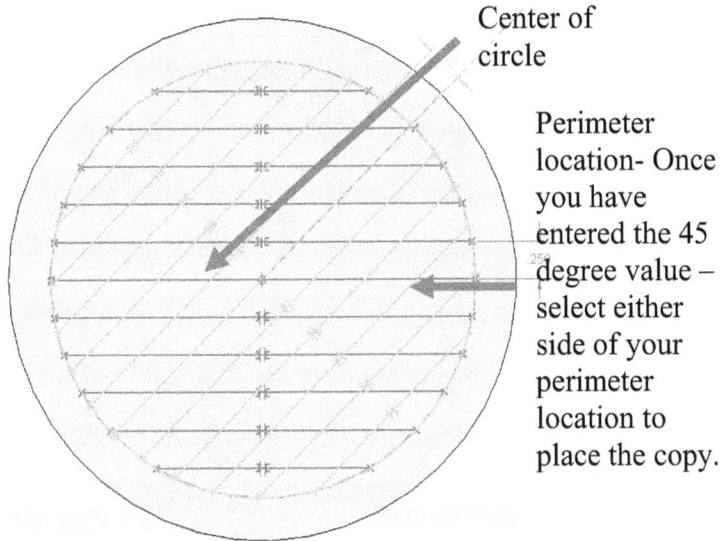

Center of circle

Perimeter location- Once you have entered the 45 degree value – select either side of your perimeter location to place the copy.

Right click on the screen to exit **Rotate** and **Close Sketch.**

Click on the **Vent** tool under the Thin Wall dropdown menu in the Solids options box.

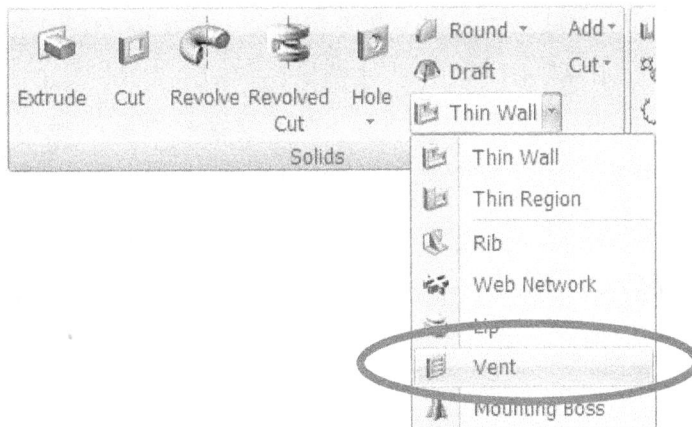

An options box that looks like this should appear:

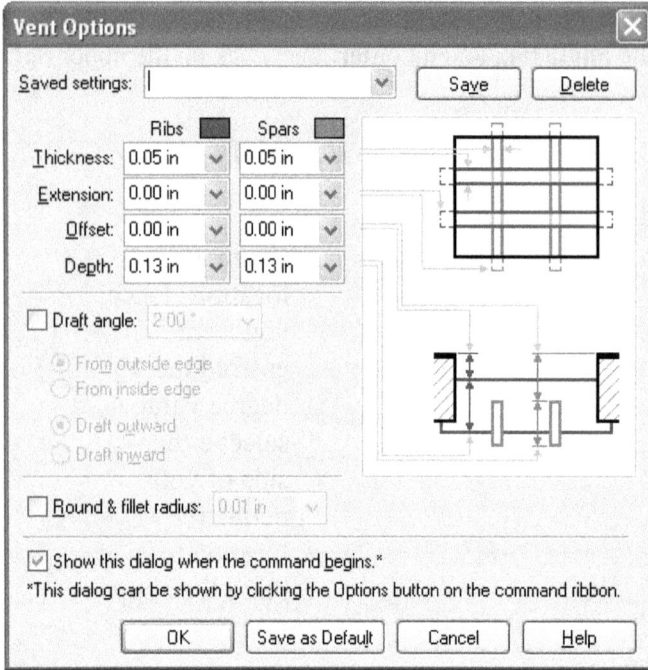

Enter **0.05 in** for the **Thickness** of both the **Ribs** and the **Spars**

Enter **0.125 in** for the **Depth** of both the **Ribs** and the **Spars**

Click OK.

In the **Select Boundary** Step click on the outer circle and click **Accept**.

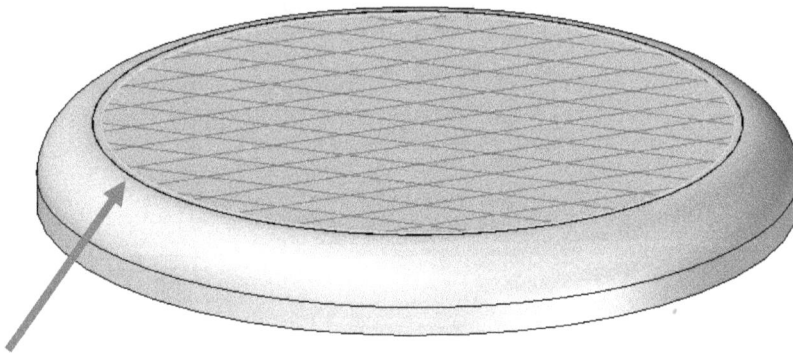

Select outer circle

In the **Select Ribs** Step select all the **horizontal lines** and click **Accept.**

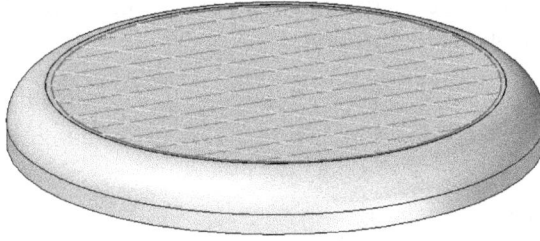

In the **Select Spars Step** select all the **angled lines** you created and click **Accept**.

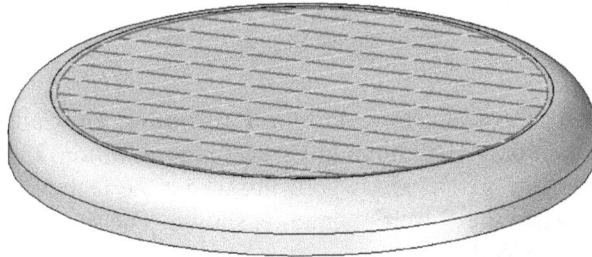

Select **Through All** in the **Extent Step** and click below the part to specify the direction.

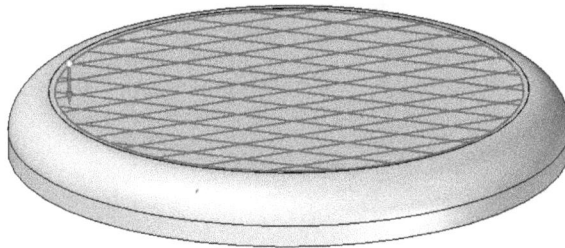

Click **Finish** to exit the Vent tool. The part will now look like the following figure.

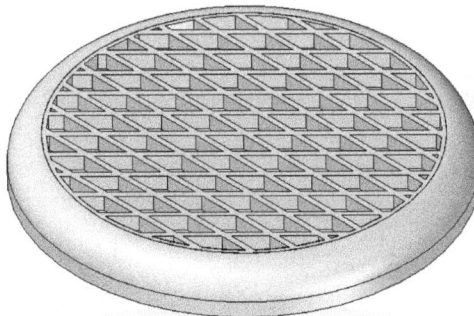

Note: Look at the underside of the part and notice that the round did NOT create a universal thickness on the inside. Notice the sharp edge in the following figure. This is due to the round being created before the Thin Wall operation.

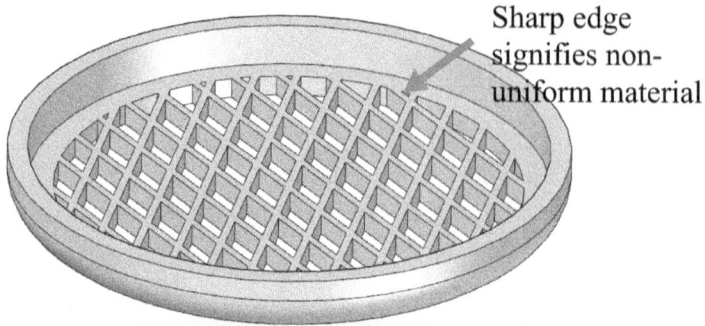

Sharp edge signifies non-uniform material

Here is a great trick to fix this problem. Go to the **Pathfinder**, click and hold the round with the left mouse button, then drag (still holding the left mouse button) the Round feature up to the Protrusion. Notice a little 'Insert' arrow showing where the feature will be positioned. Let go of the mouse button when the 'Insert' symbol is shown. The feature's time of creation has been moved before the Thin Wall.

Insert symbol

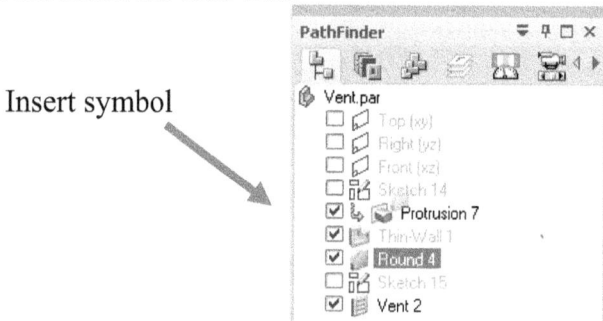

The model will look like the following figure (now with uniform wall thickness i.e. a Round on the inside too!).

Round on inside

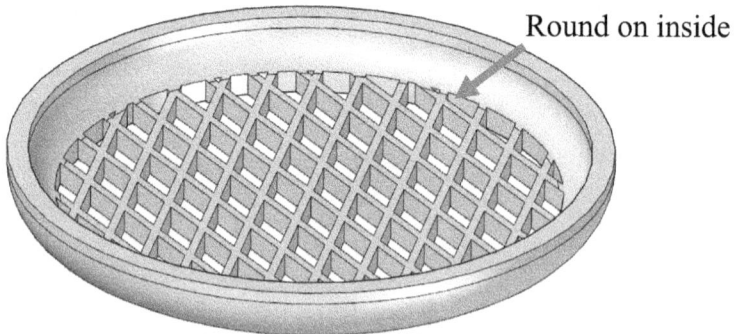

Exercise Complete

Exercise 27: Rectangular Hole Pattern by Grouping Faces

In this exercise we are going to group a number of features from the pathfinder to create our 'unique feature', then instance that grouped feature in a rectangular pattern. Start a New ANSI part file.

Click on a datum plane and **sketch** a **rectangle** that is 10in x 6.5in. **Extrude** this by 1.5in to create a block. Create another **sketch** on the top face of that block. Draw the following sketch in the bottom left hand corner of that face.

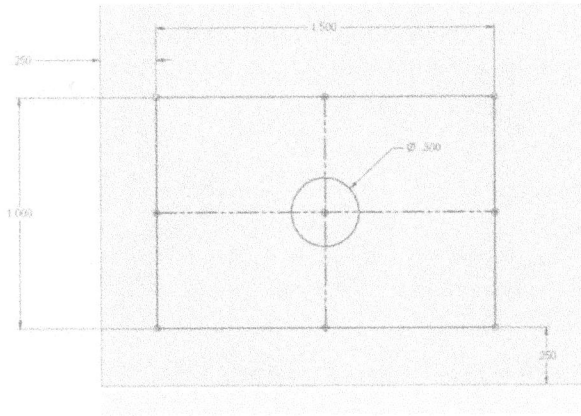

Select the **Cut** tool. Pick the rectangle from the sketch and cut a distance of 0.5in into the block that was created, as shown in the following figure.

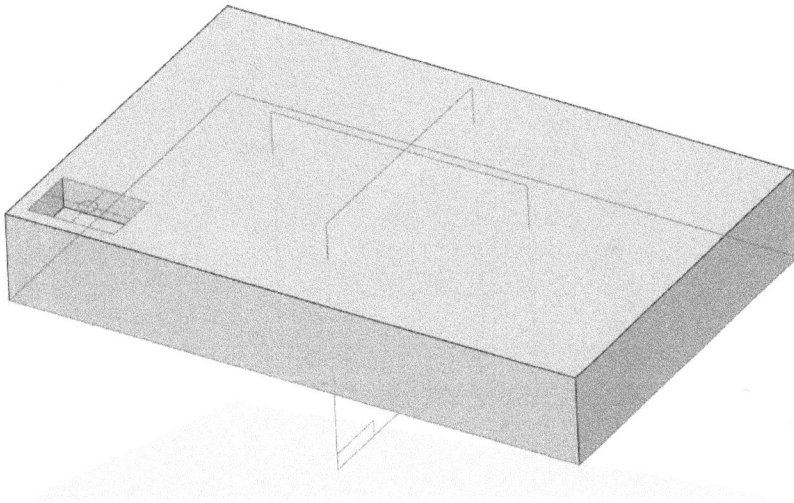

Now click on the **extrude** tool and extrude the circle from the sketch symmetrically 1.1in, as shown in the following figure. Add 0.15in **rounds** to all the inside edges of the unique feature.

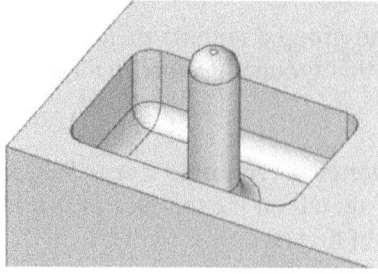

In the **pathfinder,** highlight all the features used to create our **'unique feature'** by holding the **shift key** and click on **group**, as shown below.

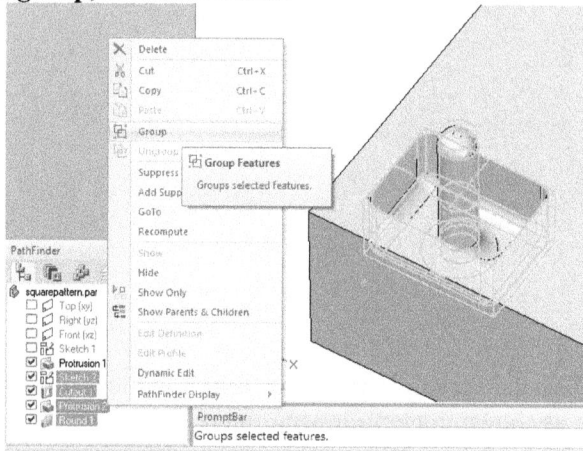

The **pathfinder** will now show the original features contained under one grouped feature heading, as shown below.

Select the **pattern** tool under the Home tab. In the **Select Step** choose the feature option then pick our grouped feature in the pathfinder, as shown below. **Accept** when done.

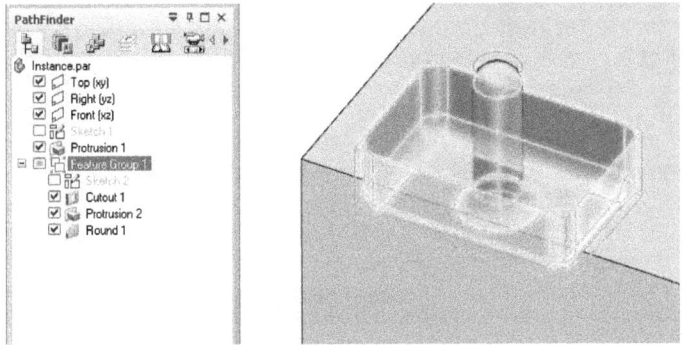

The next step is to create the pattern. In the **Sketch Step**, select the top face of the part where the unique feature exists. This action will open the sketch mode. Use the **Rectangular Pattern** button and draw the rectangle by clicking the center of unique feature and then click near the opposite corner of the block. **Dimension** the sketch as shown in the following figure. In the **edit definition box** type in the **X** and **Y** number as 4, this is how many copies of the feature we want.

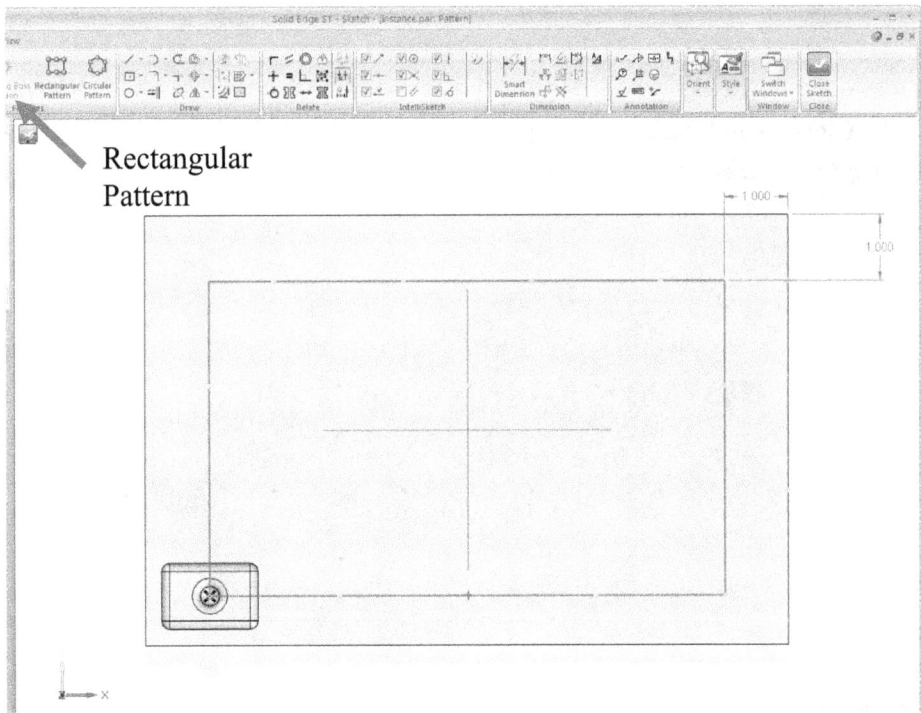

Rectangular
Pattern

Click on the **stagger options**, also found in the **edit definition box**, and select **'Row'** as the stagger option and **1/2 offset** as shown in the following figure.

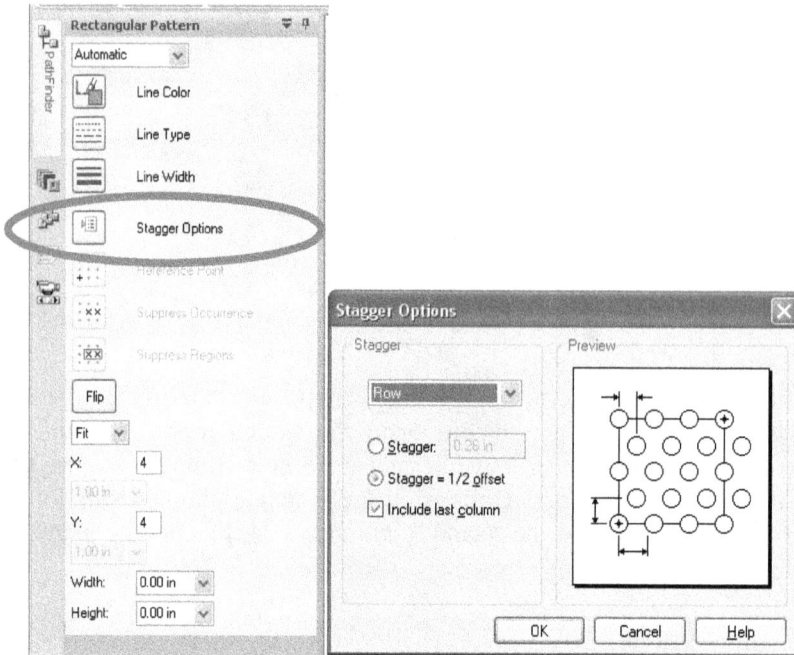

Click on the **Close Sketch** button to complete. *Note: If the pattern goes outside of the solid, SE will prompt with an error message. Click 'OK' on the error message and those particular copies will not be created once the operation is complete.*

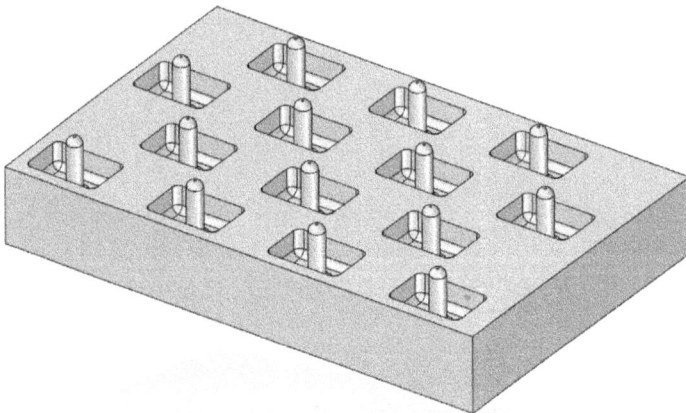

Exercise Complete

Exercise 28: Circular Hole Pattern

Create a **new ANSI part file**. Click the **Extrude** button and select the **XY plane** to sketch on.

Create a **sketch** with two concentric circles with diameter of 4in and 8in shown below. **Close Sketch** and **Extrude** above the sketch plane 1.25 in.

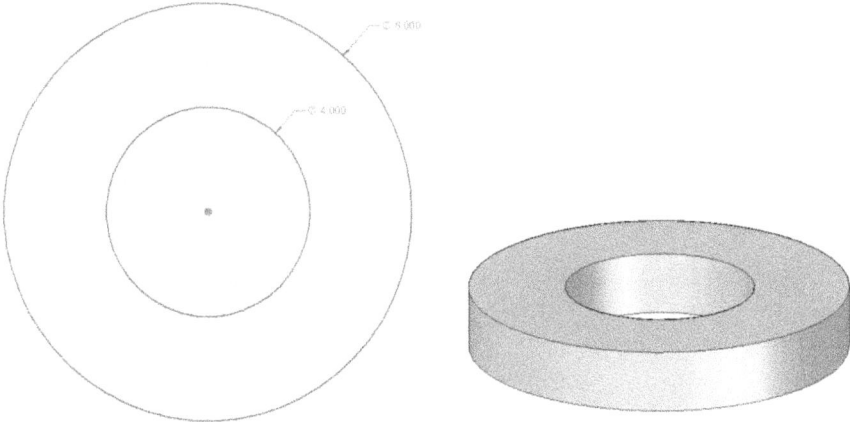

Click on the **Hole** tool in the Solids options box.

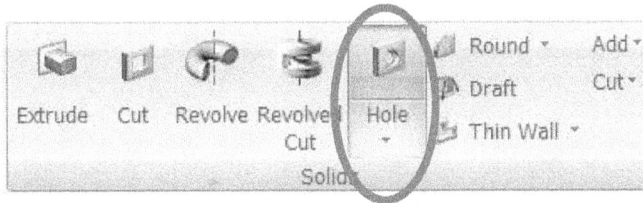

In the **Plane Step** create a tangent plane tangent to the outside cylinder at an angle of 90°.

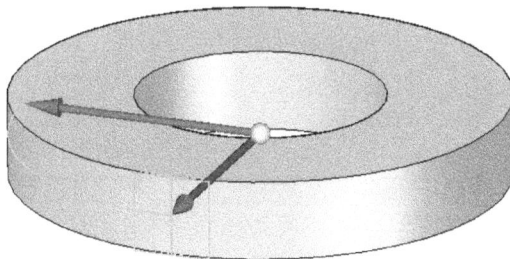

In the **Hole Step** place a hole as shown on the plane you created.

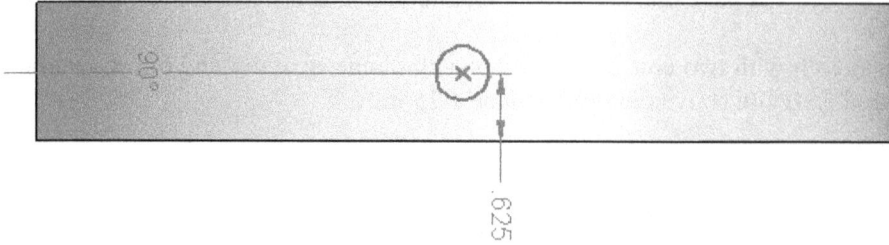

Close Sketch.

Click on the **Options** button in the upper left corner of the left menu. Set the Diameter to 0.5 in and the Hole Depth to 0.5in.

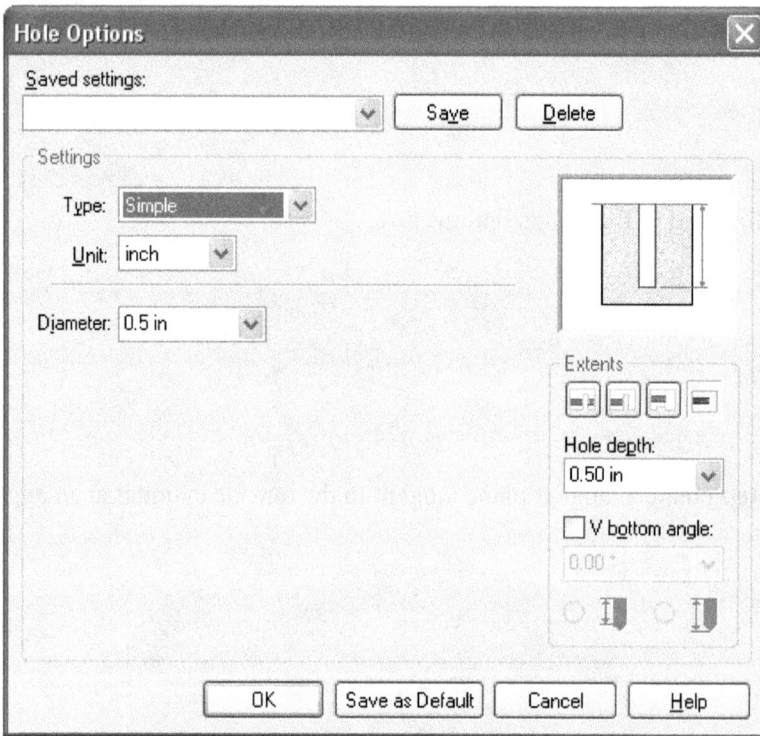

Click **OK.**

In the **Extent Step** click **Finite Extent** and enter a distance of 0.5in then click the screen so the hole goes into the solid rather than away from it.

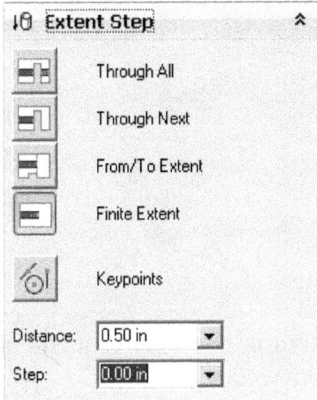

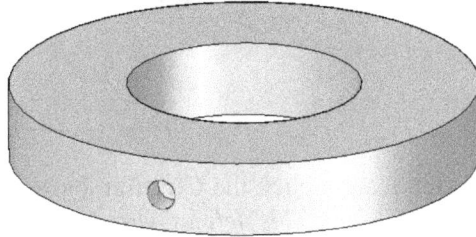

Click **Finish**.

Click on the **Pattern** tool in the Pattern options box.

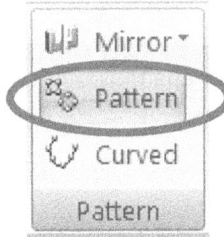

In the **Select Step** make sure **Feature** is selected and click on the hole feature then, **Accept.**

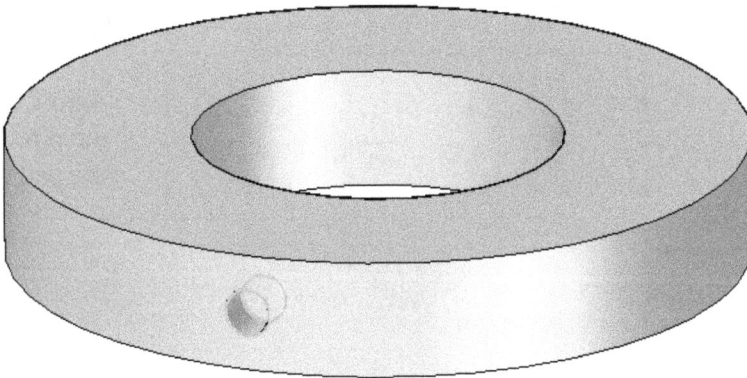

In the **Sketch Step** make sure coincident plane is selected and click on the top of the cylinder.

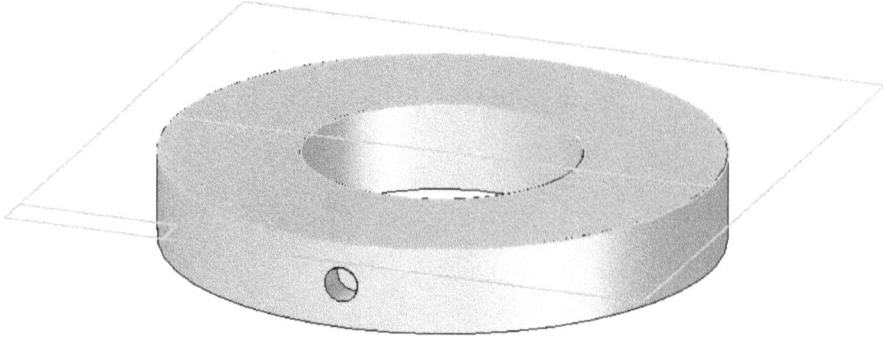

Once in the sketch feature click the **Circular Pattern** tool in the features menu on the Home tab, as shown below. *Note: This defines what type of pattern to create.*

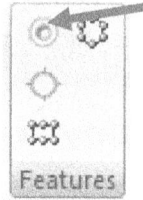

Click on the center mark of the cylinder and the outer edge then click to specify the direction.

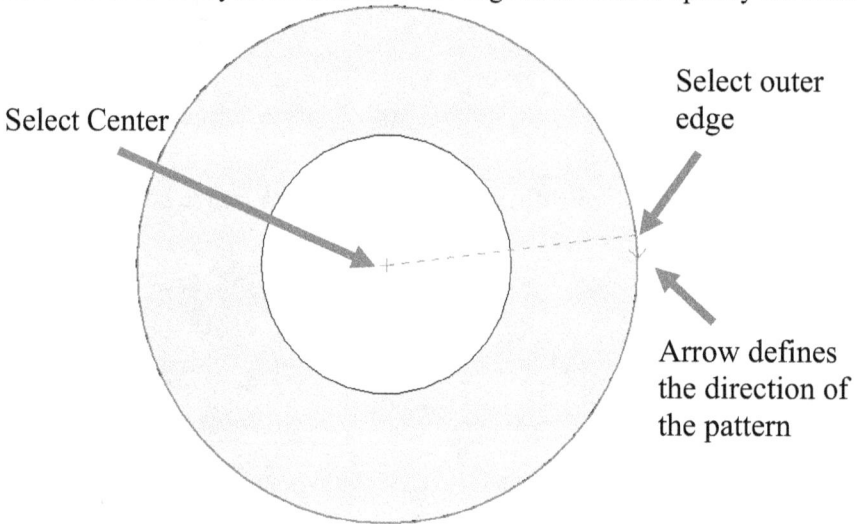

Select Center

Select outer edge

Arrow defines the direction of the pattern

Enter 12 into the **Count box** on the left and hit **Enter** on the keyboard.

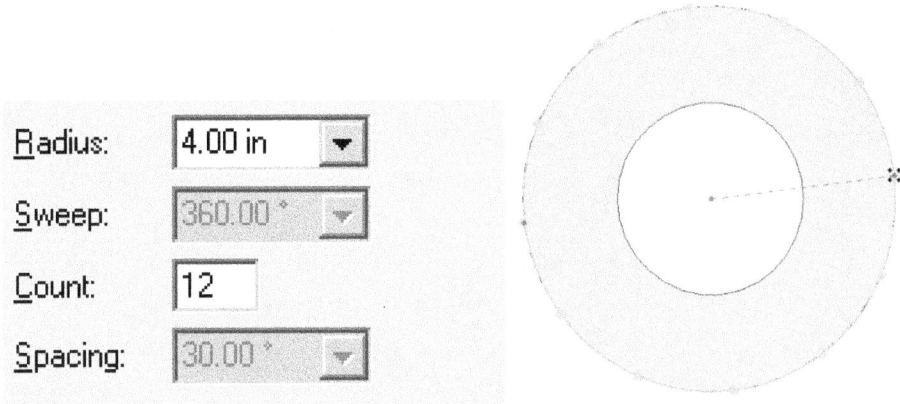

Close **Sketch.**
In the **Draw Profile Step** click on the **Smart button** then click **Finish.**

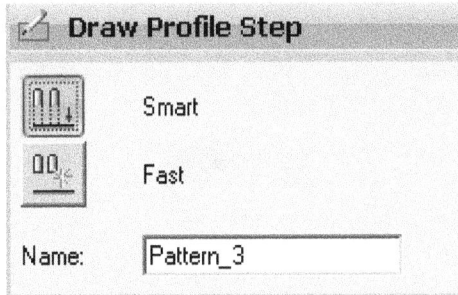

The model will now look like the following figure.

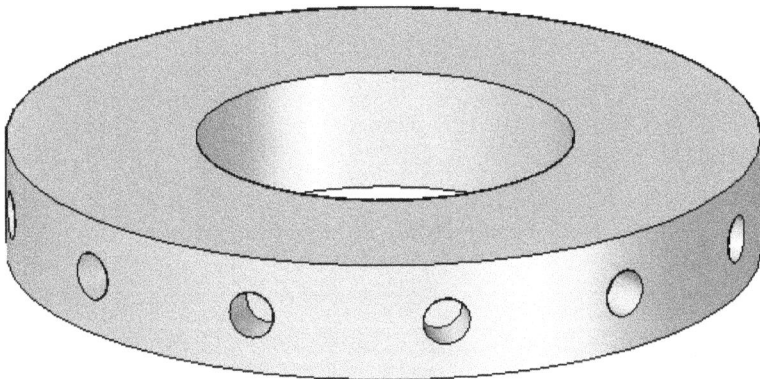

Click the **Extrude** button and select on the bottom face of the part to sketch on.

Draw a circle with a diameter of 6in that looks like the figure shown below.

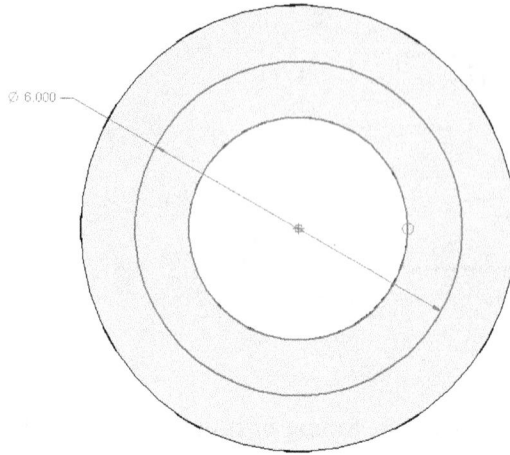

Ø 6.000

Close Sketch.

In the **Extent Step** enter a distance of 1.5 in and extrude below the sketch plane, click **Finish.**

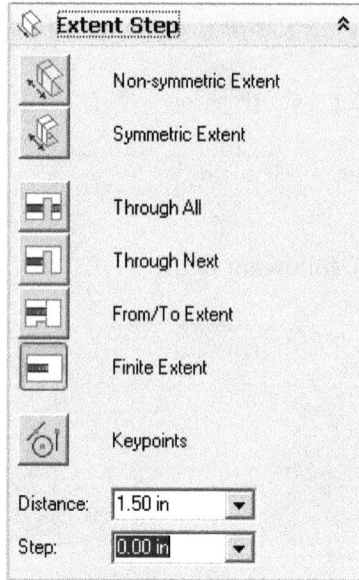

Extent Step	☆
Non-symmetric Extent	
Symmetric Extent	
Through All	
Through Next	
From/To Extent	
Finite Extent	
Keypoints	
Distance:	1.50 in
Step:	0.00 in

The part should look like the following figure.

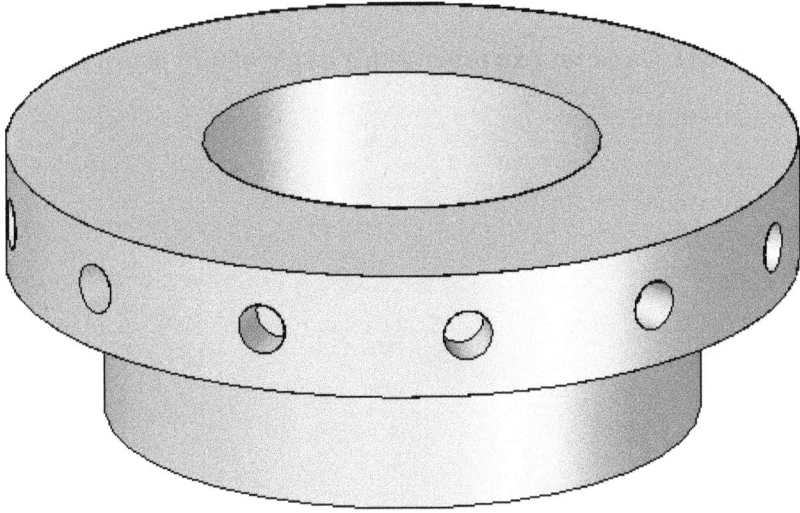

Exercise Complete

Exercise 29: Curved Pattern

Start a **new ANSI part**. Create an **Extrusion** with a height of 0.25 in that looks like the figure shown below.

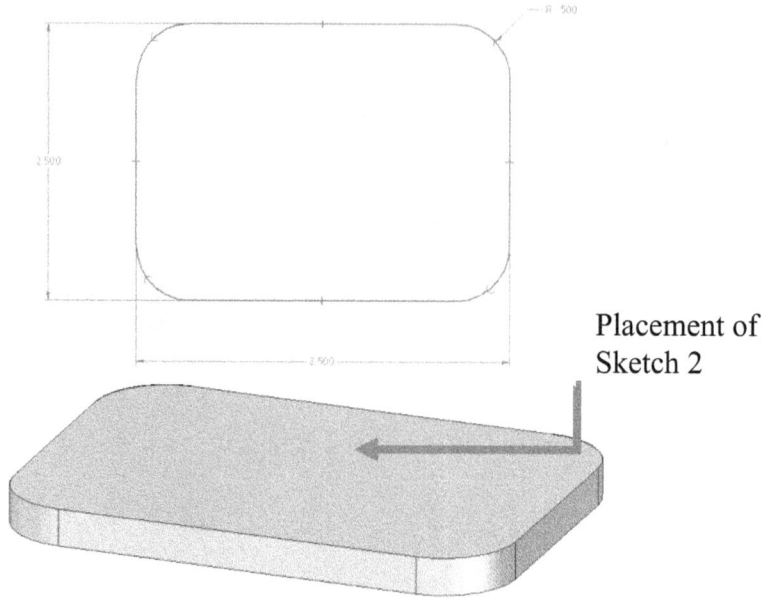

Placement of Sketch 2

We are going to pattern a hole along a curve. To create the curve let's make use of the already existing geometry and simply offset the edges. *Note: This is quicker technique than recreating the lines and constraints.*

Create a **sketch** on the top surface of the solid as shown above.

Click on the **Include** tool in the Draw Tool bar, as shown in the following figure. *Note: The Offset button is only for 2D elements, not edges.*

Offset - for 2D elements only - Great for quick offsets of existing curves.

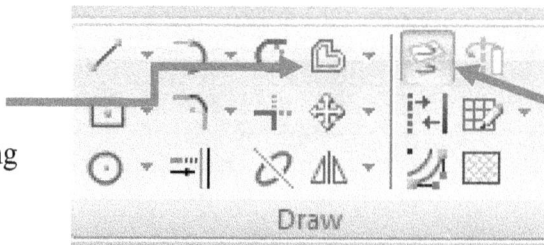

Include - allows use of 3D elements in the sketch environment

SE will open the **'Include Options'** window as shown in the following figure. Check the box 'include with offset' so we can offset as we include the edges. *Note: this saves us an additional step of offsetting the included curves.*

After clicking **OK**, select the edges to include (select all edges) Type the distance of the **offset** (0.3in) and click to specify the direction of the offset. Select the direction by clicking either side of the included edges - choose towards the center of the part as shown in the following figure.

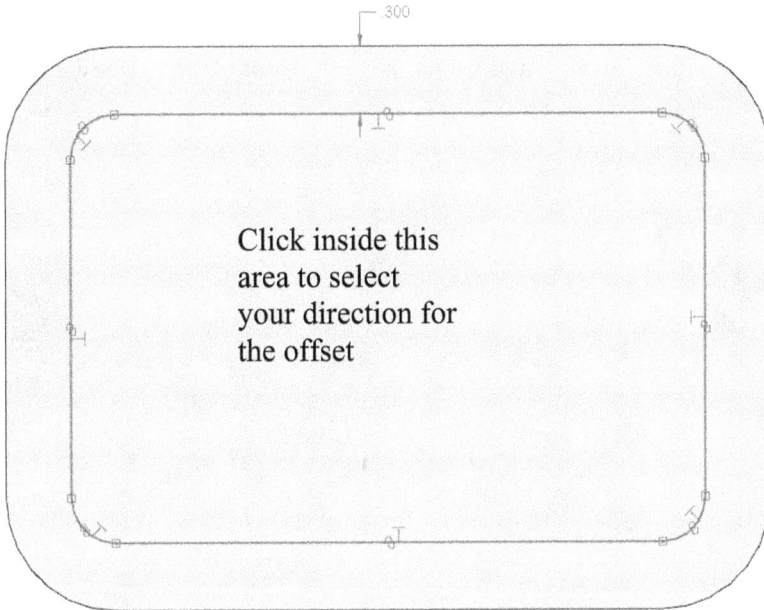

Click inside this area to select your direction for the offset

Curved patterns behave more predictably when the pattern feature is placed at the beginning of a line entity. **Delete** one of the lines as shown in the following figure by simply selecting the line and clicking 'delete' on the keyboard.

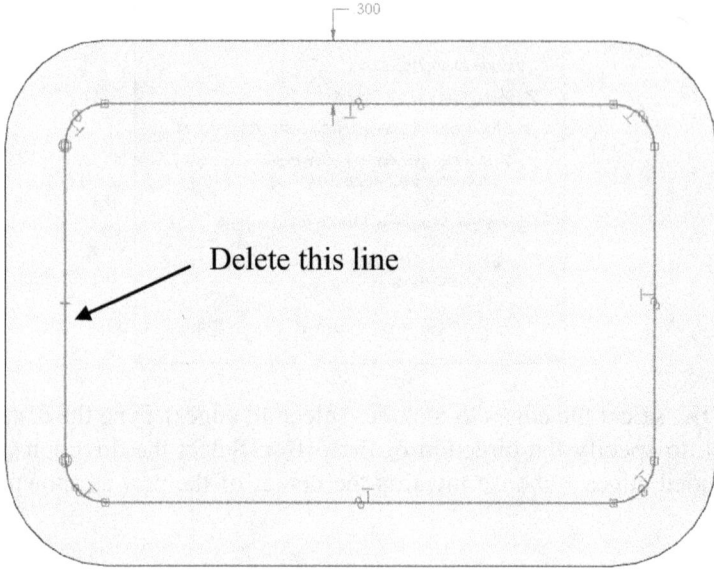

Delete this line

Create two lines and make them equal length using the **equal** tool, as shown below.

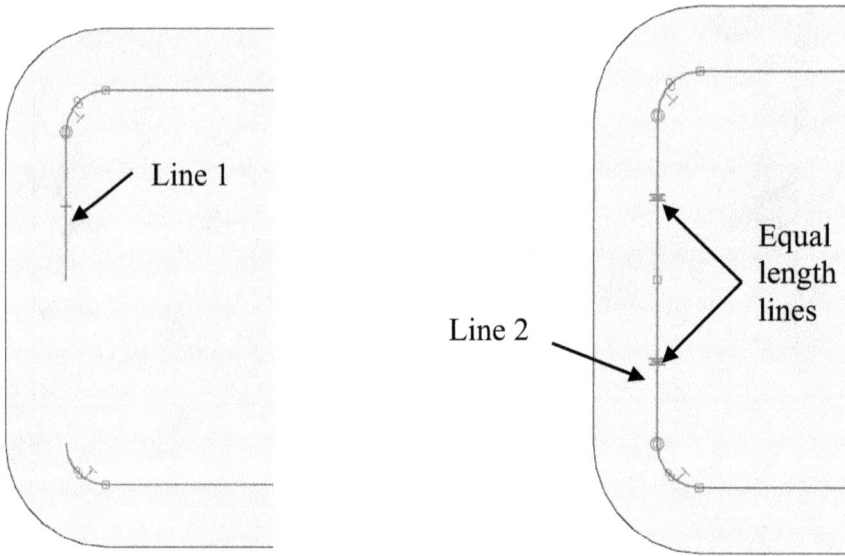

Line 1

Line 2

Equal length lines

The model will now have the following offset. **Close Sketch.**

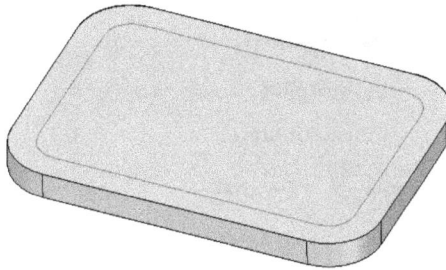

Click on the **Hole** button in the Solids options box and choose the top face of the solid.

Click on the **Options** button in the upper left corner of the left menu. Enter a Diameter of 0.125in. Click **OK.**

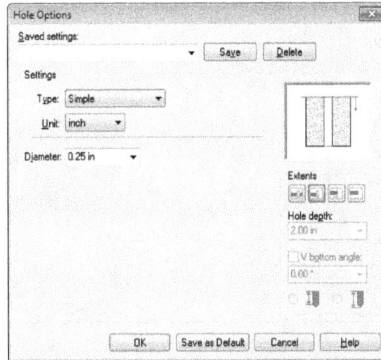

Place the hole randomly on the face and then use the **connect** tool to join the center of the hole to the end point of one of the lines that you just created. The hole should be placed as shown in the following figure.

Close Sketch.

In the **Extent Step** make sure **Through All** is selected and click below the solid to define the direction, click **Finish** to complete.

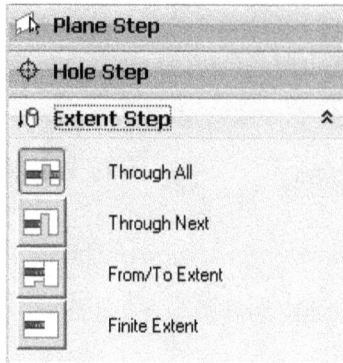

Choose the direction of the hole, as shown in the following figure.

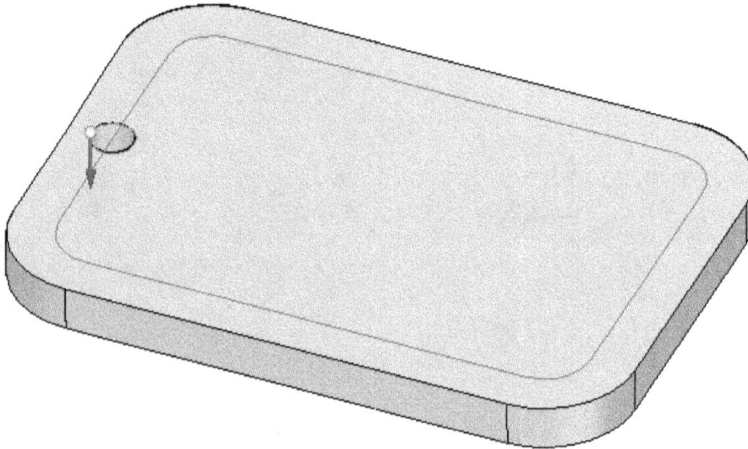

Click on the **Curved** tool in the **Pattern** options box, as shown below.

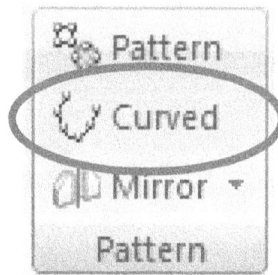

In the **Select Step** make sure 'Feature' is selected, click on the hole then, **Accept.**

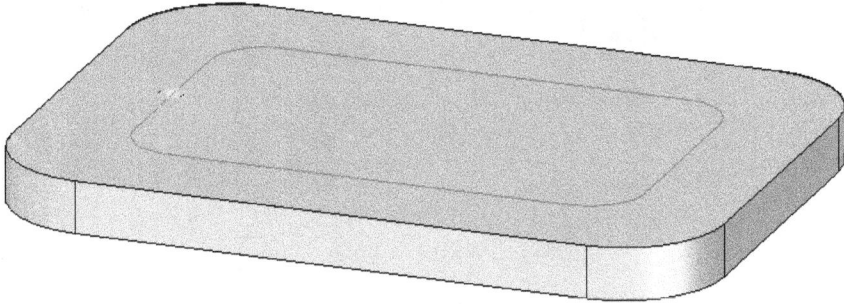

In the Select Curve Step, under **Pattern** Curve, click on all the curves in the sketch then enter 25 into the Count box.

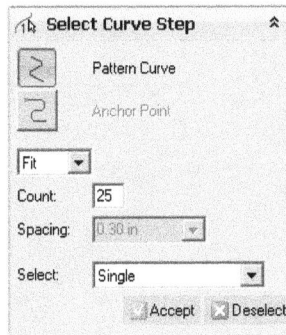

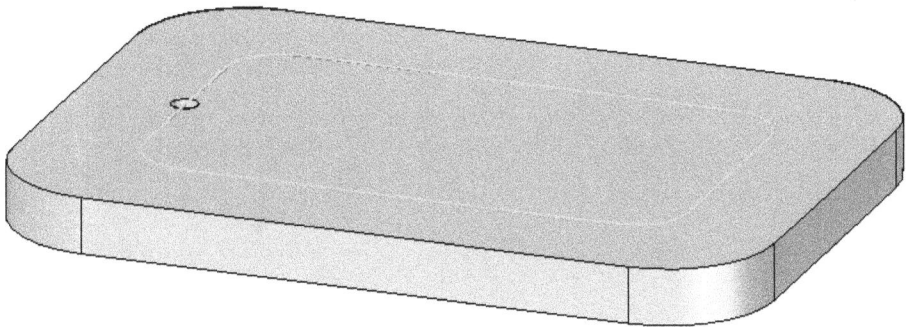

Click Accept.

Click on the center mark of the hole to select an **Anchor Point** and click to specify the direction (clockwise or anti-clockwise of the hole).

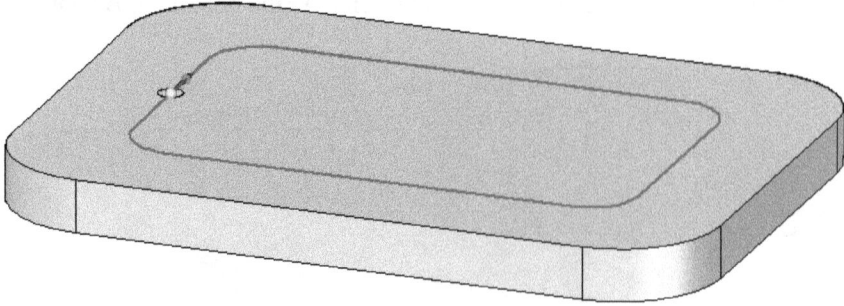

In the **Advanced** definition Step make sure **Full** and **Curve Position** are selected and select on the **Smart** button.

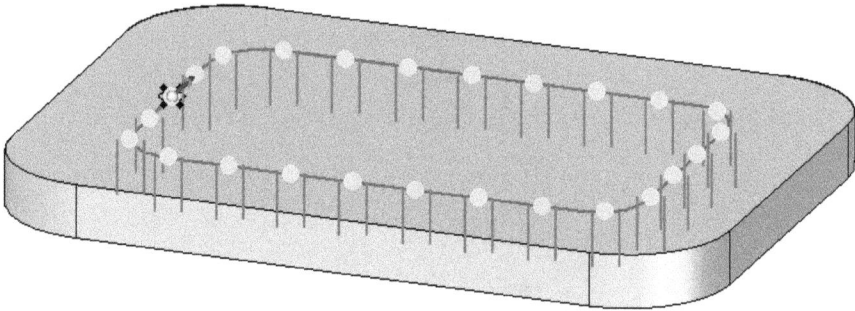

Click **Preview** then **Finish.**

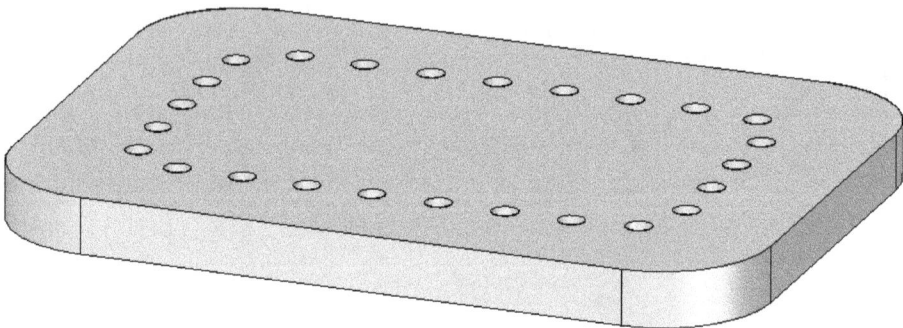

Exercise Complete

Exercise 30: Helical Cut Neural

Click the **Extrude** button and select the **XY plane** to sketch on.

Sketch a 3in diameter circle as shown below.

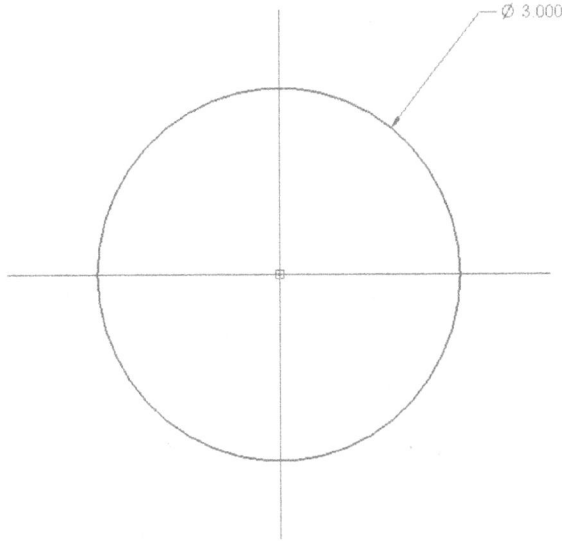

Ø 3.000

Close **Sketch** and **Extrude** symmetrically about the sketch plane a distance of 3.00in.

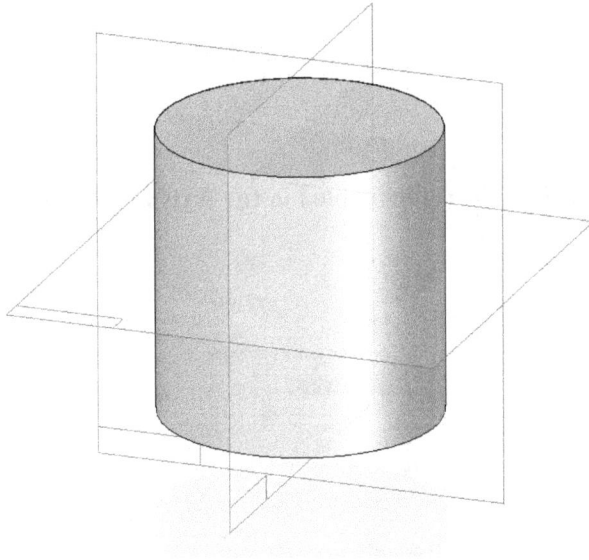

Sketch a triangle with equal length sides of 0.1in and a separate vertical line that spans the cylinder as shown below. *Note: Make sure the triangle goes just outside the area of the cylinder and the end point of the line touches the bottom edge of the cylinder, as shown below.*

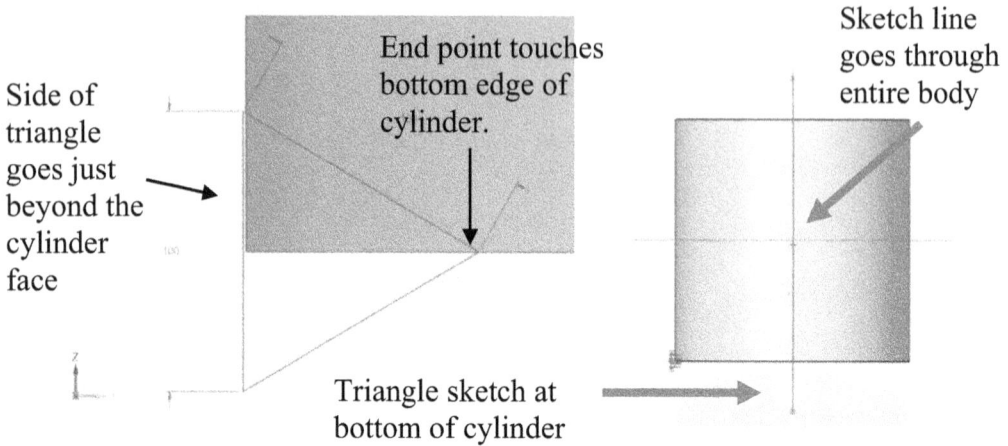

End point touches bottom edge of cylinder.

Side of triangle goes just beyond the cylinder face

Sketch line goes through entire body

Triangle sketch at bottom of cylinder

Close Sketch.

Click on the **Helix** tool under the Cut dropdown menu in the Solids options box.

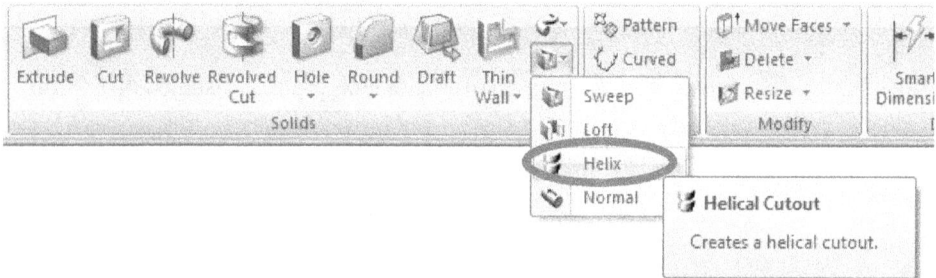

Make sure **'Select from Sketch'** is highlighted in **the Axis and Cross Section** Step in the left menu.

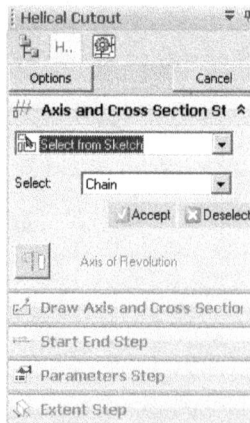

Select the Triangle as the **Cross Section,** and the Line as the **Axis of Revolution**, shown in the following figure.

In the **Start End Step** select the lower point of the **Axis of Revolution** (line).

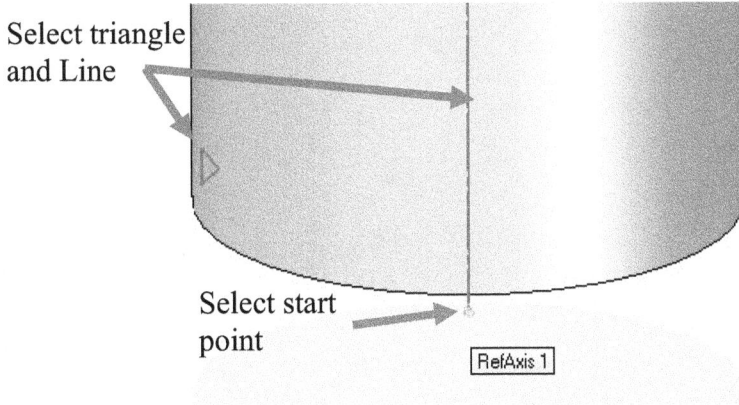

Select triangle
and Line

Select start
point

RefAxis 1

Enter 12in as the **Pitch**.

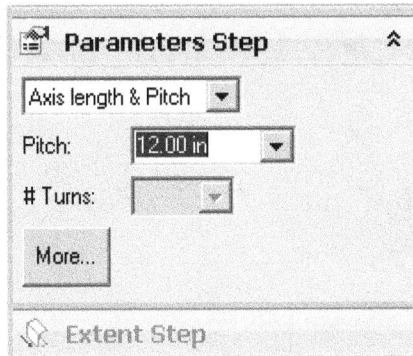

Parameters Step

Axis length & Pitch

Pitch: 12.00 in

Turns:

More...

Extent Step

Hit **Enter** on The keyboard to confirm.

In the **Extent Step**, click the **'From/To Extent'** button.

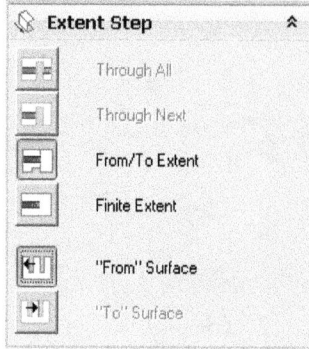

Select the bottom of the cylinder as the **'From Surface'** and select the top as the **'To Surface'**.

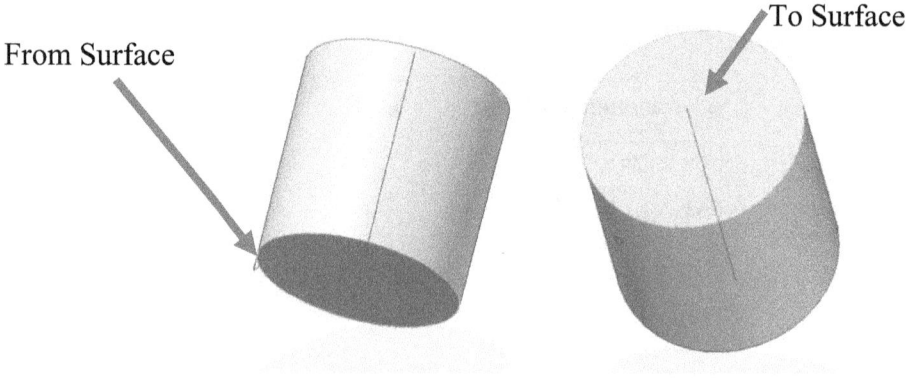

From Surface

To Surface

Hit **Enter** on The keyboard twice. The **Helical Cut** should look like the following figure.

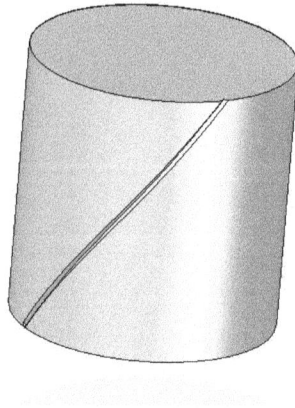

Click the **Mirror Copy Feature** tool under the Mirror dropdown menu.

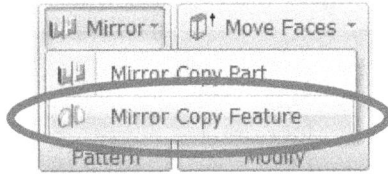

In the **Select Features Step** in the left menu, click on the helical cut and, **Accept.**

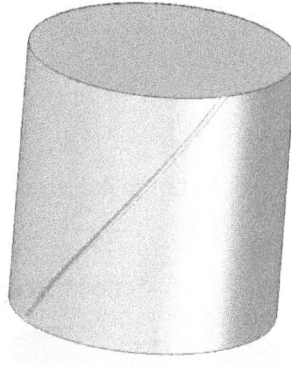

In the **Plane Step** make sure **Coincident Plane** is selected in the dropdown menu and select the Smart button.

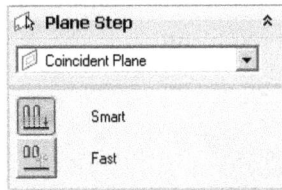

Click on the sketch plane then select the plane to mirror about.

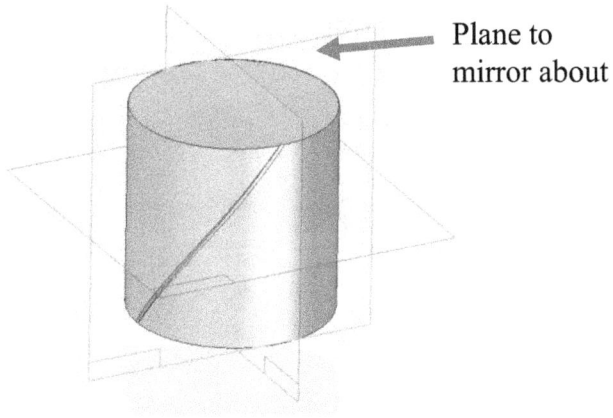

Plane to mirror about

The Mirror should look like the following figure.

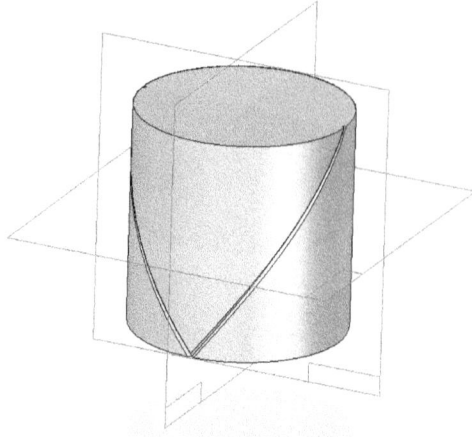

Click on the **Pattern** button. In the **Select Step** make sure Feature is selected in the drop down menu.

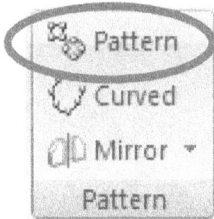

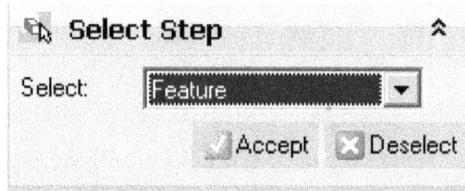

Click both the original and the mirrored feature.

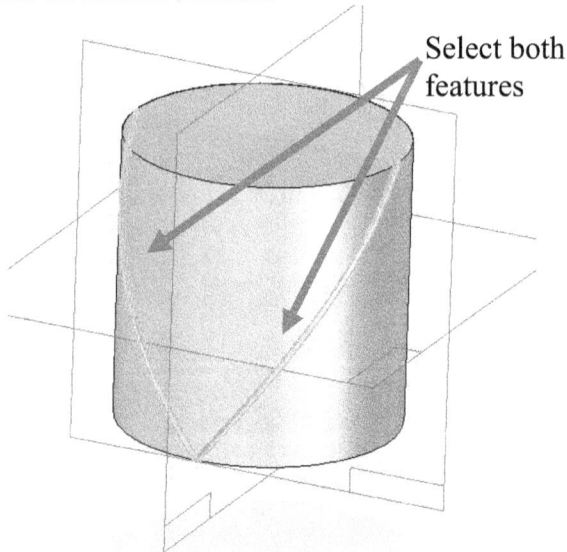

Select both features

Click Accept.

In the Sketch Step make sure **Coincident Plane** is selected in the dropdown menu and select the Smart button.

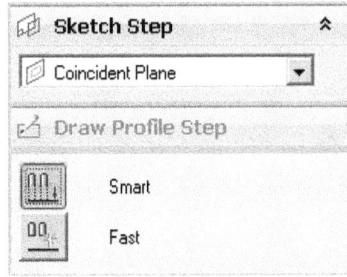

Click on the top surface of the cylinder to sketch on.

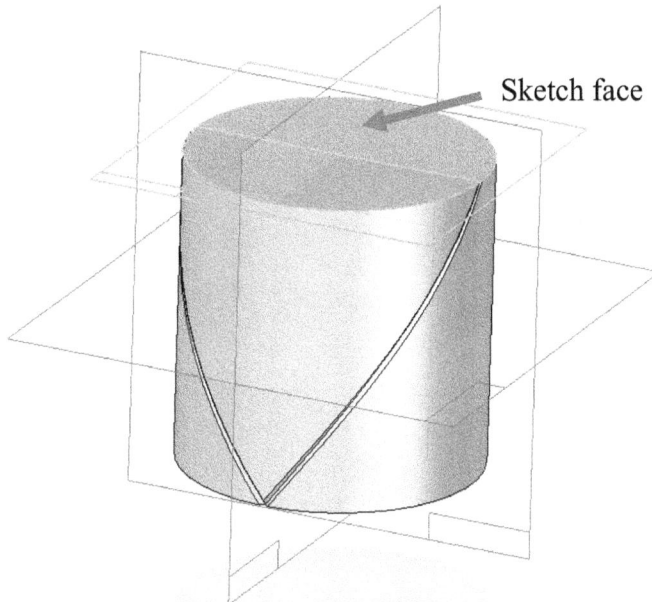

Sketch face

Once in the sketch environment select **Circular Pattern**.

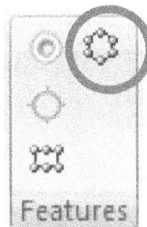

Click on (A) Center of the axis and (B) the outside of the circle in a similar way to sketching a normal circle

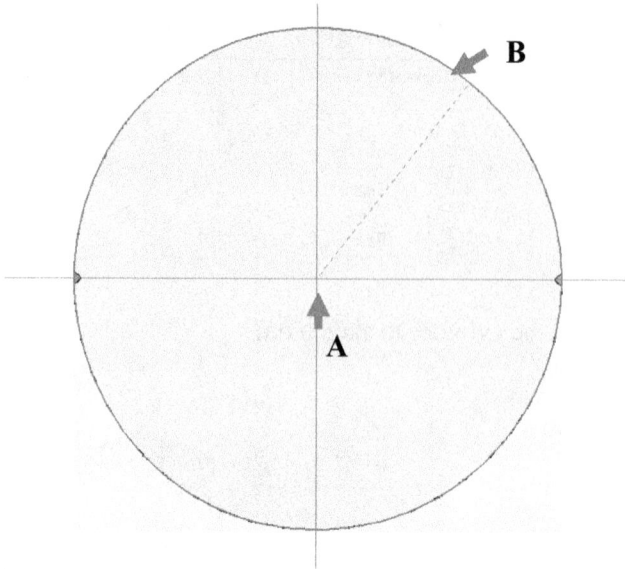

Click to the left of the dotted line to specify the direction of the pattern.

Enter 45 in the Count box.

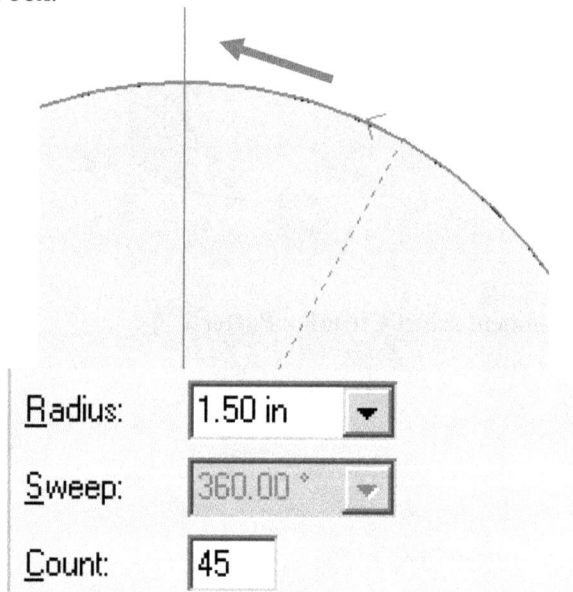

Radius:	1.50 in
Sweep:	360.00 °
Count:	45

The graphics display should now look like the following figure.

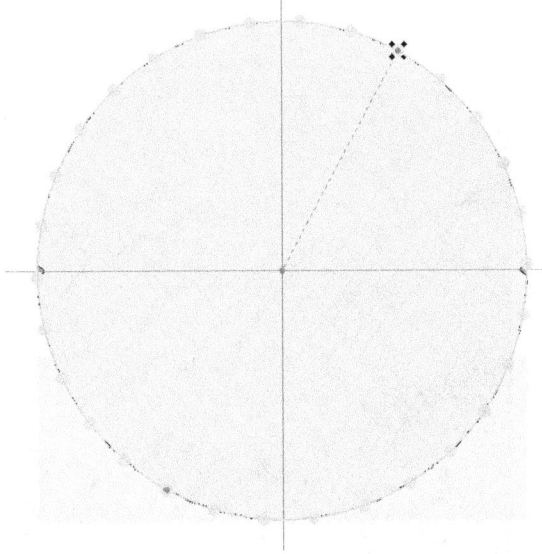

Close Sketch.

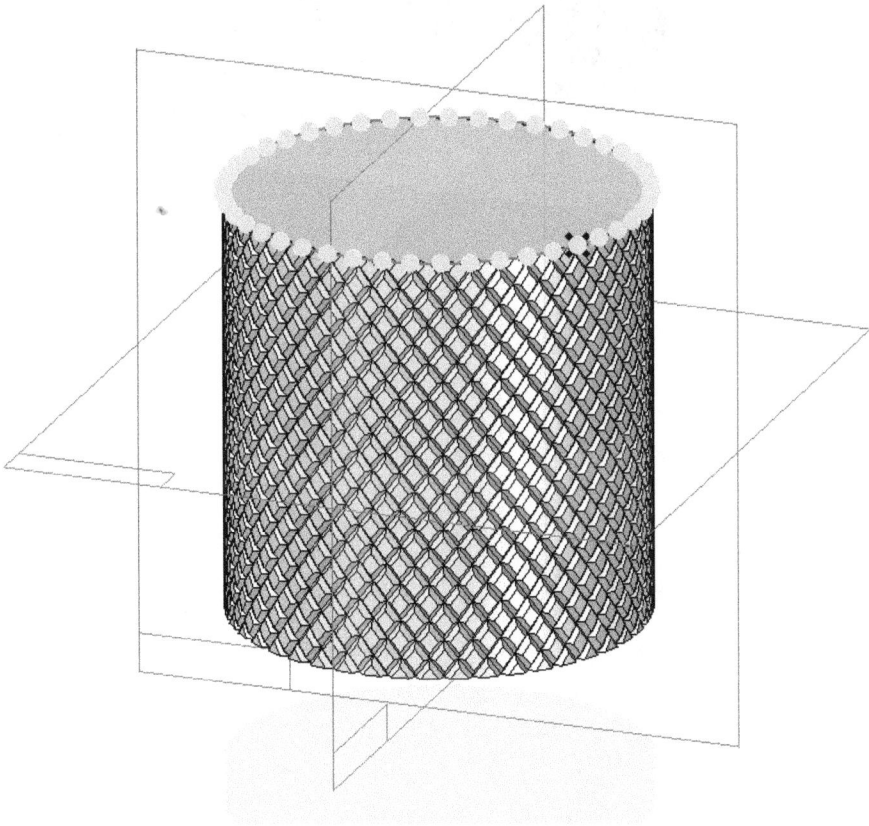

Click Finish.

The Neural will now be complete.

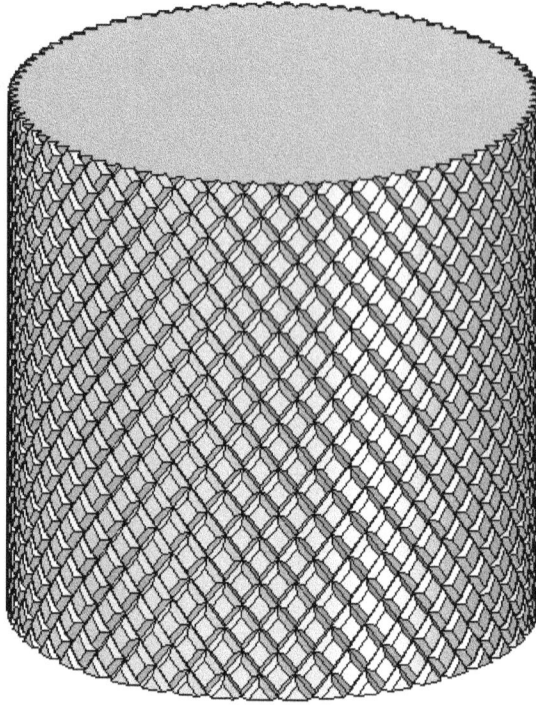

Exercise Complete

Exercise 31: Text tool, Wrapping, Normal cut

Open a **New ANSI Part File**. Right click anywhere in the design window and click "Transition to Synchronous". We will discuss the difference between the two different modeling types later in the book.

Create a **sketch** that is 3in X 1.5in and looks like the figure shown below.

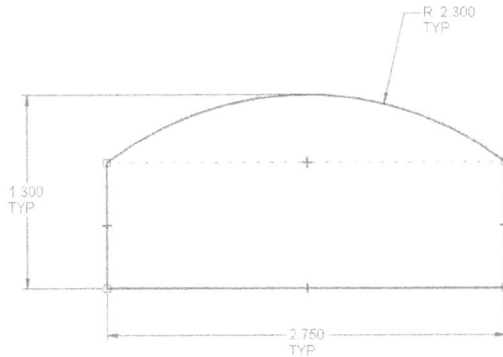

Extrude about the sketch plane a Distance of 5in.

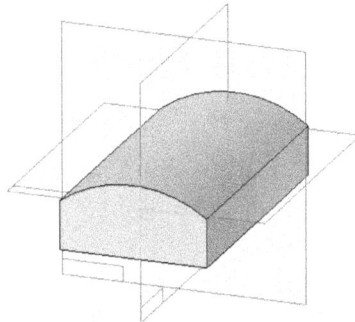

Create a **sketch** on a plane **Tangent** to the curved surface.

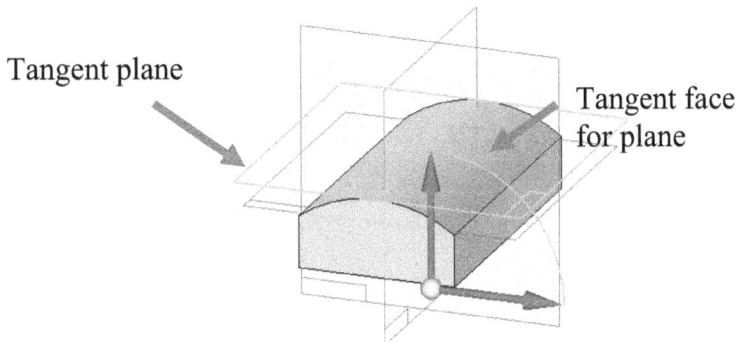

Under the **Sketching tab**, click on the **Text Profile** tool in the Insert options box.

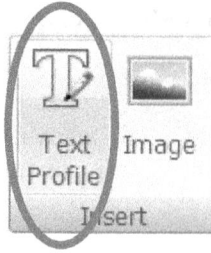

Within the **Text tool,** type 'SOLID' into the Text box, and change the Size to 1.5 in. *Note: the Size refers to text height.*

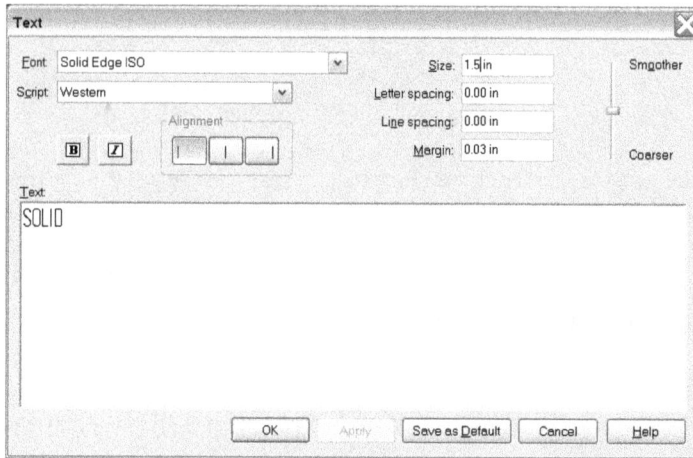

Click OK.

Click the **Anchor** button on the left menu and select the **Center** option.

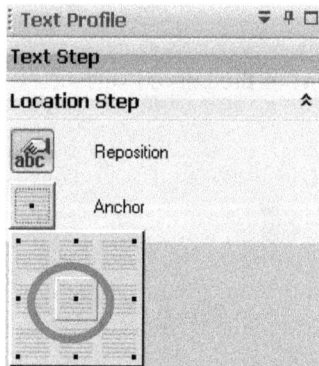

Place the **Text** in the center of the part, as shown in the following figure.

Close **Sketch**
Under the **Surfacing tab** select the **Wrap Sketch** tool in the Curves options box.

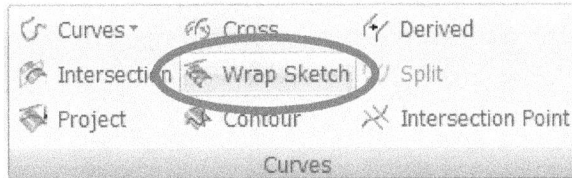

In the **Select Surface** Step in the left menu choose the curved surface as shown below.

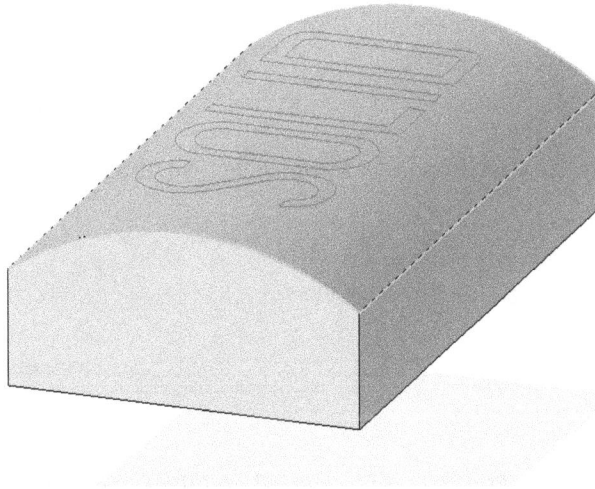

Click Accept.

In the **Select Sketch Step** select the sketch as shown below and, **Accept.**

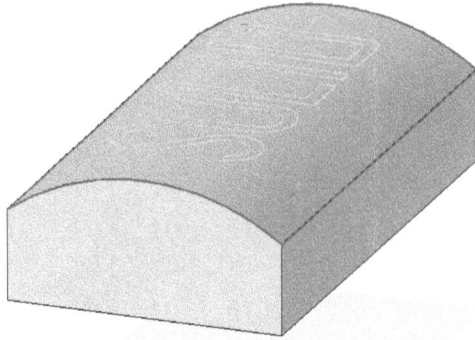

Click **Finish**.

Right click on the screen and click **Hide All/Sketches**. *Note: This is for clarity purposes only*

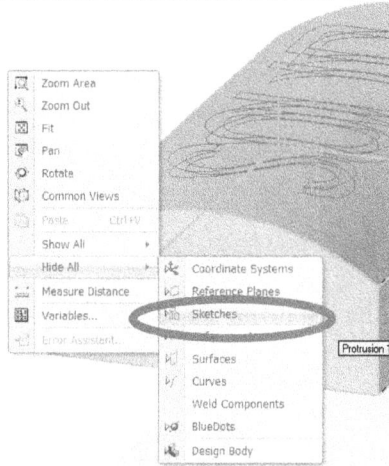

Under the Home tab in the **Cut** dropdown menu, select **Normal.**

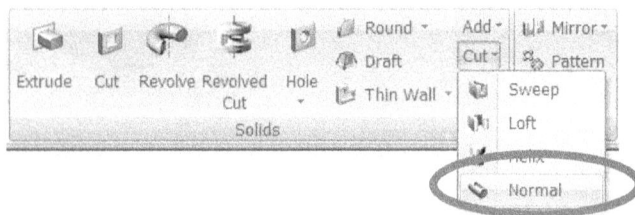

Note: There is a similar tool under the ADD button that adds geometry normal to a surface.

In the **Select Curve Step,** with the **'Faces Touching Curves'** on, click all the edges of the sketch. Then **Accept**.

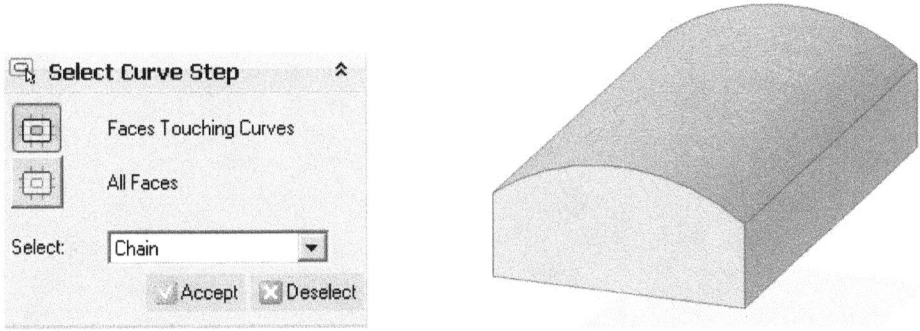

In the **Side Step** enter .25 in as the Depth

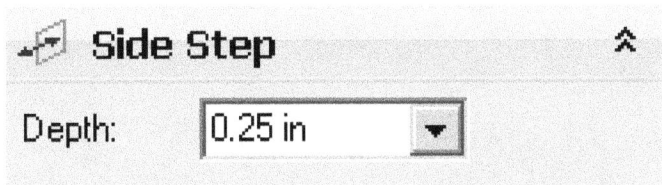

Select the direction to cut by changing the vector using the cursor location.

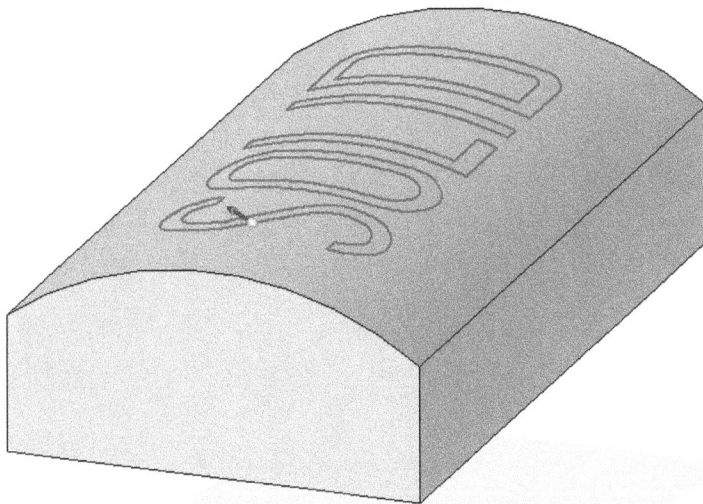

Click **Finish.**

Exercise Complete

Exercise 32: Blue Surface through 3 sketches

In a **tradition ANSI part file** create a **sketch** on a plane **Parallel** to the XZ plane that has an offset Distance of 2.5in.

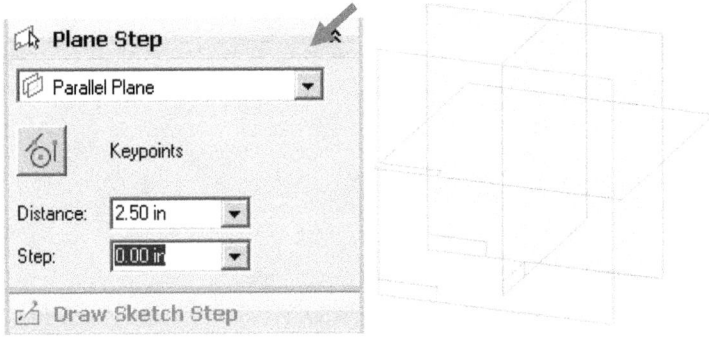

Using the **Arc by 3 Points** tool create a sketch that looks like the following figure.

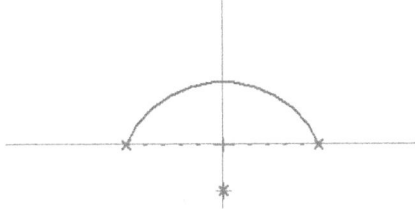

Close Sketch

Create a **sketch** on the original **XZ plane** that looks like the following figure. *Notice: only draw larger arc.*

Sketch on plane

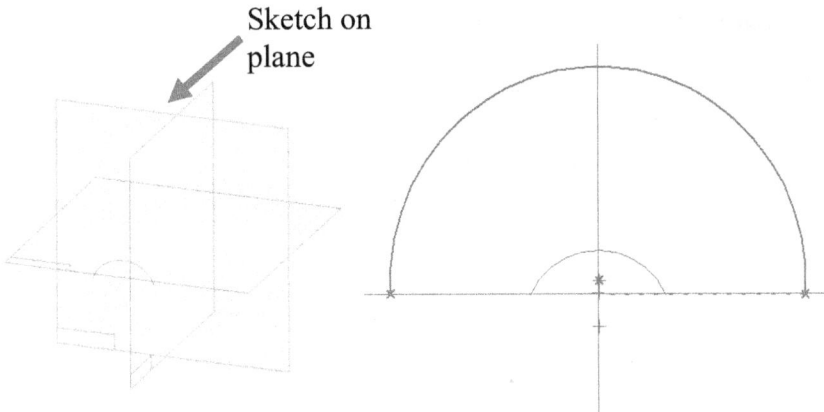

Close Sketch.

Create another **sketch** on a **Parallel Plane** that has an offset Distance of 2.50in the opposite direction as the first sketch plane.

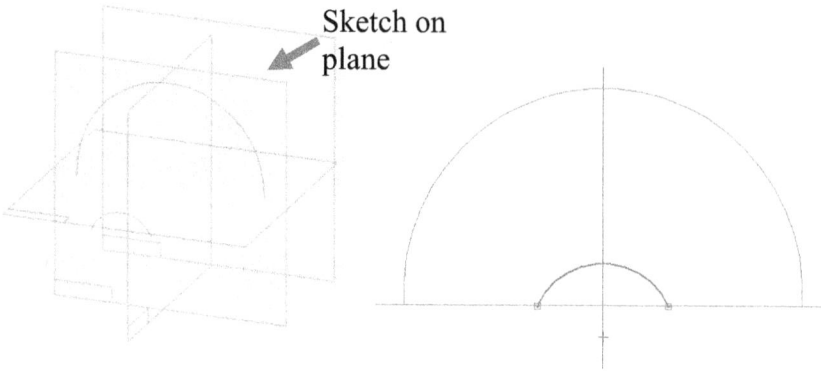

Sketch on plane

Copy the original arc.

Close **Sketch**

Under the Surfacing tab click the **BlueSurf** tool in the Surfaces options box.

In the '**Cross Section Step**' click the first small arc as shown below and, **Accept.**

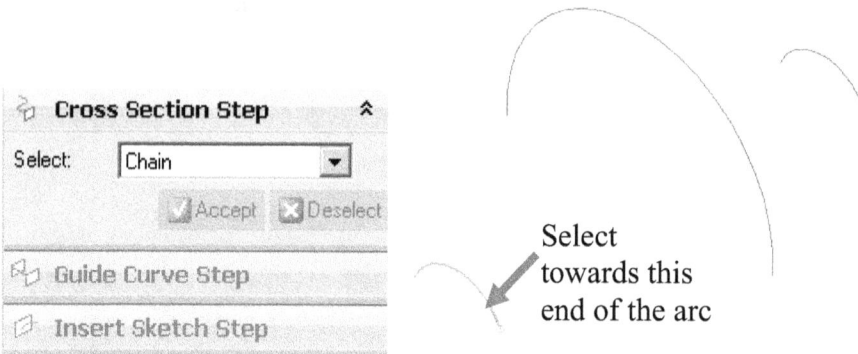

Select towards this end of the arc

Remaining in the '**Cross Section Step**' click the large arc as shown below and click Accept.
Note: Make sure the green construction line is not crossed over to the other end point.

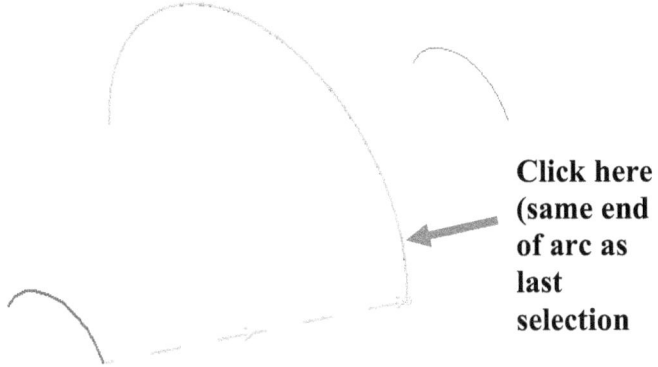

**Click here
(same end
of arc as
last
selection**

Click the last small arc as shown below.

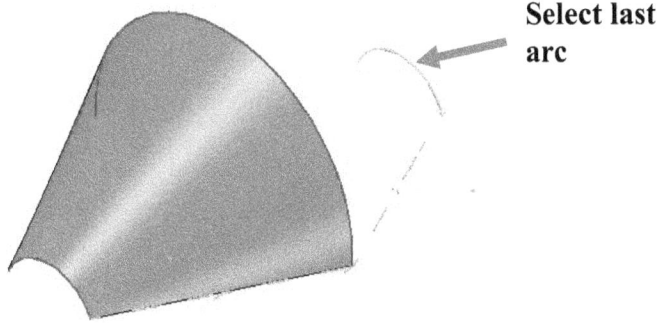

**Select last
arc**

Click the **Preview** button then the **Finish** button.

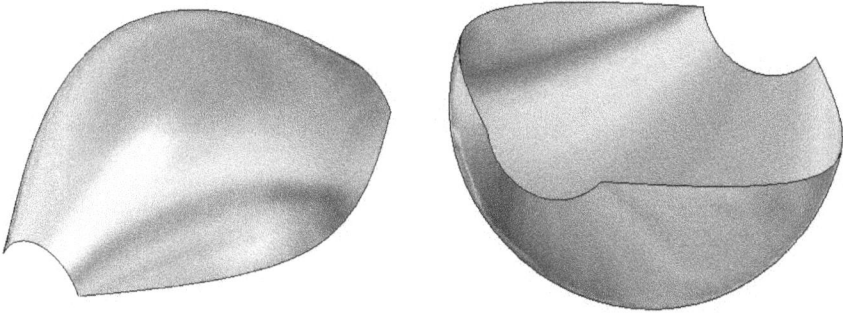

Exercise Complete

Exercise 33: Blue Surface with guide Curves

In a **New ANSI part file** create a **Sketch** on the **XY Plane**.

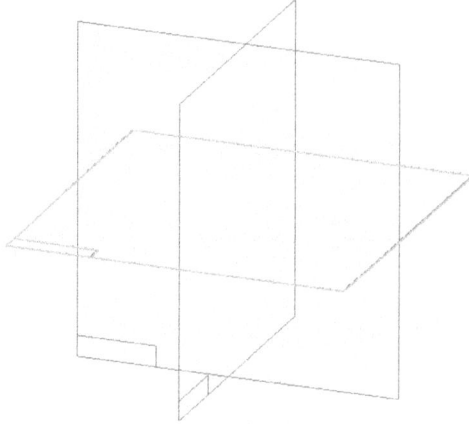

Using the **Ellipse** tool located in the Circle dropdown menu create a sketch that looks like the following figure.

Note: create two lines and use the trim tool to remove ¾ of ellipse.

Trim line 1

Trim Line 2

2.150 TYP

.700 TYP

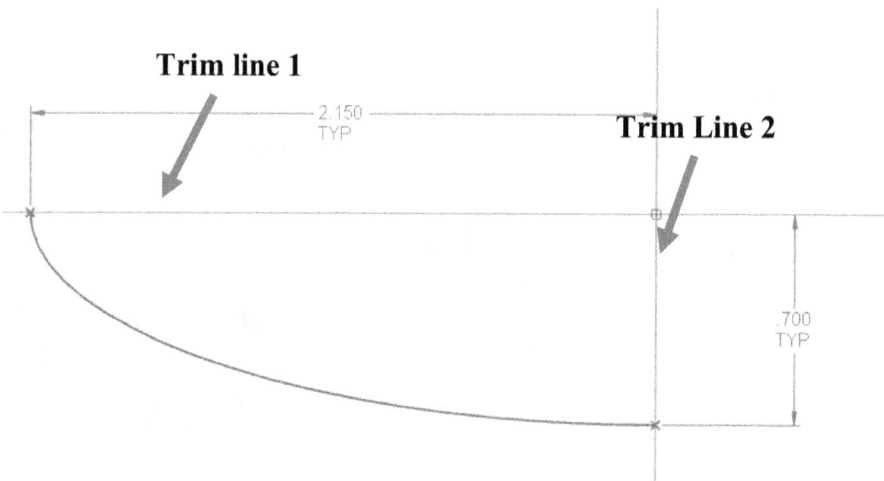

Close Sketch.

Create a **Sketch** on the **ZX Plane**.

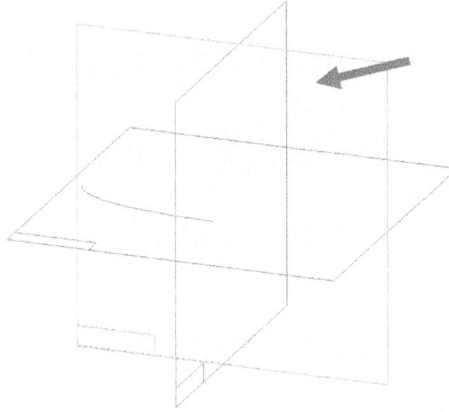

Once again with the **Ellipse tool** create a sketch that looks like the figure below, remember to use the trim tool to trim off ¾ of the ellipse.

Match end point
with end point of
other ellipse

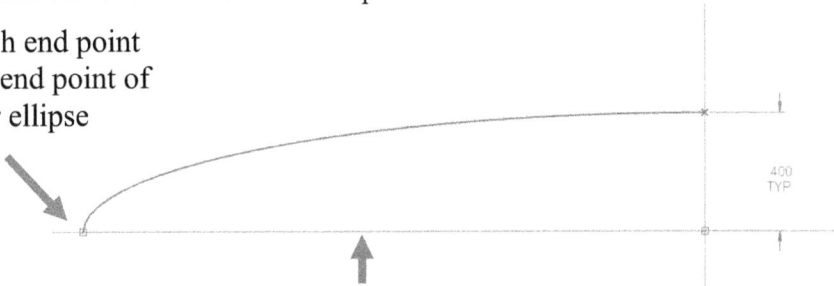

400
TYP

Notice: This line is not part of this sketch it is the ellipse from the other sketch.

Close Sketch.

Create another **Sketch** on the **XY Plane** and draw a ¼ just like the previous step.

Match end point
with end point of
other ellipse

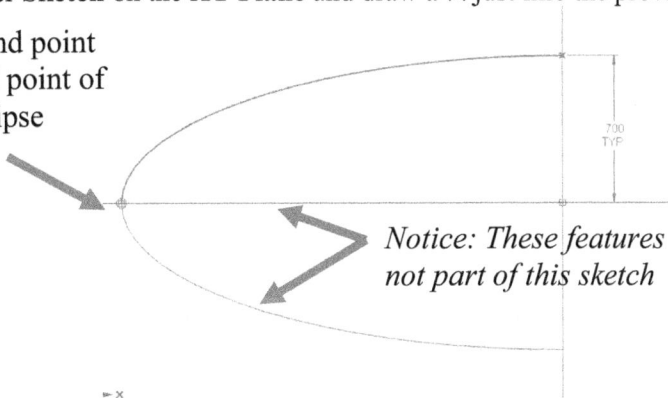

700
TYP

Notice: These features are not part of this sketch

Close Sketch. *Note: If we had made the first ellipse a half ellipse it would make line selection more difficult in the BlueSurf operation.*

Create a **Sketch** on the **ZY Plane.**

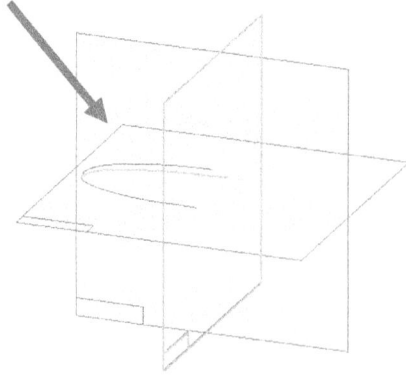

Draw a ½ **ellipse** and make sure to connect the ellipse to the end points of the other sketches using the Connect tool in the Relate options box.

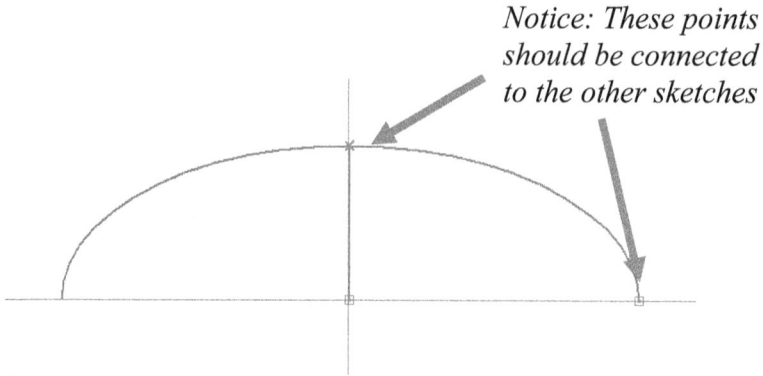

Notice: These points should be connected to the other sketches

Close Sketch.

The four sketches shown together should now look like the following figure.

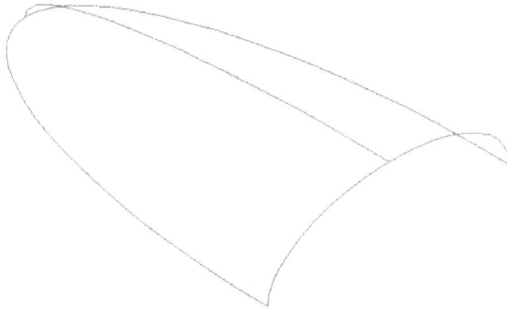

Under the Surfacing tab click the **BlueSurf** tool.

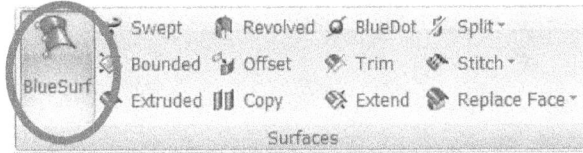

In the **Cross Section Step** click the curve shown below and click **Accept**. *Note: Pick the curve towards the side shown in the following figure. Each cross curve will have a starting location (shown by a little dot during selection). Imagine a surface being made up of X and Y locations we want all the X locations and Y locations to be acting in the positive. Therefore if all the start points of the cross curves are located on the same side then all the X's will be acting in the positive. So be sure to pick the other cross curves with a similar starting location so that the positive is acting in the same direction as the previous cross curves.*

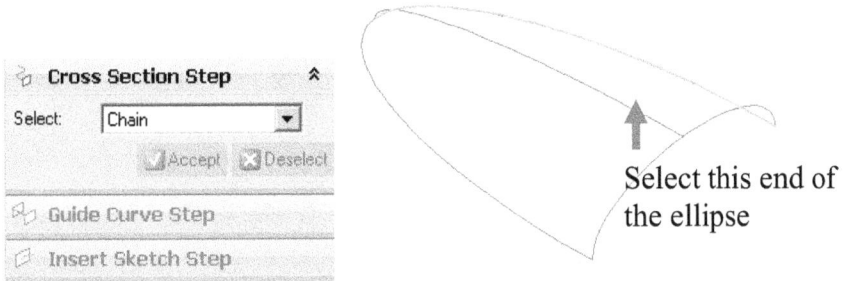

Select this end of
the ellipse

Click the middle curve then **Accept** and the left curve and **Accept.** *Reminder: Remember to pick the curves towards the same side as the initial curve, shown below. If failed to do so the surface will not work.*

Select this end of
the ellipse

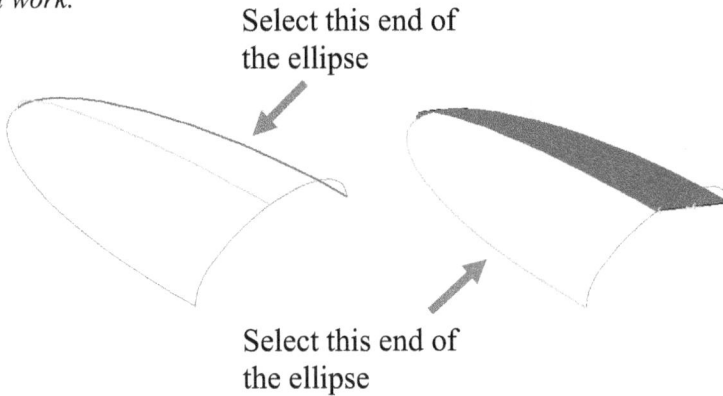

Select this end of
the ellipse

In the **Guide Curve Step** click on the last curve as shown below and, **Accept.** *Note: the same rule applies to guide curves' that was discussed earlier in the exercise, whereby the starting locations can have to start from a corresponding side.*

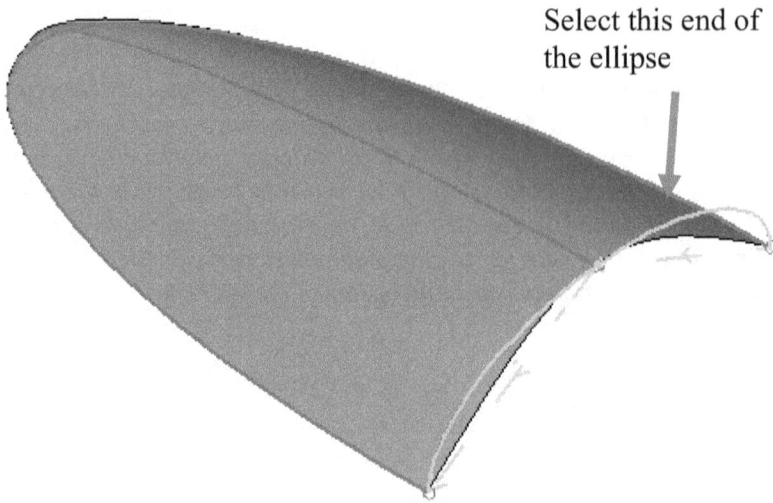

Select this end of the ellipse

Click the **Preview** button then **Finish.**

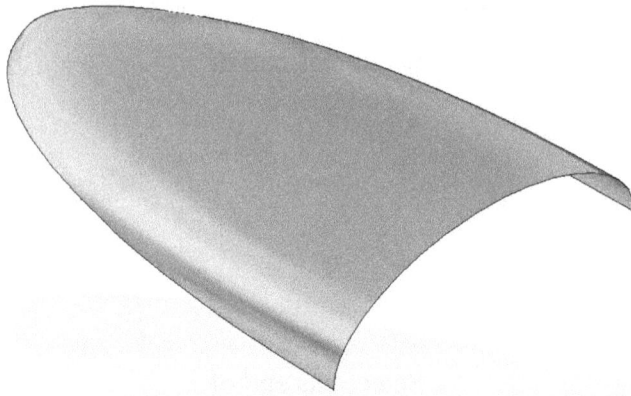

This is one of the main tools used to create industrial design quality surfaces.

Exercise Complete

Exercise 34: Changing Surfaces into a Solid

In a **ANSI part file** create a **sketch** on a **Parallel Plane** that has an offset Distance of 5in.

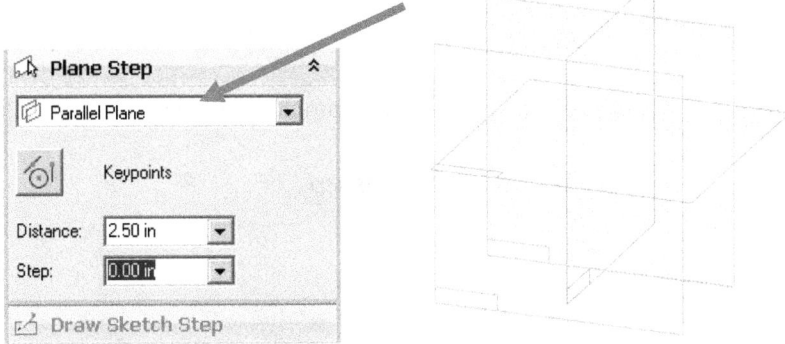

Using the **Circle** tool create a sketch that looks like the following.

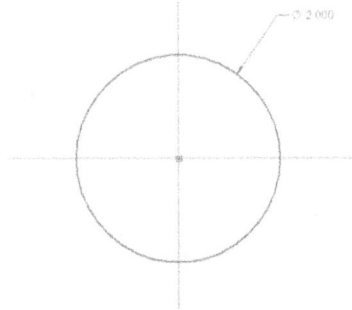

Close Sketch

Sketch an **Ellipse** on the **XZ Plane** that looks like the following figure.

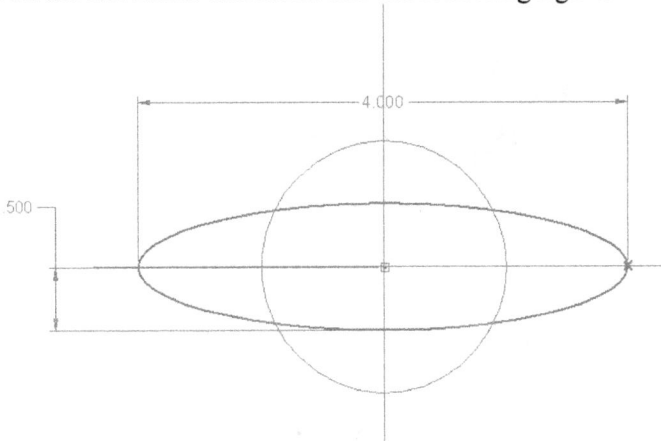

Close Sketch.

Under the Surfacing tab click the **BlueSurf** tool.

In the **Cross Section Step** click the circle then **Accept** and then click the ellipse and **Accept**.

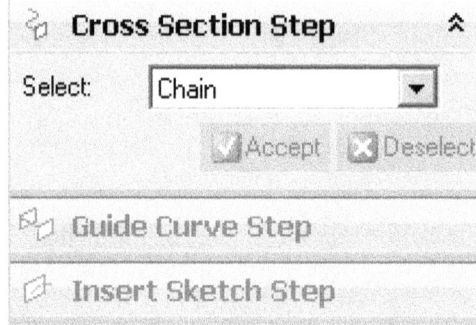

Click the **Preview** button then the **Finish** button. Your model will look like the following figure.

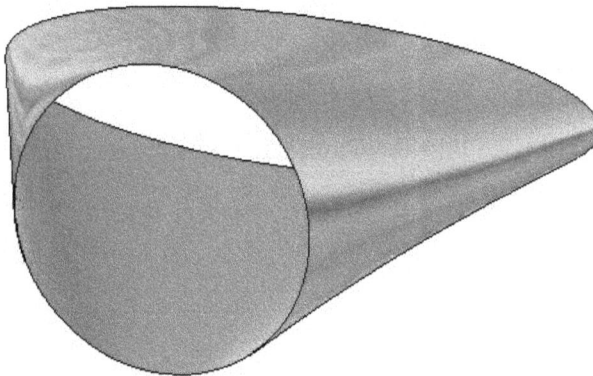

Click the **Bounded tool** in the Surfaces options box.

In the **Select Edges Step** select the edge of the circle and **Accept.**

Select edge

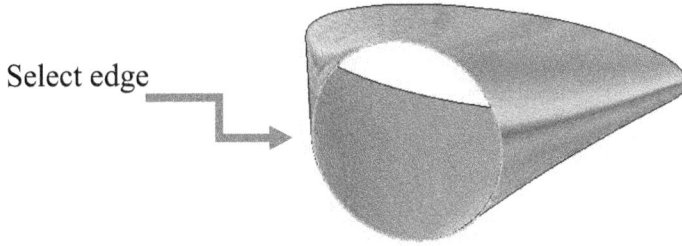

Click the **Preview** button then the **Finish** button.

Create another **Bounded Surface** on the other side of the shell by clicking on the ellipse. Click the Preview button then the Finish button.

Bounded Planes

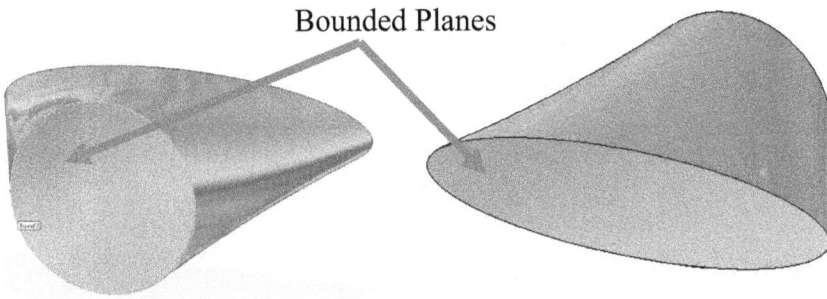

Click the **Stitched tool** located in the Stitch dropdown menu in the Surfaces options box.

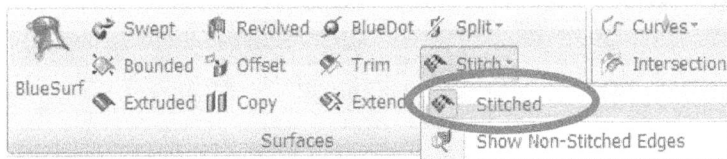

A box called **'Stitched Surface Options'** should appear. Click **OK.**

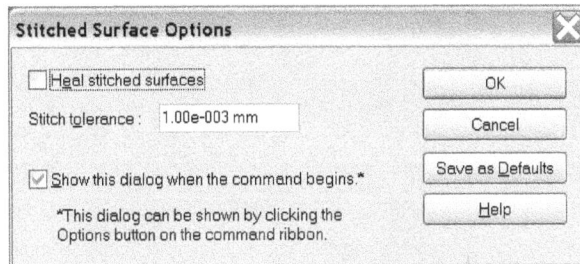

In the **Select Surfaces Step** select all three surfaces and click **Accept.**

The following options box should appear:

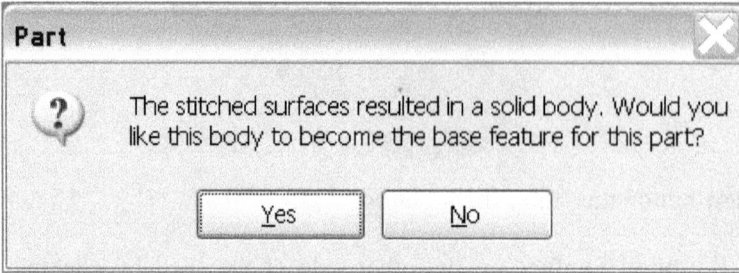

Click **Yes.**

The series of surfaces should turn into a solid part and appear like the following figure.

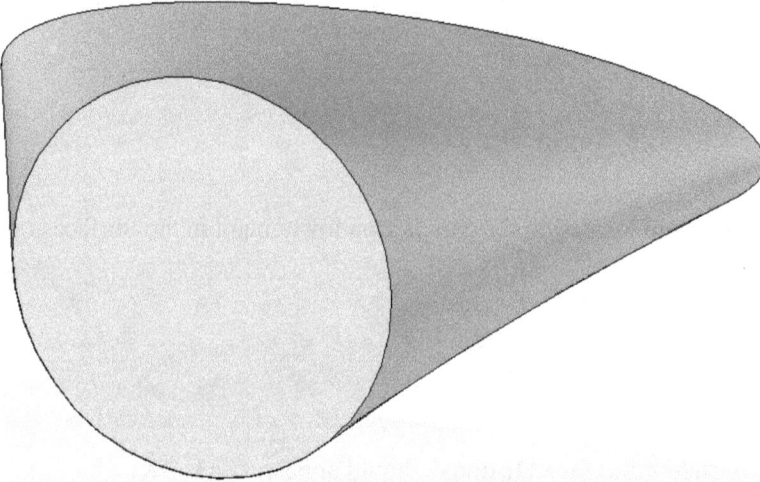

Exercise Complete

Exercise 35: Extrude Surfaces that are Tangentially Connected

In a New ANSI part file under the **Surfacing tab** click the **Extruded tool.**

In the **Sketch Step** create a **Parallel Plane** that is offset by 3in.

Offset Plane

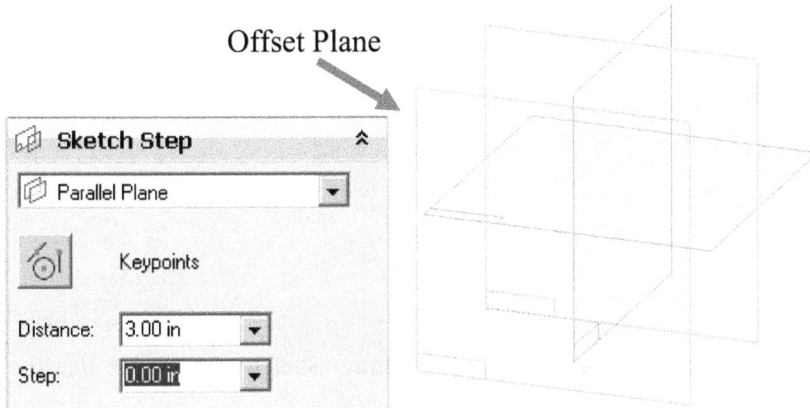

Draw a **rectangle** as shown below.

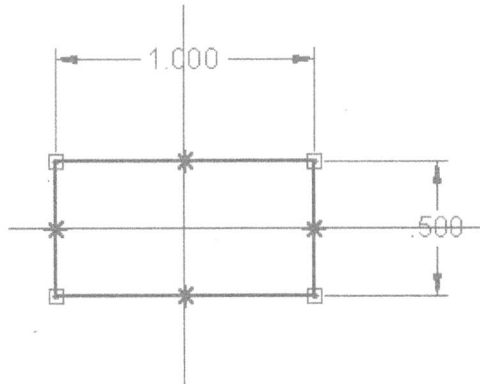

Close Sketch.

In the **Extent Step** extrude forward a Distance of 1.25in.

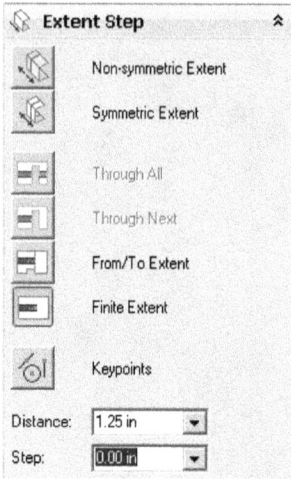

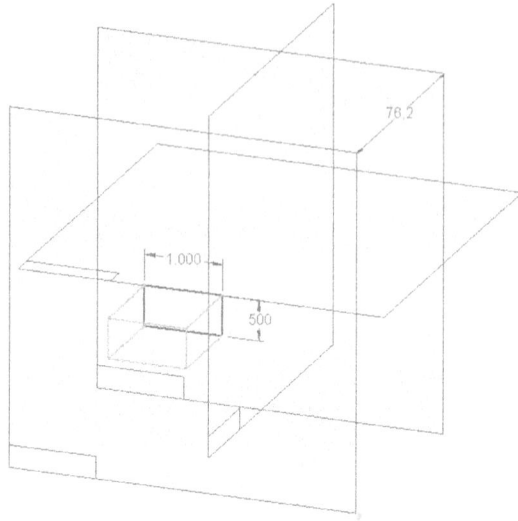

Click **Finish.**

Create another **Extruded Surface** on the **ZX Plane**. Sketch a rectangle like the one shown below.

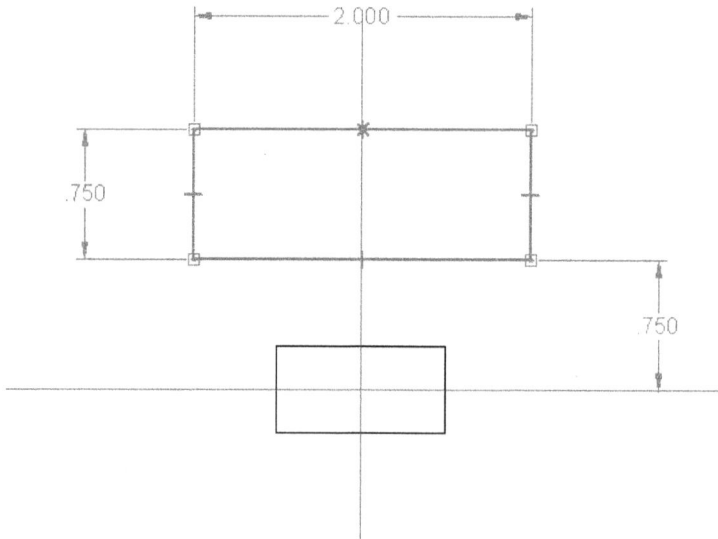

Close Sketch.

In the **Extent Step** extrude behind the sketch plane a Distance of 3in, as shown below.

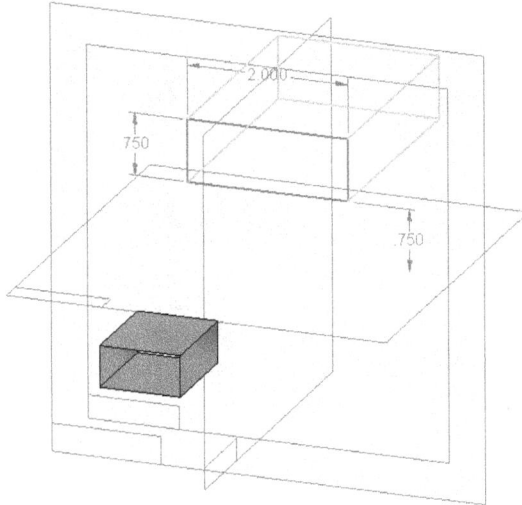

Click **Finish.**

Select the **BlueSurf** tool in the Surfaces options box.

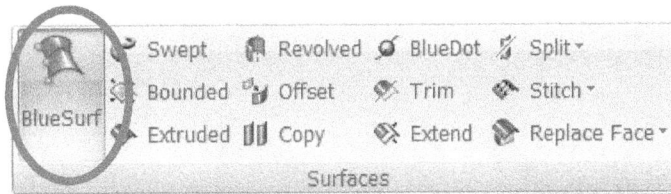

In the **Cross Section Step** select the edges on the back of the small rectangle as shown below.

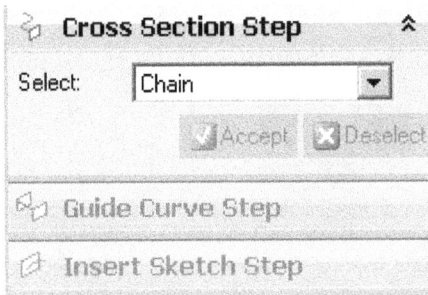

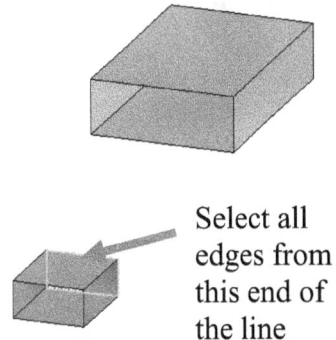

Select all edges from this end of the line

Click **Accept** and select the corresponding sides on the front of the large rectangle.

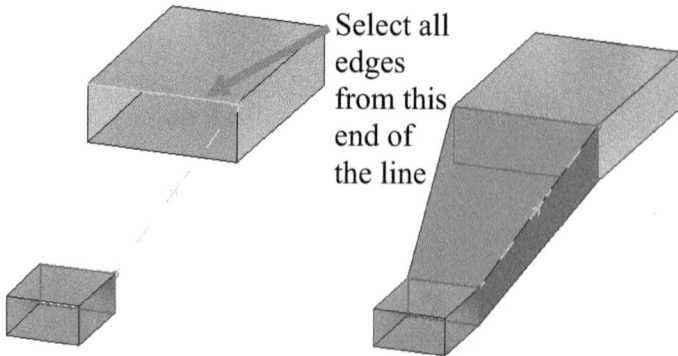

Select all edges from this end of the line

Click the **Options** button at the top of the left menu.

Under **Start Section** and **End Section** select **Tangent continuous** and click **OK.**

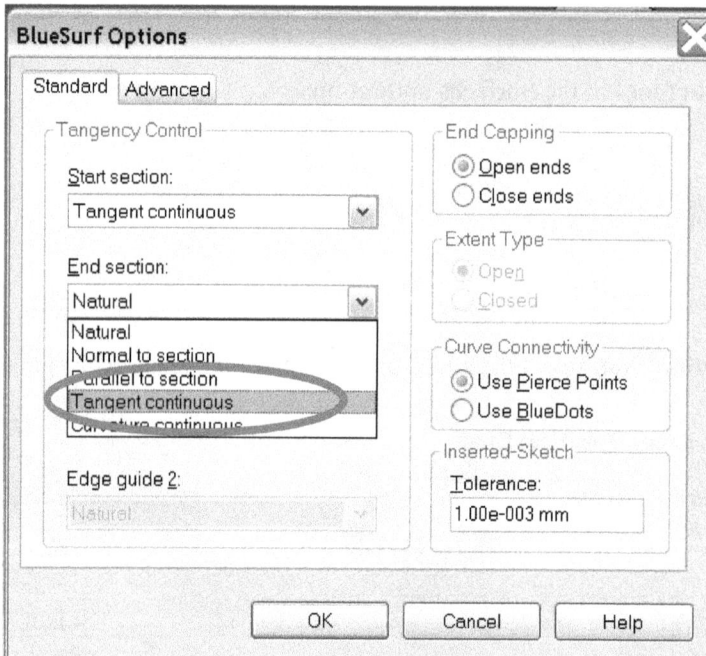

Click the **Preview** button then the **Finish** button, the model should look like the following figure.

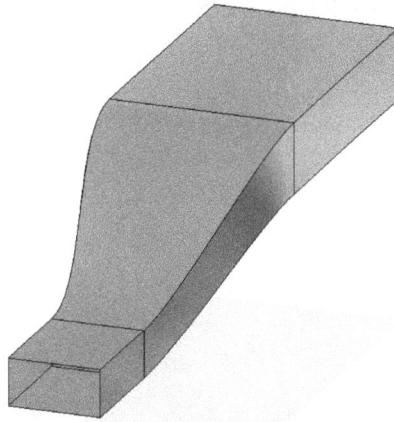

In the Add dropdown menu select the **Thicken** tool.

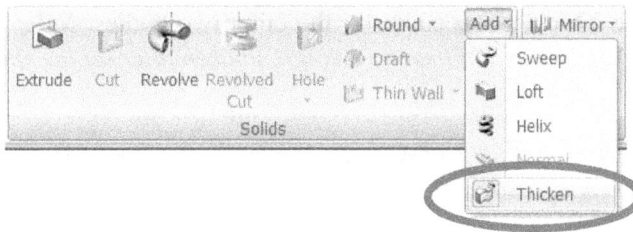

To Thicken, In the Select Step, click on the small rectangle.

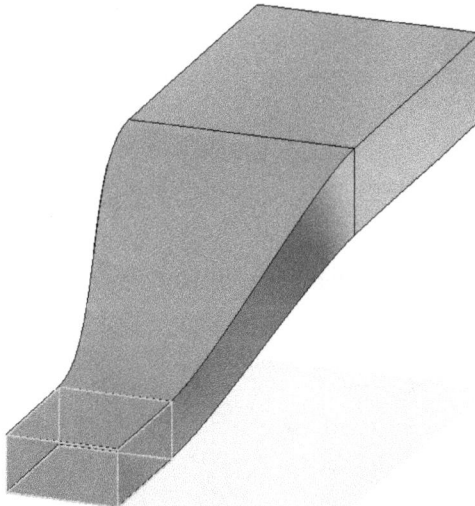

In the Offset Step enter 0.125in into the Distance box then click on the little yellow sphere connected to the red arrow to specify that the operation will thicken symmetrically. This is a similar process that was achieved in a previous exercise.

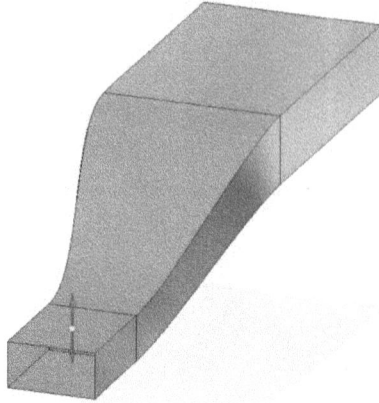

Click **Finish**

Go through the same **Thicken** procedure for the other two bodies. *Note: It is also possible to stitch all the surfaces together and then do a single thicken after all the sheet bodies are one.*

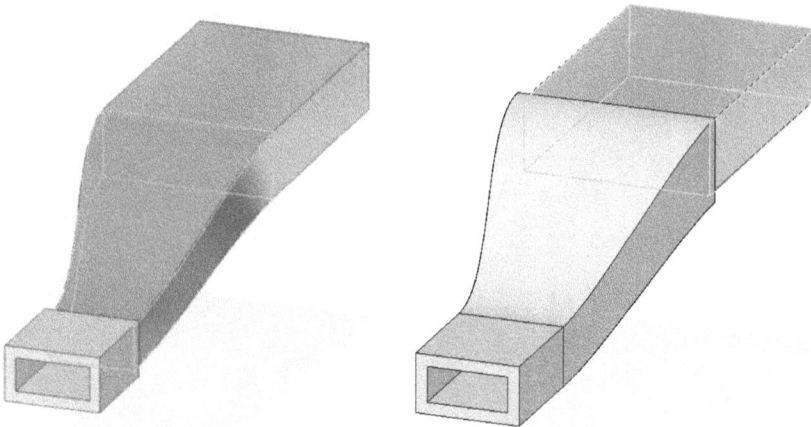

As the solids are added they will automatically unite to each other, becoming one solid body.

The Duct is now complete!

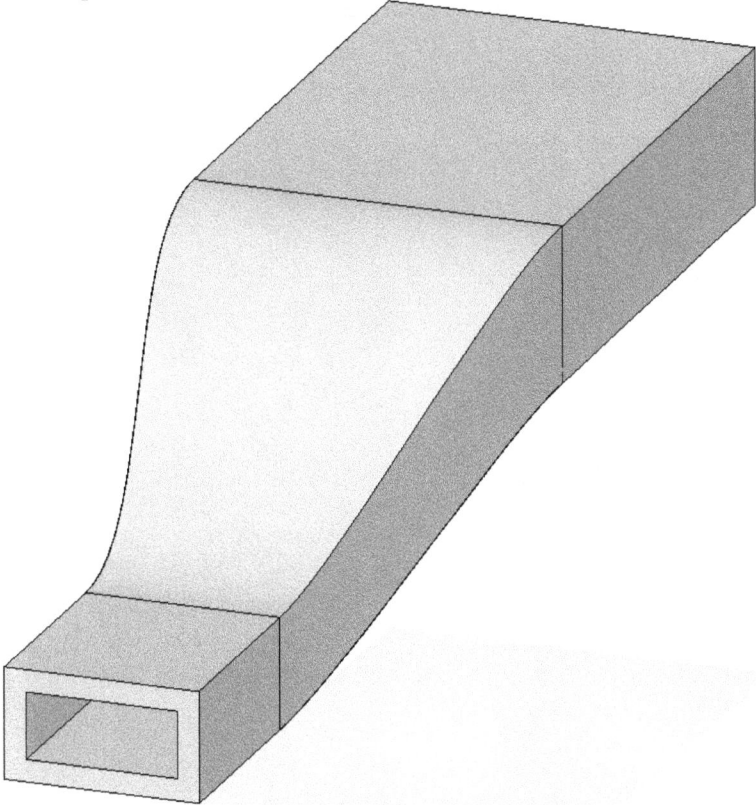

Exercise Complete

Exercise 36: Surfacing, Trim, Sweep, Extend, Split

In **a New ANSI part** file create a **sketch** on the **XZ Plane** shown in the following figure.

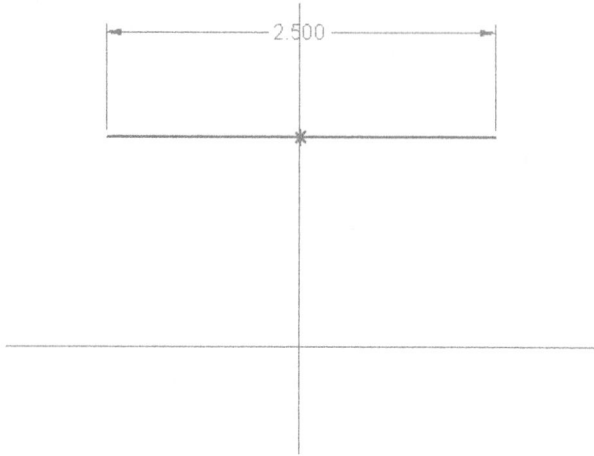

Close Sketch

Now create a sketch on the **YZ Plane** that looks similar to the following figure. Use the **Curve** tool on the Line drop-down menu.

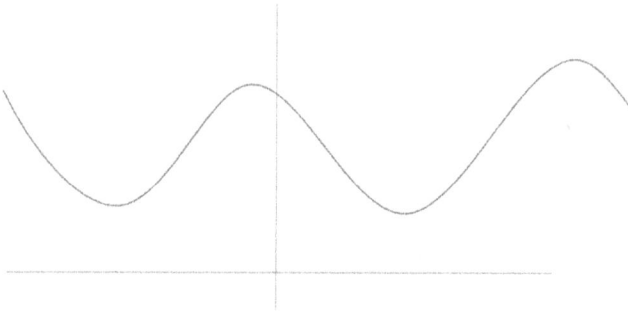

Under the **Surfacing** tab click the **Swept tool** in the Surfaces options box.

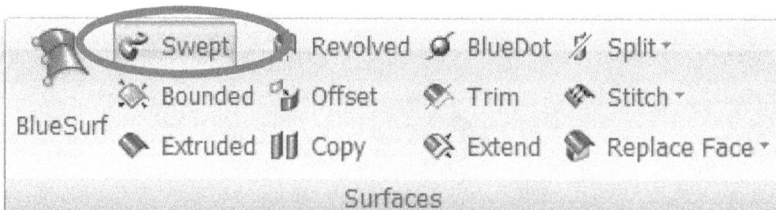

In the **Sweep Options box**, make sure **Single path** and **Cross section** is selected and click **OK**.

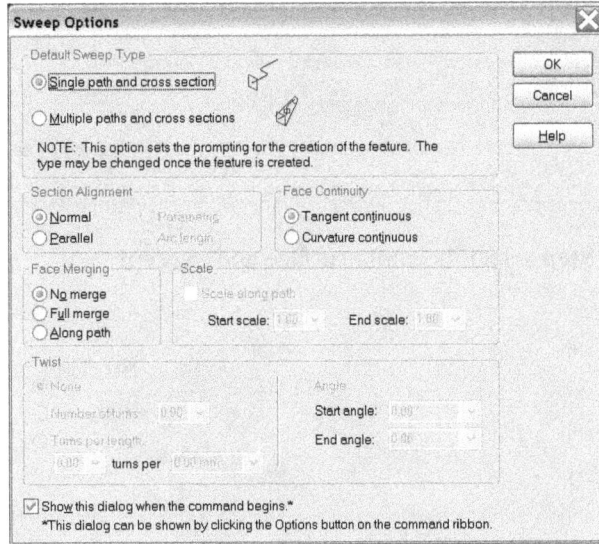

In the **Path Step** click on the curve and in the **Cross Section** Step click on the line.

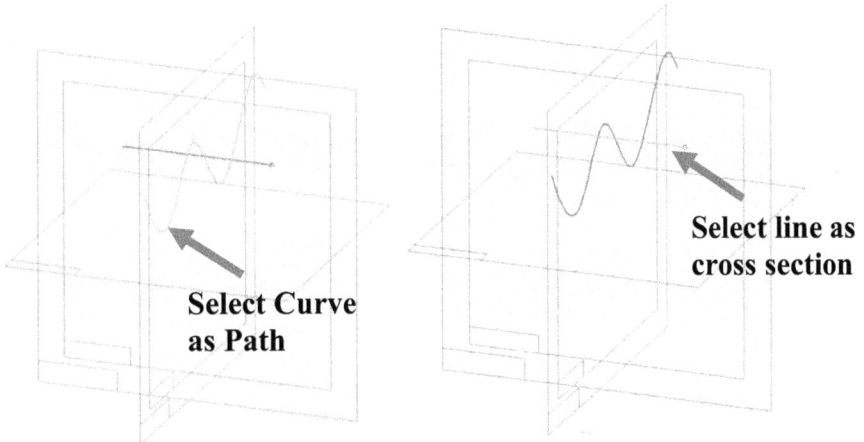

Select Curve as Path

Select line as cross section

Click the **Preview** button then the **Finish** button.

Click the **Split tool** in the Surfaces options box.

In the **Select Surface Step** select the entire surface and, **Accept.**

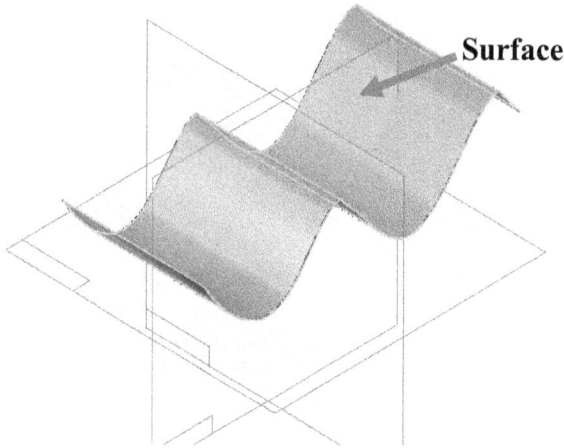

To split the surface, in the Select Splitting Geometry Step, click on the **ZX Plane**, and Accept. *Note: a plane doesn't have to be used as the splitting geometry, a separate sketch, or another surface for example could also be used.*

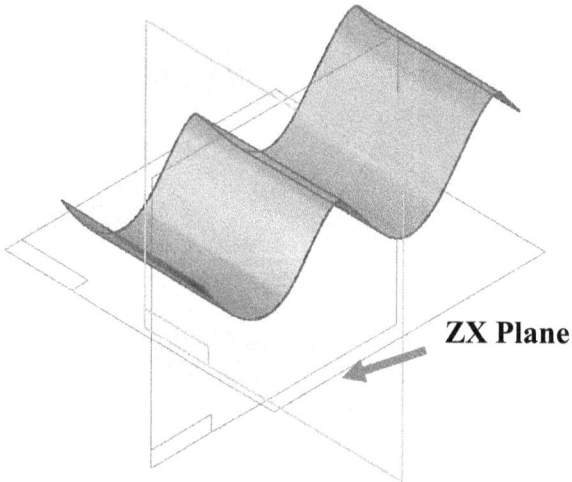

Click **Finish.**

The surfaces have been split into 2 parts that can be highlighted individually as shown below.

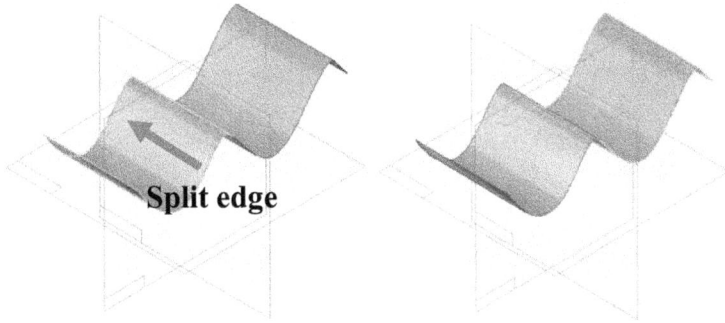

Split edge

Click on the **Extend** tool in the Surfaces options box.

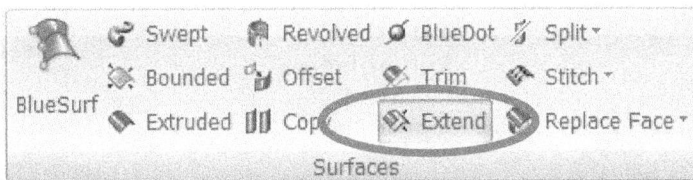

In the Select Edges Step make sure Edge is selected and click on the right edge. Make sure the **'Natural Extend'** button is clicked on as shown below.

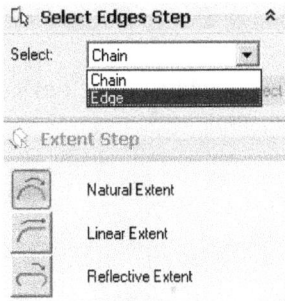

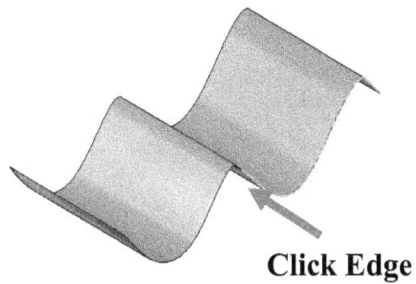

Click Edge

Click **Accept.**

In the **Extent Step** enter a Distance or control it freehand by moving the cursor in the direction of the Extend.

In this exercise, freehand the distance and click when it is approx 1.in .

Make sure the **Natural Extent** button is chosen and click **Finish.**

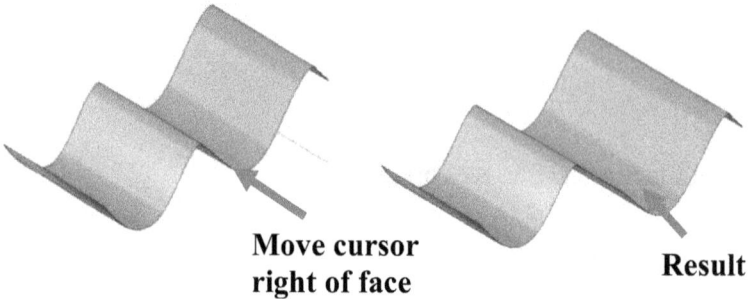

**Move cursor
right of face**

Result

Click on the **Trim tool** in the Surfaces options box.

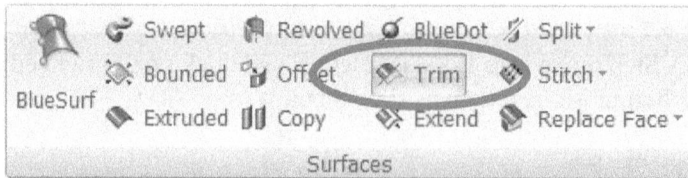

In the **Select Surface Step** select the entire surface and click **Accept.** In the **Select Trim Tool** Step select the **ZY Plane** as shown below.

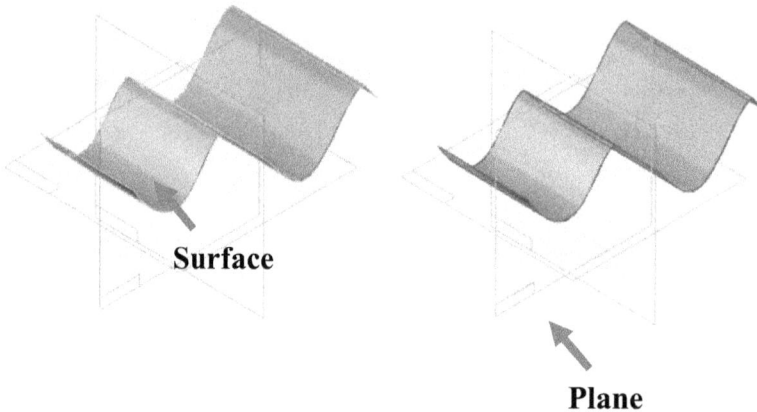

Surface

Plane

To trim, in the Side Step, click to the left of the trim plane to specify the direction to cut.

**Click to the
left of the
ZY Plane**

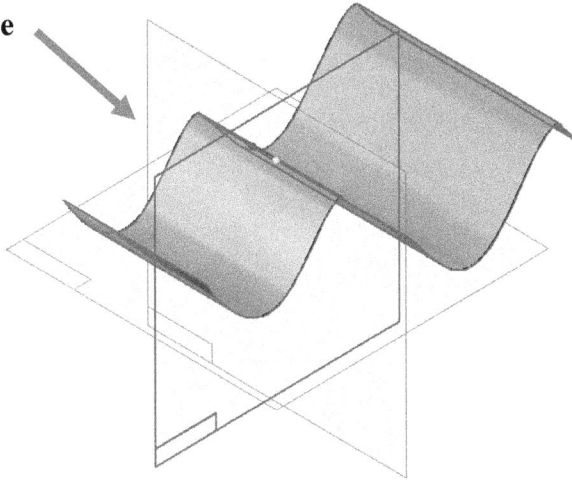

Click **Finish.**

These are some great techniques for creating and editing surfaces.

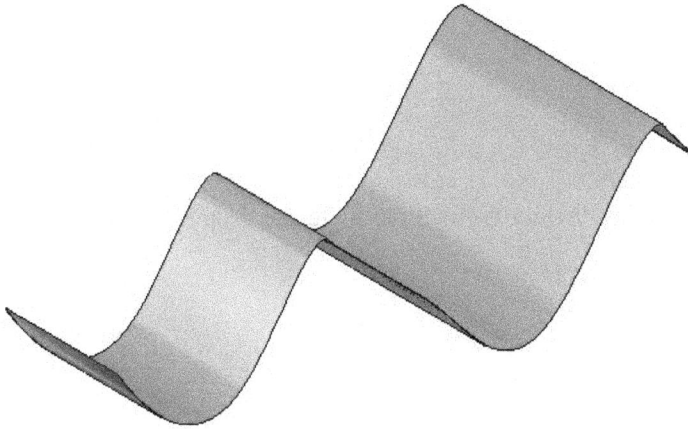

Exercise Complete

Exercise 37: Offset Surfaces

Start a **new ANSI part**. Create a sketch that looks similar to the figure shown below. Be Creative.

Close Sketch

Under the Surfacing tab click the **Revolved tool** in the Surfaces options box.

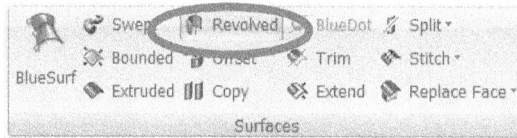

In the Sketch Step click **'Select from Sketch'** in the dropdown menu and click on the curve then, **Accept.**

Choose the vertical line as the **Axis of Revolution**.

In the **Extent Step** click the **Revolve 360** button then **Finish.**

Select the **Offset tool** in the Surfaces options box.

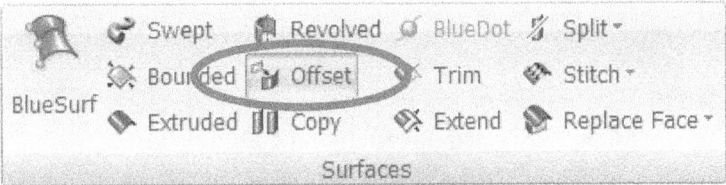

In the Select Step click on the entire surface then Accept.

In the Offset Step make sure **Remove boundaries** is highlighted and enter a distance of 0.125in.

Click below the part to specify the direction of the offset to the outside of the part.

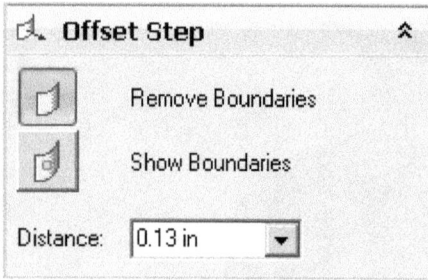

Arrow indicates offset direction

Click **Finish.**

Offset Surface

Rotate the model to view the underside. Use the **Bounded tool** under the surfacing tab and patch the two edges shown in the following figure. Simply select edge 1 and click accept through the menus. Repeat this for edge 2 shown below.

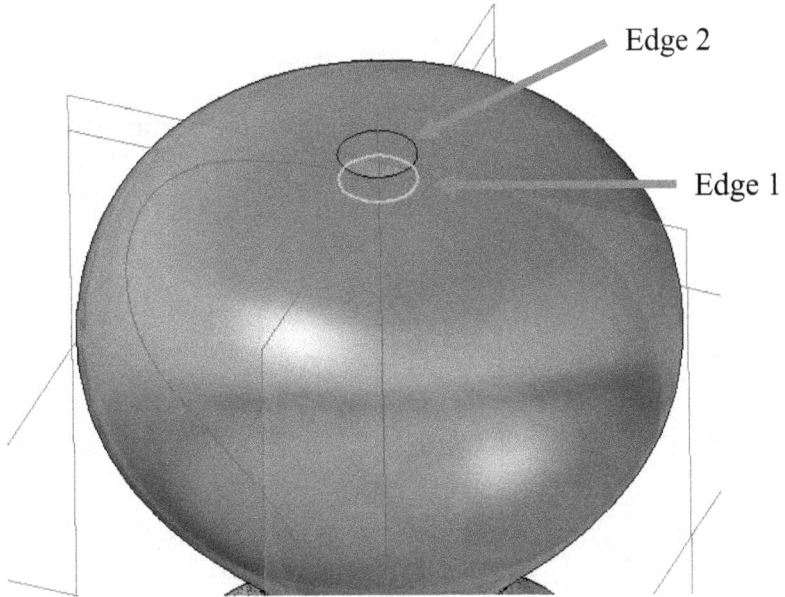

Edge 2

Edge 1

Sketch on the **ZX plane**. Sketch a line entity joining the two edges of the surfaces as shown in the following figure.

End points of lines touch each surface

Use the **Swept** tool under the surfacing tab to sweep the line entity around one of the edges. First select the path (the edge) then select the section (the line). A new surface is now effectively closing the gap between the two surfaces.

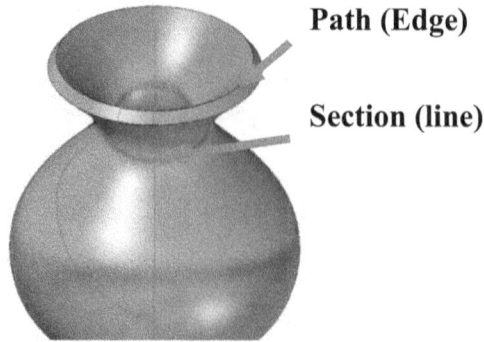

Path (Edge)

Section (line)

Finally, use the **Stitched tool** under the surfacing tab to turn the fully connected surfaces into a solid. Click **OK** through the 'Stitched Surface Options', drag a box over all the surfaces and click, **OK.**

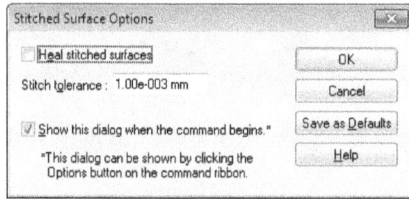

The solid is complete.
Exercise Complete

Exercise 38: Replace Face

Create a Traditional Part file. Create an **Extruded** part that looks like the figure shown below. *Sizes are not important in this exercise- be creative!*

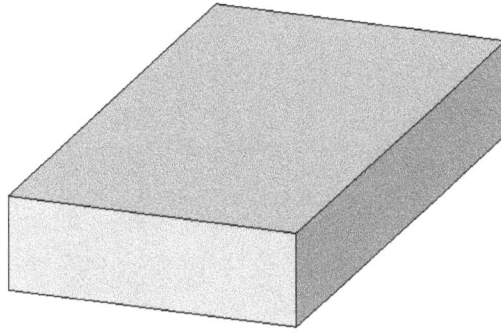

Create a **Sketch** on the surface shown below.

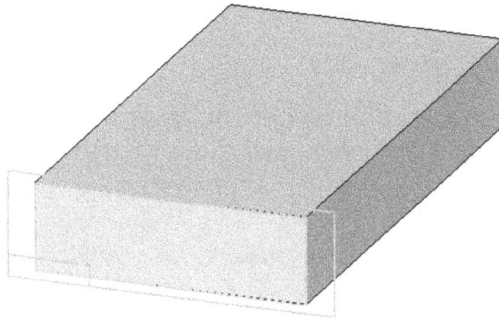

With the **Arc by 3 points** tool draw an arc similar to the one shown below and **Close Sketch.**

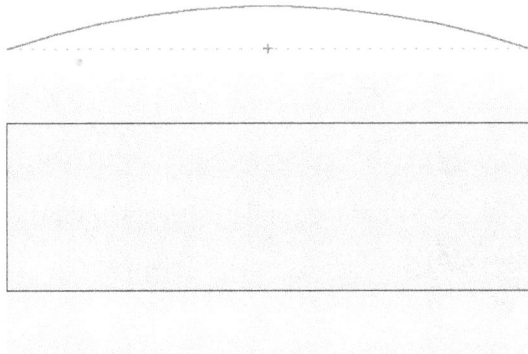

Create another **Sketch** on the face shown below, perpendicular to the last surface you started on.

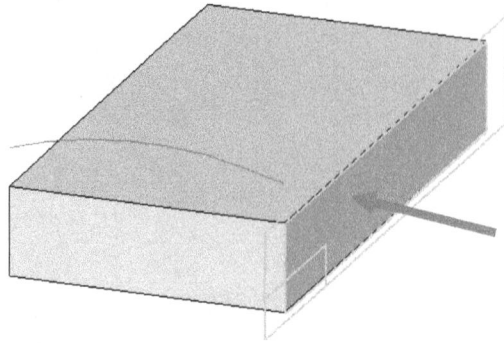

Using the **Curve tool** in the Line dropdown menu, draw a curve similar to the figure below. *Note: to finish the curve - right click after placing the last point.*

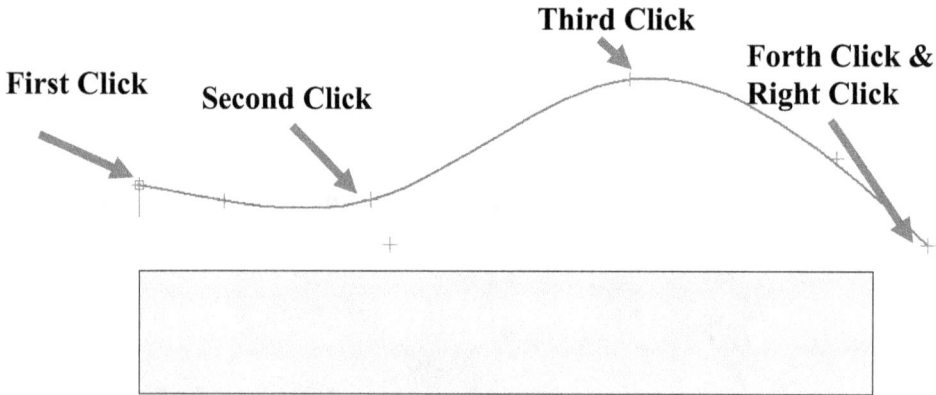

First Click **Second Click** **Third Click** **Forth Click & Right Click**

Close Sketch.

Under the Surfacing tab click the **Swept tool** in the Surfaces options box.

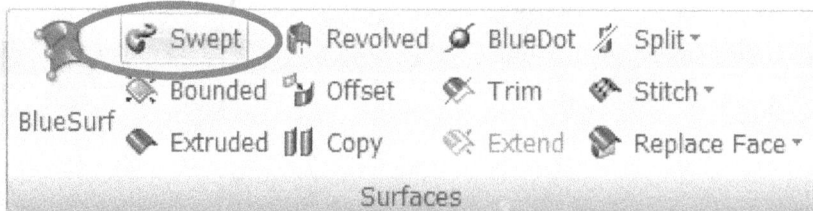

In the **Sweep Options** box, make sure **Single path** and **cross section** is selected and click **OK.**

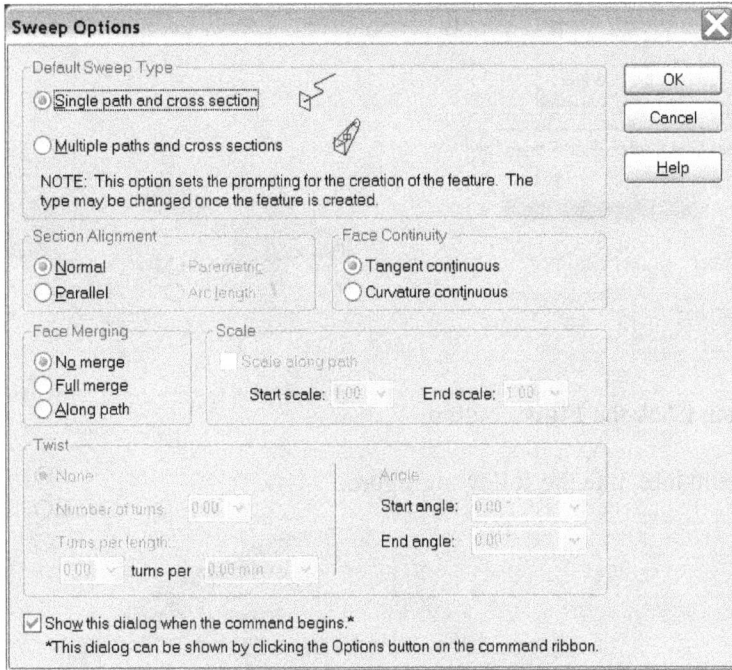

In the **Path Step** make sure **Select from Sketch/Part Edges** is selected and click on the curved path.

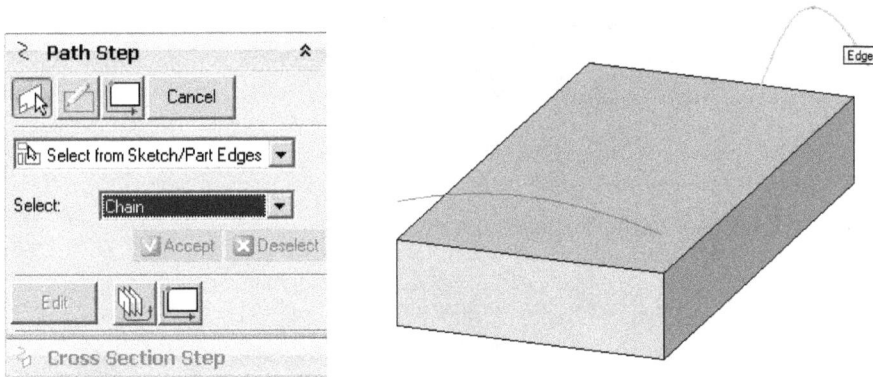

Click **Accept.**

In the **Cross Section Step** make sure **'Select from Sketch/Part Edges'** is selected and click on the arc.

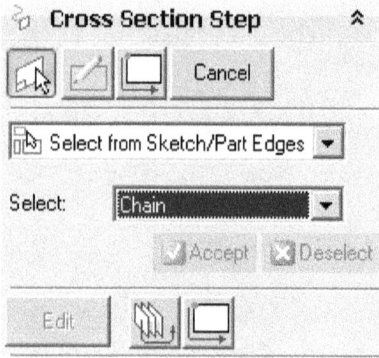

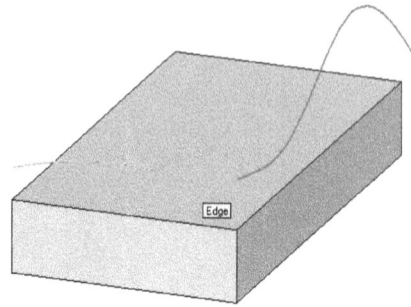

Click **Accept** then Click the **Finish** button.

The Surface should look like the following figure.

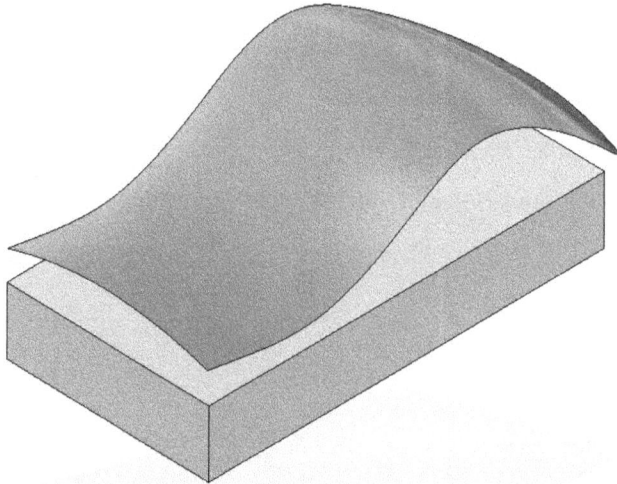

Click the **Replace Face** tool in the Surfaces options box.

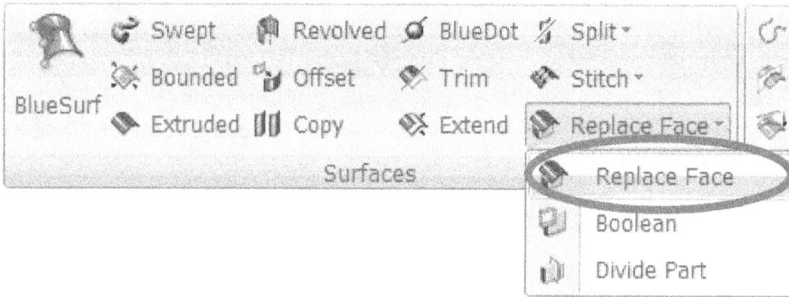

In the **Select Face Step** select the Original face and, **Accept.**

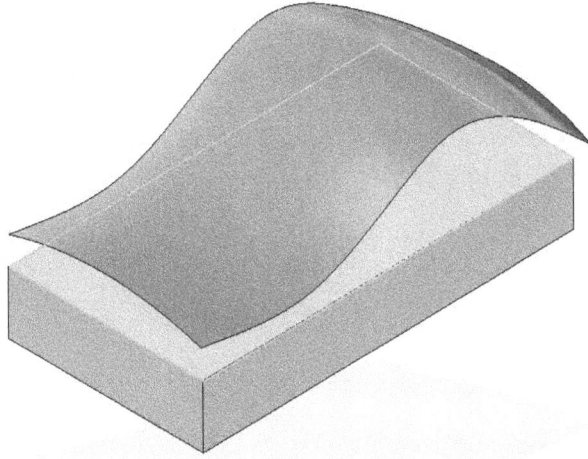

In the **Select Replacement Step** select the Curved surface and click **Finish.**

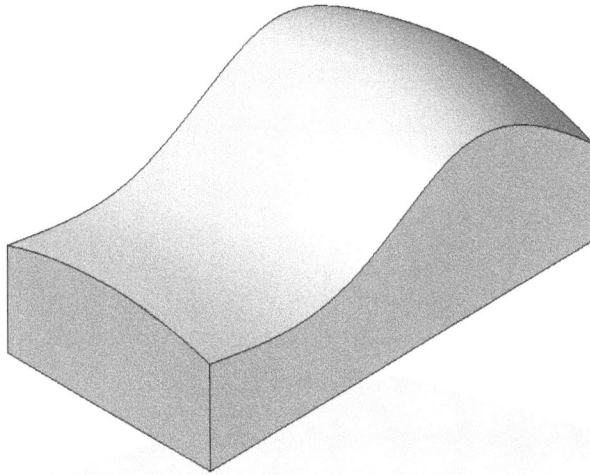

This is a very powerful tool and there are many great benefits to using it.

Exercise Complete

Exercise 39: Copying Surfaces

In a **New ANSI part** file create an **Extrusion** that looks like the figure shown below.

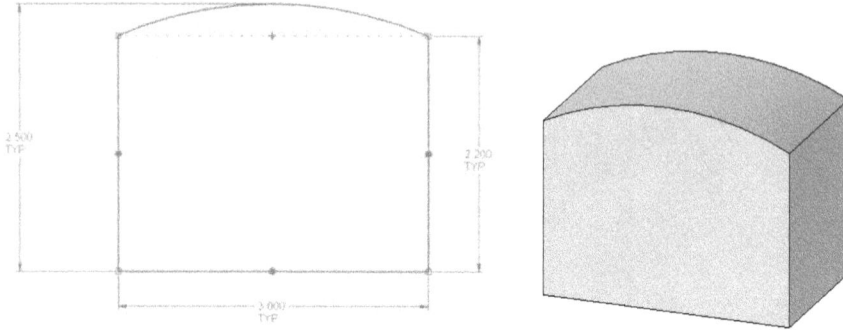

Click on the **Thin Wall** tool in the Solids options box.

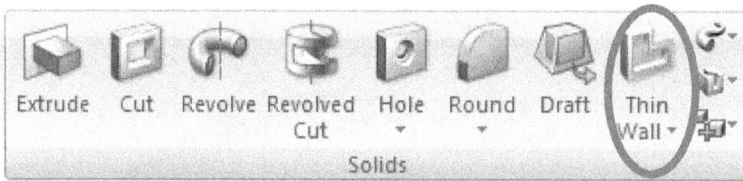

Select **Offset** Inside and type 0.125in into the **Common Thickness** box.

In the **Open Faces Step** click on the back face of the part by rotating it around and **Accept.**

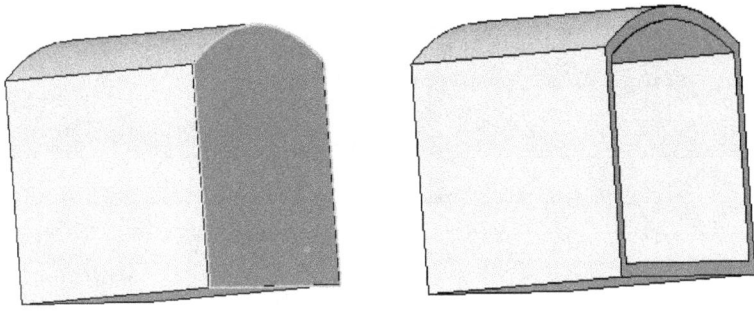

Click the **Preview** button then **Finish.**

Using the **Cut** tool in the Solids options box create a plane on the front face of the part and draw an **Ellipse** as shown below.

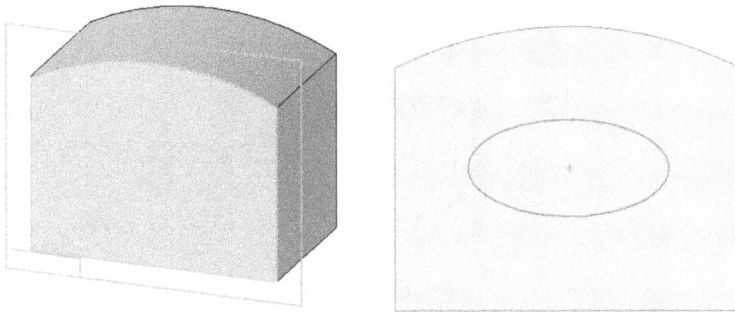

In the **Extent Step** select **Through All** and cut through the thin wall.

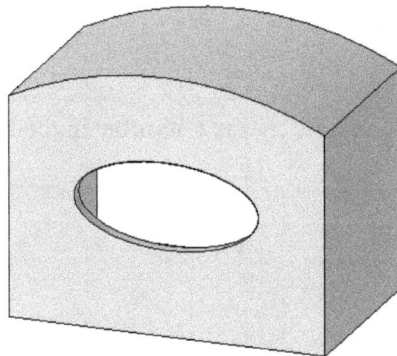

Click on the **Chamfer tool** in the **Round** dropdown menu.

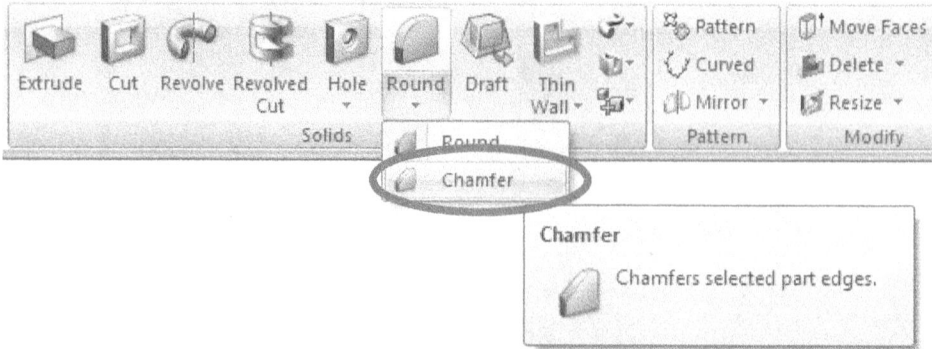

In the **Select Edge Step** click on the edge of the **ellipse** and enter a Setback of 0.125in.

Click **Accept** then **Finish.**

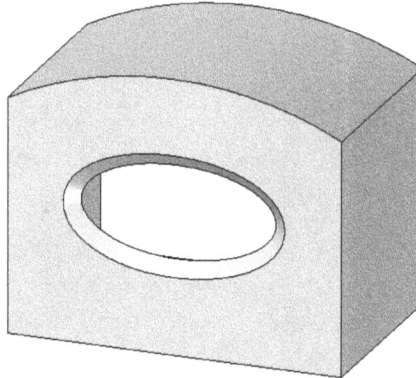

Under the **Surfacing** tab click the **Copy** tool in the Surfaces options box.

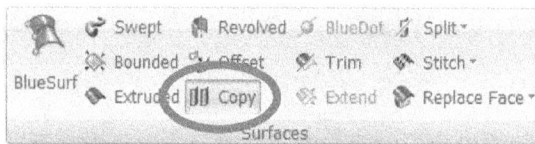

In the **Select Step** select the face create by the **Chamfer** in the last step and click **Accept** then **Finish.**

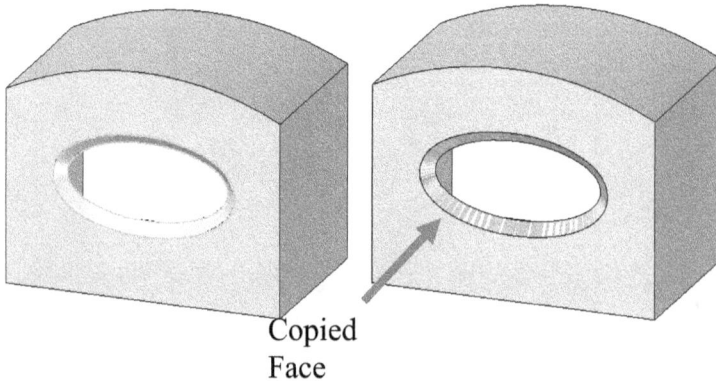

Copied
Face

Back in the Home tab click the **Move Faces** tool in the Modify options box.

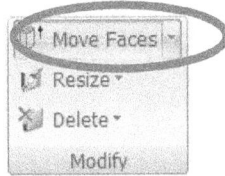

In the Select Faces Step select the face made by the **Copy** operation and **Accept.** *Hint: Hover over the face then right click to choose from a list of options for your selection.*

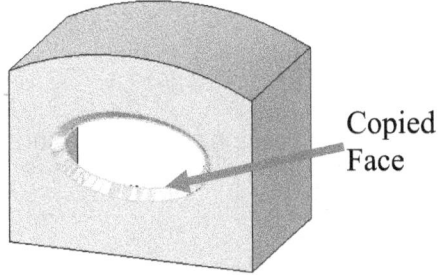

Copied
Face

In the **Movement Step** choose the option **Along Edge** and click the edge shown below. *Note: This step specifies which vector to move the surface along.*

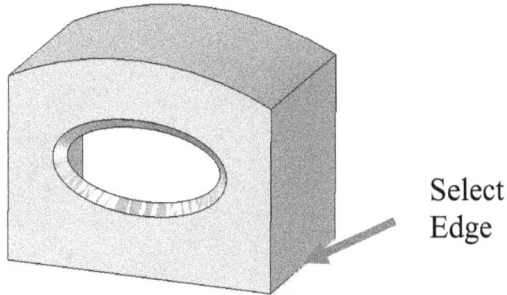

Select
Edge

In the **From Point** Step choose a point on the surface to move the face with. *Note: The easiest way to do this is using "silhouette" to choose a point.*

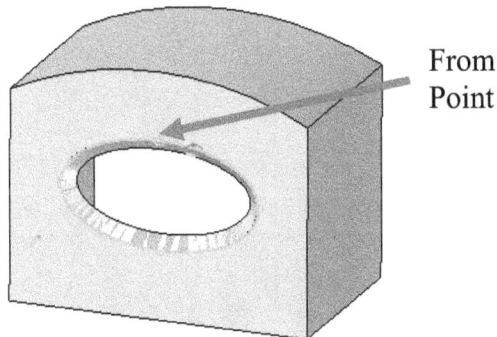

From
Point

In the **To Point Step** enter 3in into the distance box and click **Finish.**

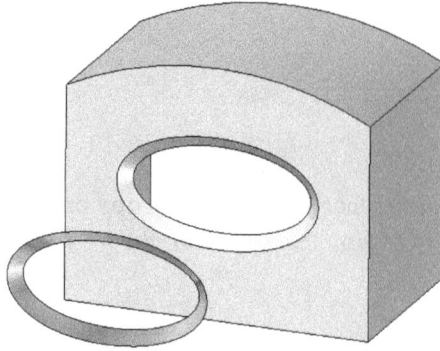

Now use the **rotate tool** in similar fashion to rotate the surface 90°, as shown in the following figure. *Note: Instead of choosing an edge to move along, choose an edge to rotate around. Wireframe view is used for clarity.*

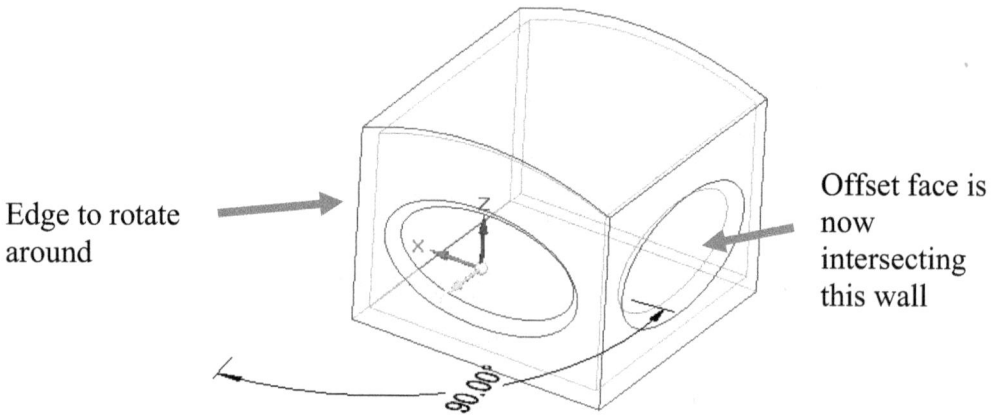

Edge to rotate around

Offset face is now intersecting this wall

90.00°

We are going to use this surface to trim the body. *Note: Boolean operations that are line on line have the tendency to fail or not work, therefore we will be enlarging the surface to prohibit that tendency.*

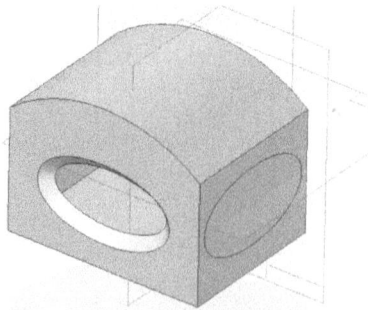

Under the Surfacing tab, choose **'Extend'** and make the two edges of the surface larger, as shown in the following figure.

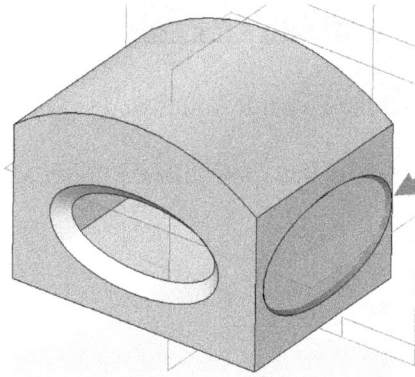

Click to extend edge – choose distance - then repeat for other side of copied face

Now use the **Subtract** tool from the replace face drop down. First select the **Tool** (surface) and then the Side (body).

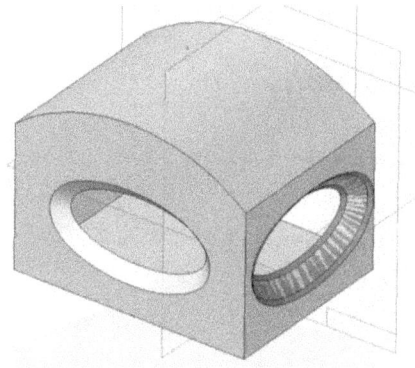

Exercise complete

Exercise 40: Boolean, and Curve Projection

Create a **New ANSI part**. Click the **Extrude** button and select a plane to sketch on.

Create a sketch using the Ellipse tool that looks like the figure shown below.

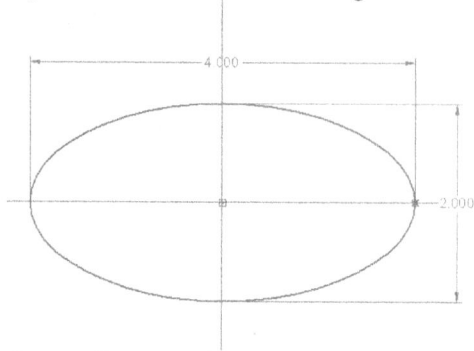

Extrude above the sketch plane a distance of 1in.

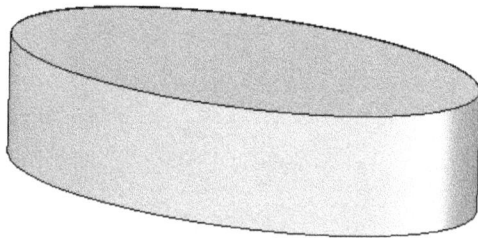

Create a **sketch** on a plane **Parallel** to the **ZY Plane** that is offset by a distance of 2.00 in.

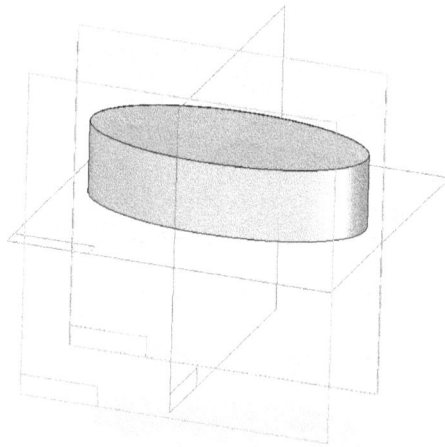

Using the **Curve** tool sketch a curve that looks similar to the following figure. Be creative.

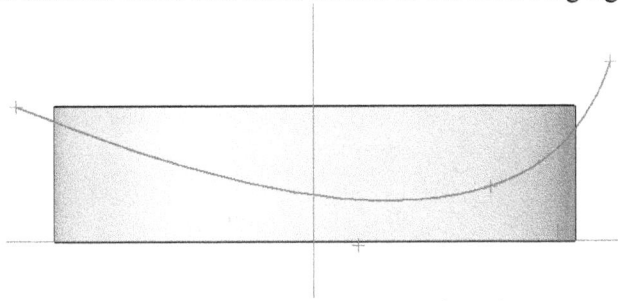

In the Surfacing tab select **'Extruded in the Surfaces'** options box.

In the Sketch Step make sure **'Select from Sketch'** is chosen and select the curve.

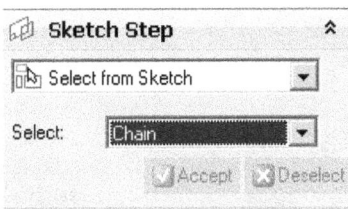

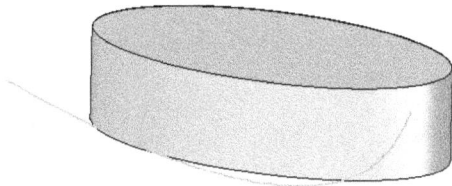

Click **Accept.**
In the **Extent Step** extrude a distance of 4in, as shown below.

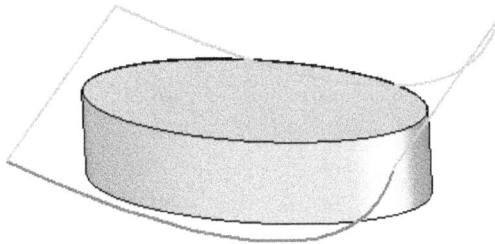

Click **Finish** and view the following surface.

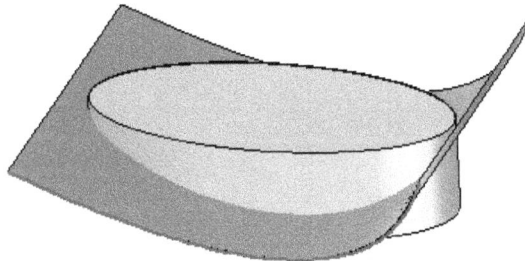

Under the Surfacing tab select the **Subtract tool** under the **Replace Face** dropdown menu in the Surfaces options box.

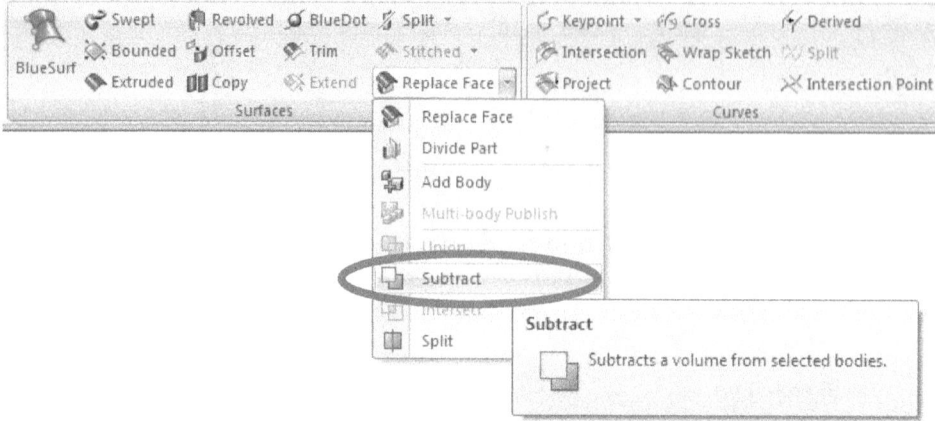

In the **'Target Step'** select the Body and then select the surface.

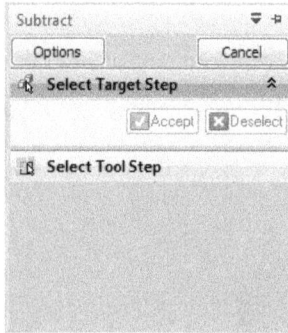

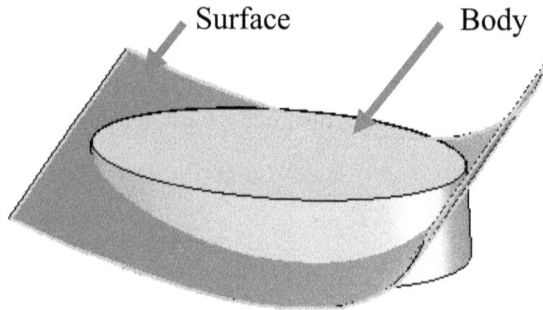

In the **'Side Step'** click a direction to subtract the material as shown in the following figure.

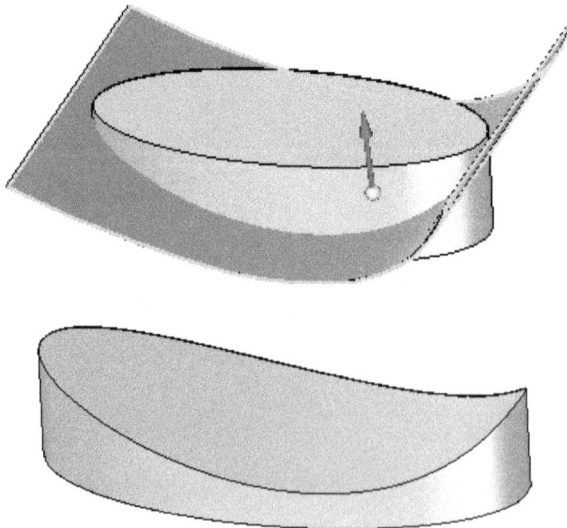

Exercise Complete

Exercise 41: Adding Threads

Create a **New ANSI part.** Create a **sketch** (in the extrude tool) that looks like the figure below. Using the **Polygon** tool, make the hexagon shown below. *Notice: The Polygon Tool can be found in the drop down menu created by clicking the arrow next to the Rectangle Tool.*

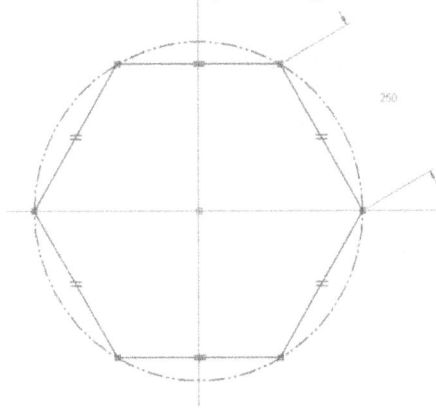

Extrude in front of the sketch plane a distance of 0.125in. Add a **Round** of 0.025in to all edges on the front and back of the hexagon and a Round of 0.01 in to the edges on the sides as shown in the following figure.

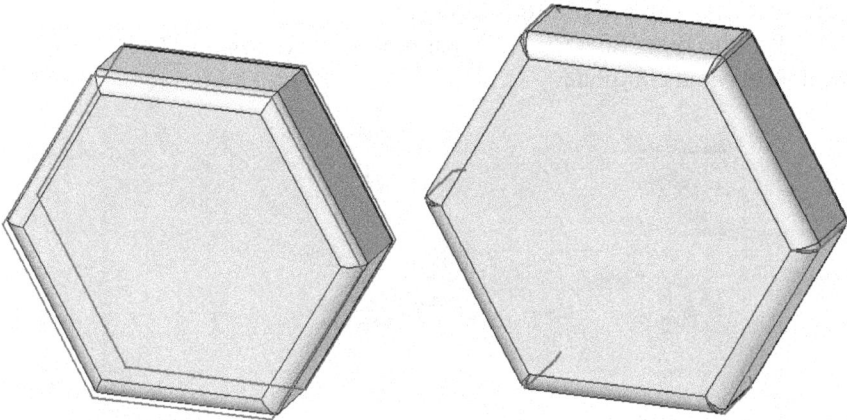

Save the part under the name Bolt to use in a later assembly.

Click on the **Hole** tool in the **Solids Options Box**.

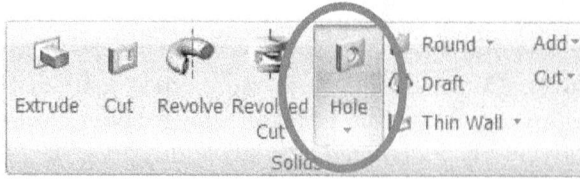

In the **Plane Step** select the front face of the part and place a hole in the center.

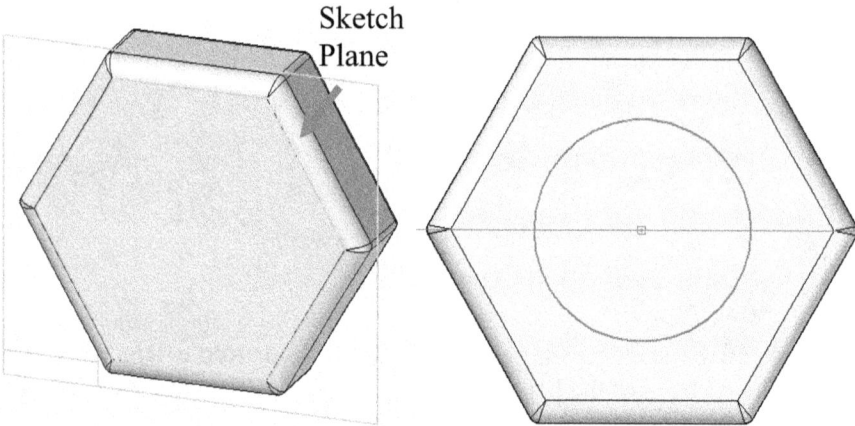

Sketch
Plane

Close Sketch and click on the **Options** button at the top of the left menu
Select Threaded as the type. Specify 0.25in in the distance box and ¼-20 UNC as the thread size, then click **OK** when complete.

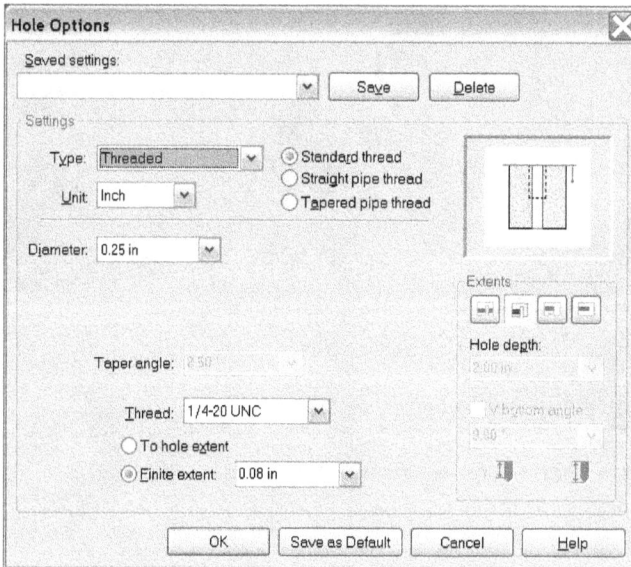

Click behind the part to specify the direction and click Finish. Add a **Round** of 0.025 in to the front of the hole as shown in the following figure.

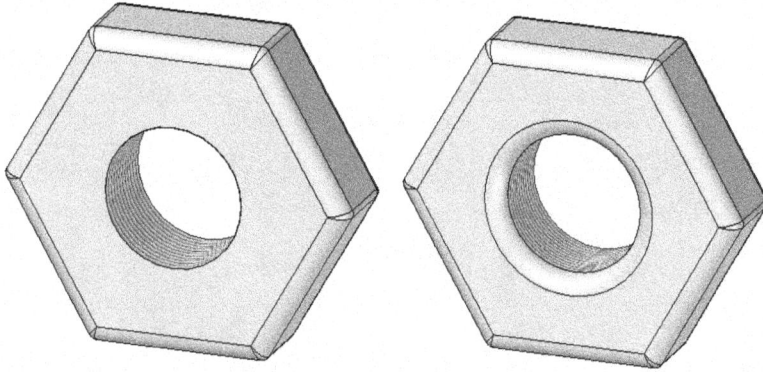

Using the **Save As** option, save the part again under the name Nut.

Open the original part named Bolt.par.

Create a **sketch** on the original sketch plane that looks like the figure below.

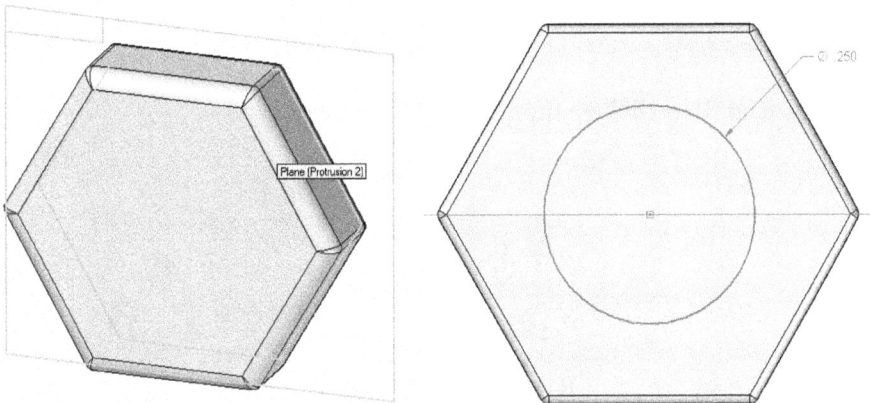

Close Sketch and extrude the sketch behind the part a distance of 0.75 in.

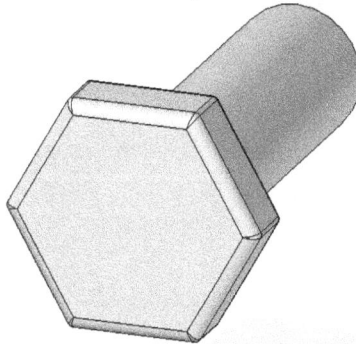

Click on the **Thread tool** under the Hole dropdown menu in the Solid options box.

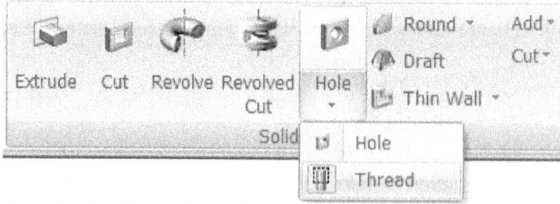

An options window that looks like the figure below will appear. Make sure 'Straight' is selected and click **OK.**

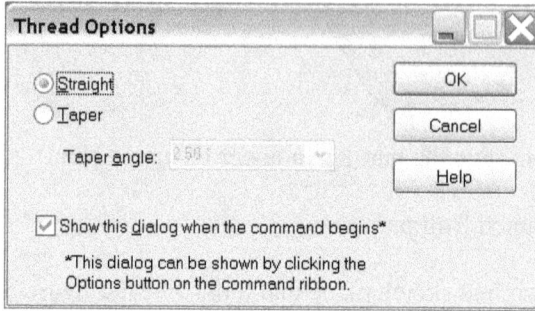

In the **Select Cylinder Step** click on the extrusion.

In the **Cylinder End Step** click on the end of the cylinder.

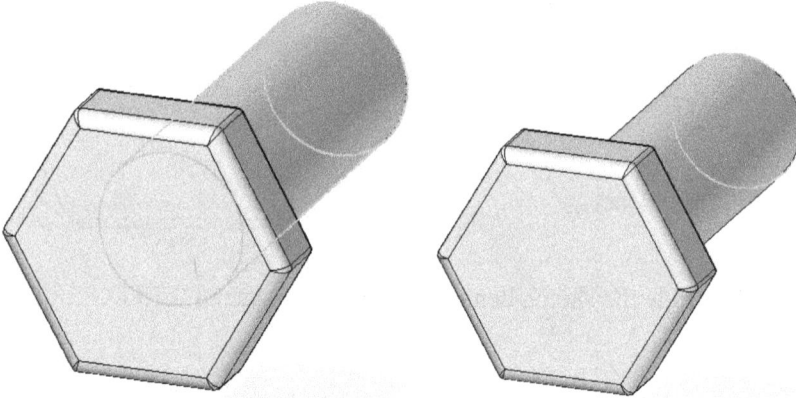

In the Parameters Step make sure the depth is set to cylinder extent and set the type to ¼-20 UNC click Finish.

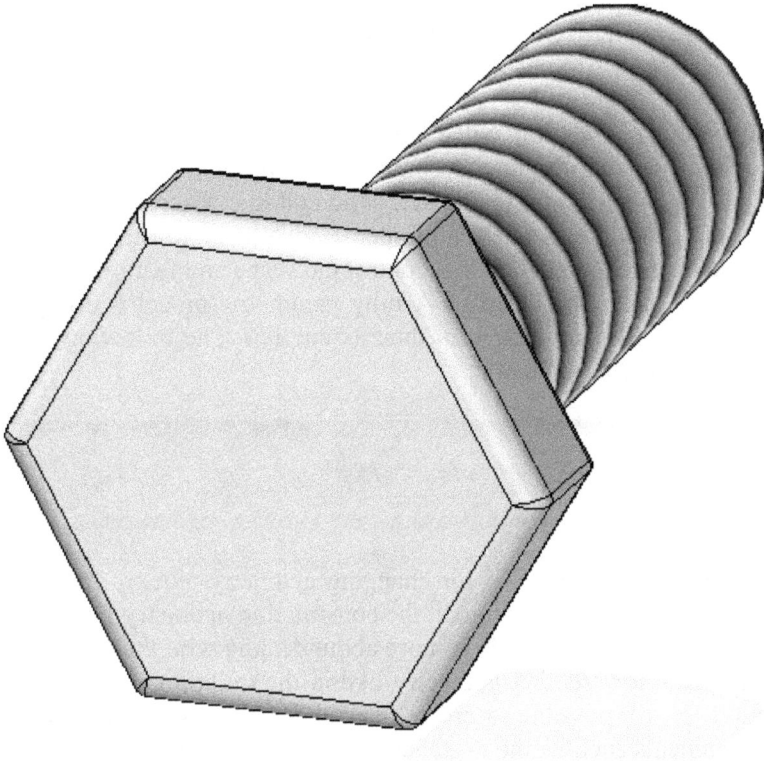

Exercise Complete

Traditional vs. Synchronous Modeling

Traditional and Synchronous are the two different modeling modes available in Solid Edge. Each have their benefits as well as their downfalls. Which mode to work in is entirely up to the user or the employer. The following is a general discussion on the pros and cons of each mode.

The **traditional mode** of modeling will come most natural to any existing CAD user as synchronous modeling is a fairly new technology. Traditional mode offers such tools as sketches, lines and arcs to define the shape of a solid and it is possible to track the creation of the model using the Pathfinder. The Pathfinder is available for the user to go back in 'modeling time' and change parameters that control the shape and features of the model. It is a great way to work and has many benefits. A fully parametric model is a huge goal at many companies and can allow full automation of their assemblies. The following screenshot shows the main toolbars in the traditional mode.

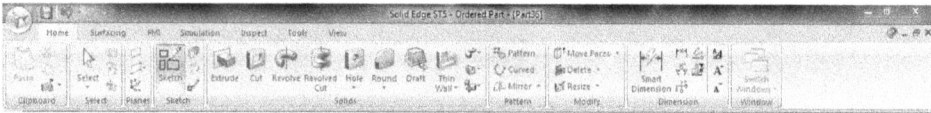

Synchronous modeling is a great tool for changing geometry quickly that otherwise would require fishing through the Pathfinder to edit the construction geometry that would be required to make any change. Synchronous mode is more about dealing with the actual geometry (faces etc) than the preparation for creating geometry like in the traditional mode. It's a more fluid modeling process. It is still possible to create lines and arcs on faces to extrude and revolve however the environments such as the sketcher and boolean are merged, in fact everything is merged to create a very unique painting palette.

The following few sections of the workbook go through both modeling environments. Each exercise will indicate type of part file will be used.

Introduction to Synchronous Modeling

Synchronous modeling is essentially a streamlined modeling environment. It is ergonomically designed to allow the user to model with maximum efficiency. Synchronous modeling does have its downsides, however: Creating organic models or models with intricate details can be difficult and these tasks are better left to the more organized Ordered mode. Also, a part in which the history and steps are important would be more suited to Ordered mode. Luckily for you, you can have the best of both worlds. You can now switch from Synchronous mode to Ordered mode from within a single part by right clicking anywhere in the design window and then choosing "Transition to Ordered".

"So how do I benefit from this?" you may be asking. Well, you can utilize both modeling environments by creating the base of a part quickly and efficiently in Synchronous mode and then switching over to Ordered mode to add detail to the model. There is, however, a catch. You can freely switch from Synchronous to Ordered mode with no problem, but if you transition back to Synchronous mode, the changes you made in Ordered mode will be effectively suppressed.

All of what has been mentioned here and more will be covered in the following lessons on Synchronous modeling.

Exercise 42: Working in a Synchronous Modeling Environment

When starting a synchronous modeling part file some of the tools that were available in the traditional environment are now gone. Like, for example, the sketcher. Fear not, most of the functionality is still present. It's hidden in the big modeling melting pot that is a synchronous part file.

Start **a new ANSI part file** and convert it to Synchronous. Under the home tab all the tools normally used in the traditional modeling environment sketcher are available. Click on the **rectangle tool** and draw a rectangle on the **XY** axis 35mm x 20mm - use the dimension tool. *Note: The dimensions stay displayed, click on one to edit.* Extrude 2.5 mm.

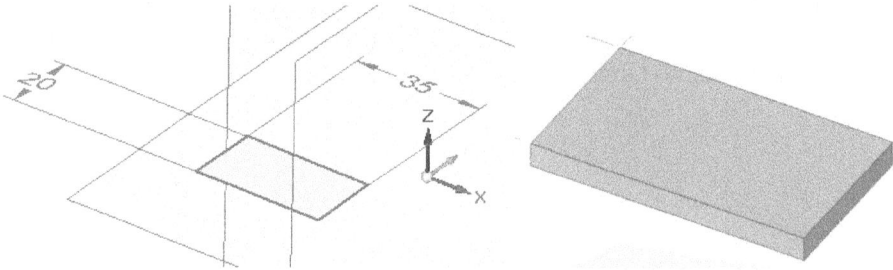

Now draw two small **rectangles** on the face of the block. Use some **construction lines** and **connect midpoint** constraints to position the rectangles equally spaced on the block, as shown in the following figure. Extrude the rectangles 30mm.

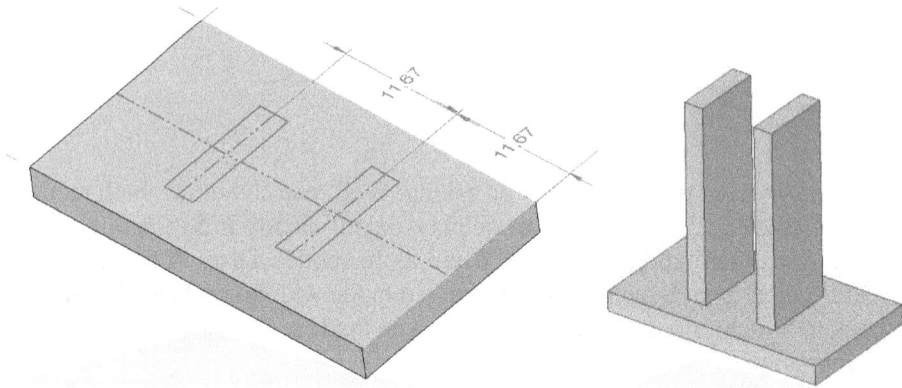

Now create a circle on the end of the last extrusion. **Extrude** using the **cut** tool located on the extrude drop down menu. This makes a cut feature.

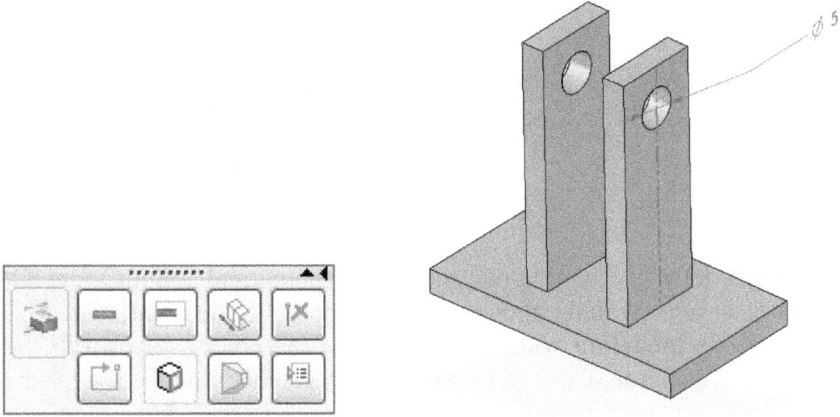

Create another extrude on the bottom face of 25mm and this time add a 5 degree draft angle using the **'Treatment'** section on the lower part of the Extrude box.

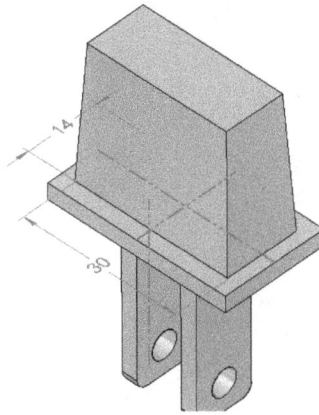

Add two **concentric circle** extrudes that match the following model.

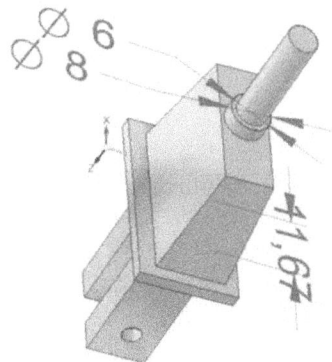

Add 0.5mm rounds to the edges shown in the following figure.

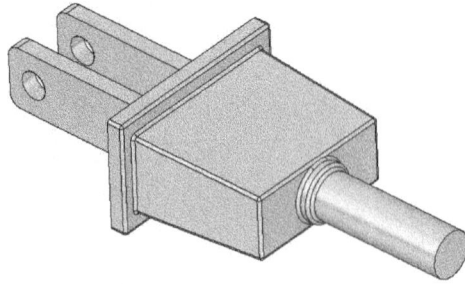

Now for the fun part, let's start editing our geometry. The normal options are not available to edit the features that were created - though they do appear in the pathfinder. Make edits by using the **synchronous modeling wheel**. Select one of the holes as shown in the following figure. Two arrows will appear. This is a tool for editing geometry on the fly.

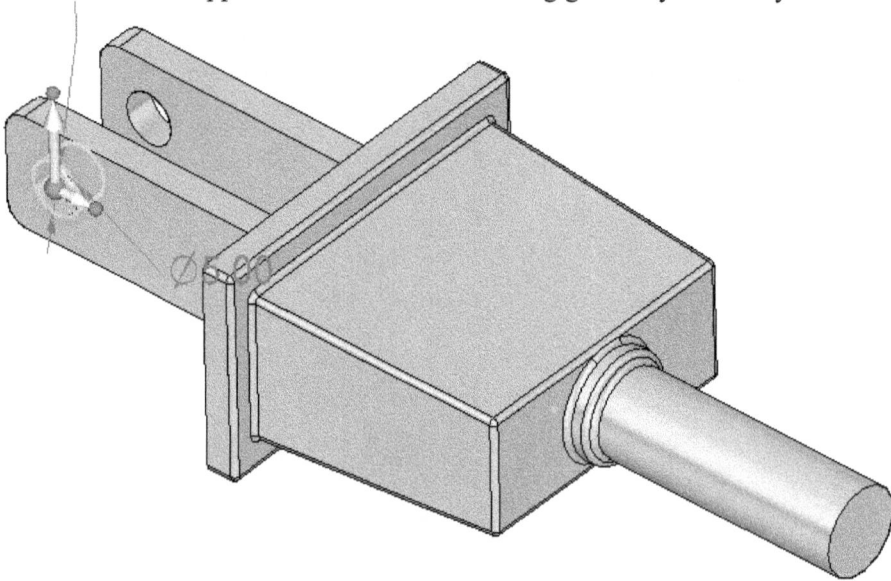

Drag the small arrow towards the main body and watch the hole move location. This is synchronous modeling in its essence. The other hole may move with the selected hole but don't be alarmed. This is just one of the amazing features in the synchronous modeling world. On the bottom of the screen, all the live rules are available. Uncheck the concentric constraint and the symmetric constraint and only the hole selected will now move.

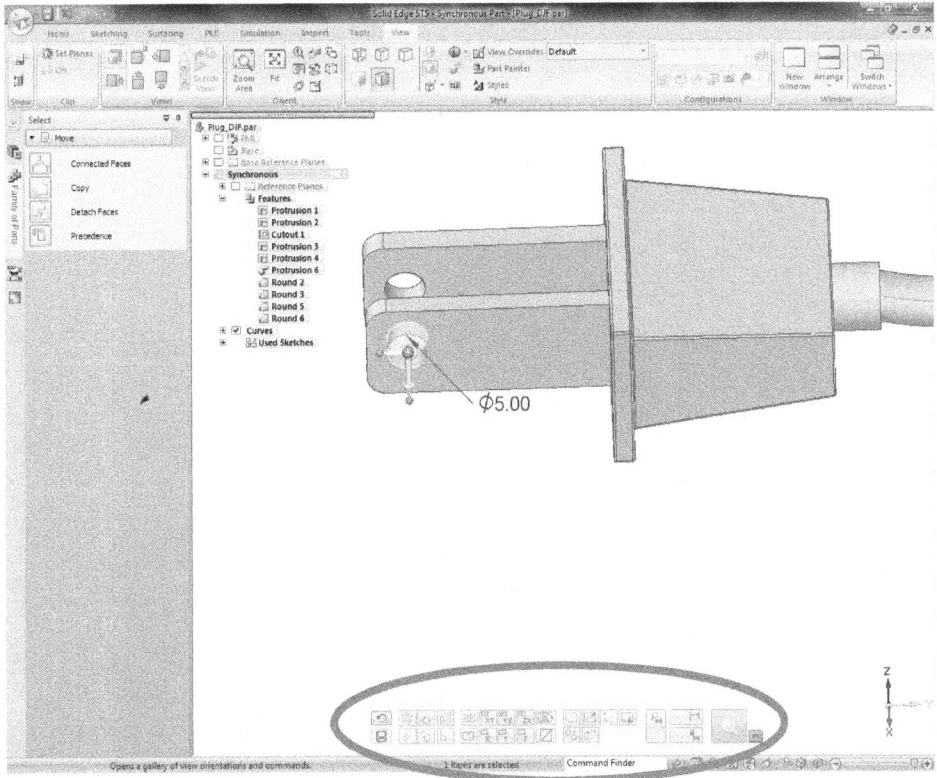

Click on a round or a series of rounded edges to quickly change the radius – change the radius to 1mm on the rounds surrounding the first extrude we create, as shown below. *Hint: Hover over one of the rounds a green plus sign will appear - click on this it will bring up a selection intent manager where it is possible to select an entire blend chain. This works with many different types of faces.*

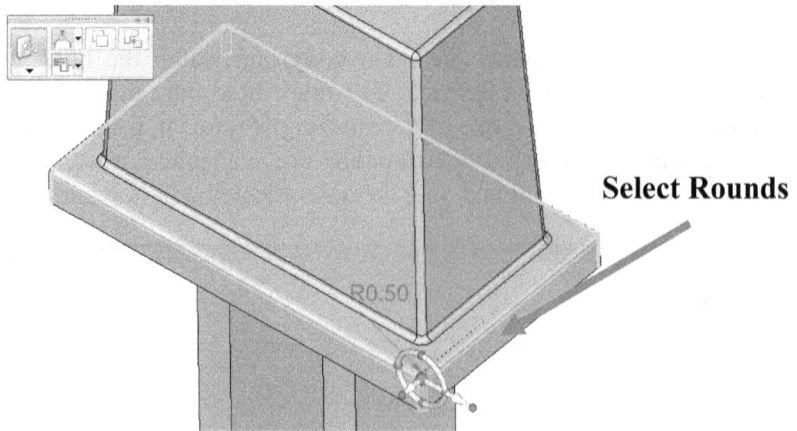

Select Rounds

In the same way select all the **rounds** shown in the following figure.

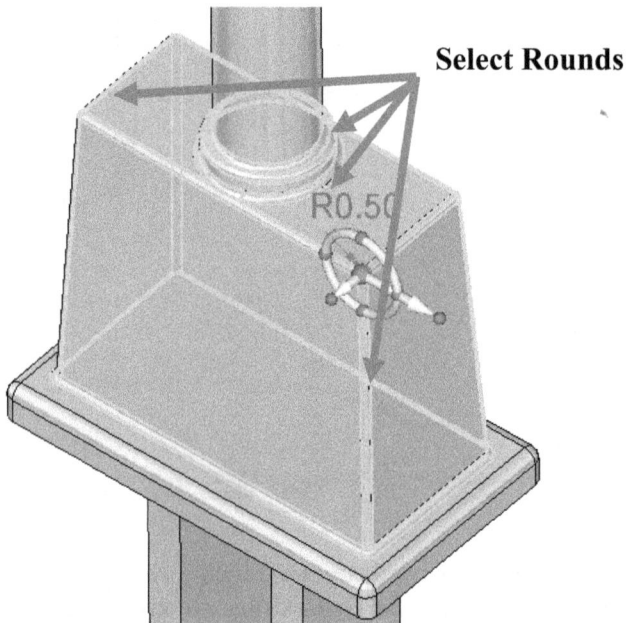

Select Rounds

Now simply click on the **delete** button on the keyboard and all the rounds will disappear, as shown below. It is possible to do this this with many different types of faces like holes, bosses, features of almost any kind.

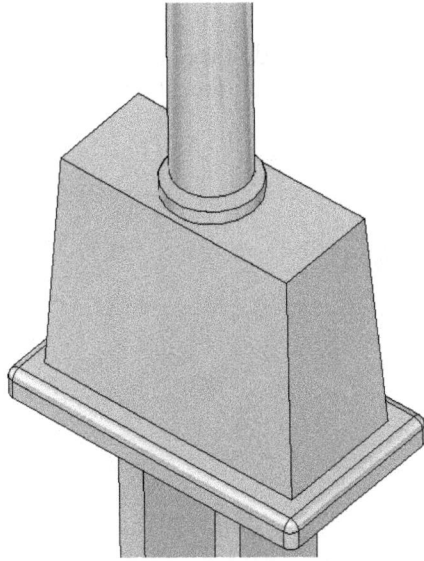

Now let's change the angle of one of the side faces. Click on the face shown below. An arrow and a **sphere** will appear, use this to move the face along the vector, don't use this.

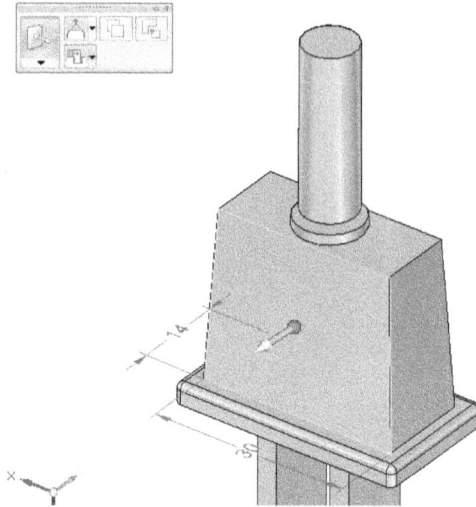

Click on the sphere to get the **SE synchronous modeling steering wheel** that helps the user manipulate face's of a model. Drag the center of the steering wheel to the top edge as shown below.

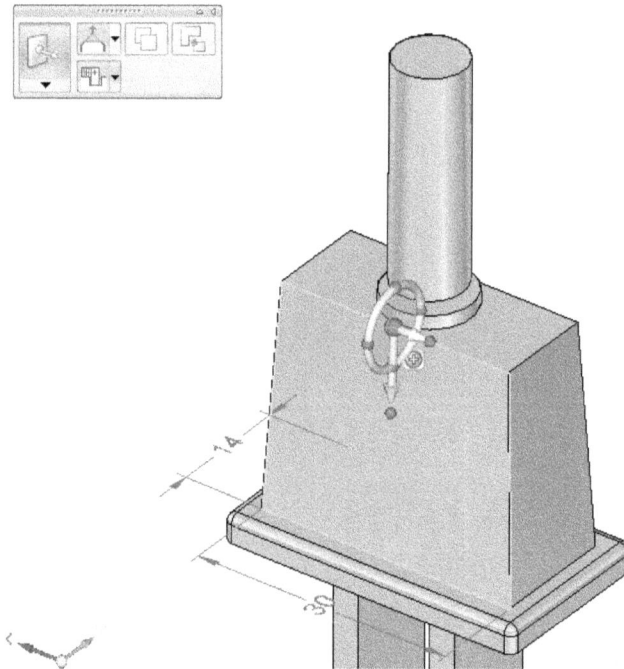

By turning the wheel using one of the 4 spheres around its perimeter the angle of the face with modify, as shown in the following figure.

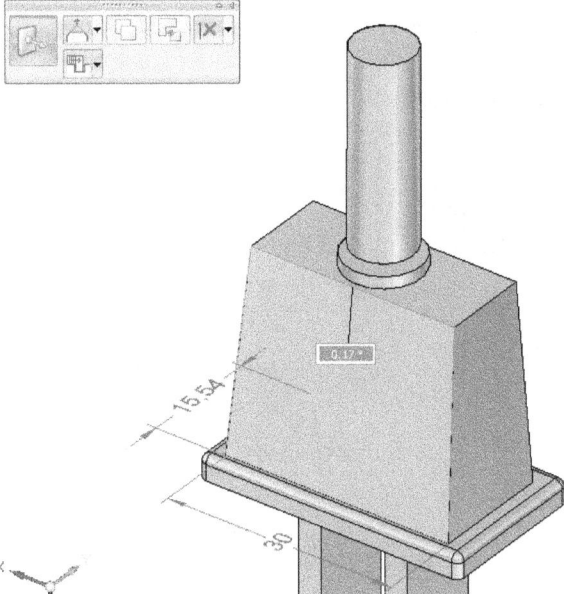

Exercise complete

Exercise 43: Live Sections and Live Rules

Live sections are exclusive to the Synchronous modeling environment and can be quite handy. They are 2D cross sections of a 3D model and can be utilized to modify the existing part or new geometry can be created using them.

Start a new **ANSI part file** and convert it to Synchronous. Create the following 5in x 5in x 5in cube.

Now create a 2.5in diameter cylinder that is 2.5in tall in the center of the cube. Then create a 1in diameter hole that goes through the center of the cylinder and all of the way through the cube.

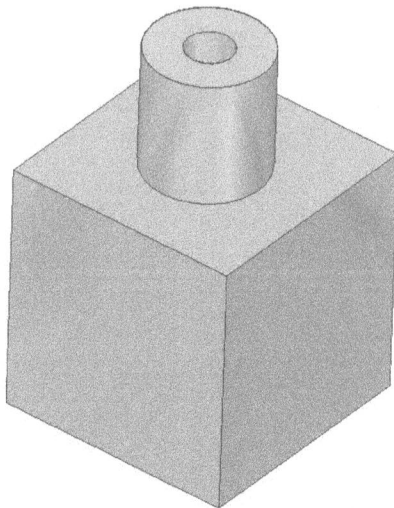

Click the **Live Section** command.

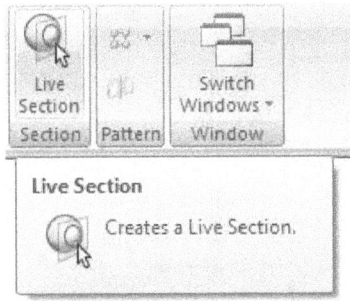

Create a live section through the center of the part as shown below. The easiest way to do this is by clicking one of the cube's sides and offsetting the live section 2.5in inward.

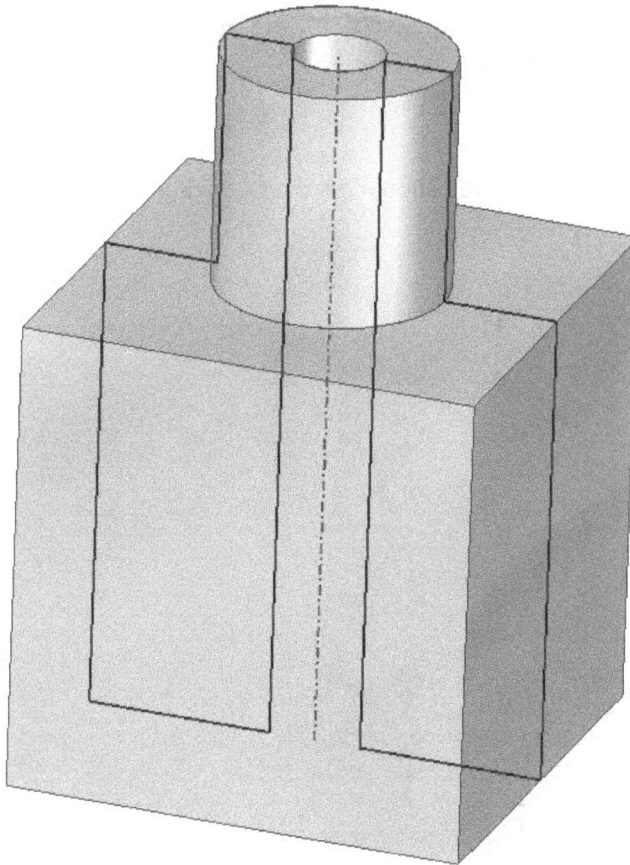

Click the edge of the live section shown below and then click the red highlighted ring.

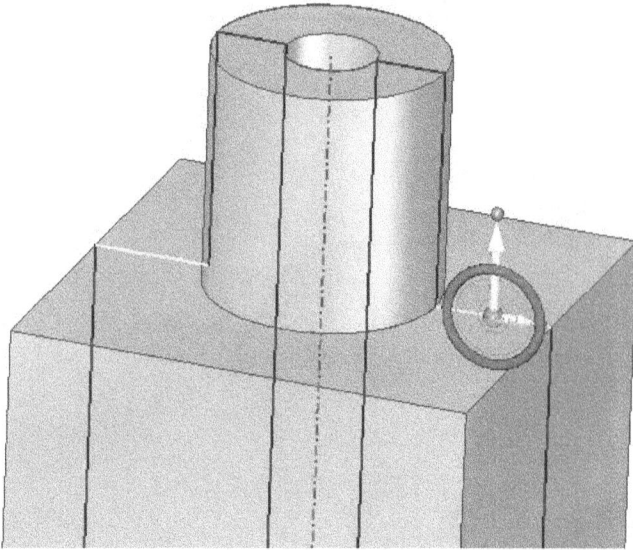

Enter a value of -30 degrees and you should get the following model.

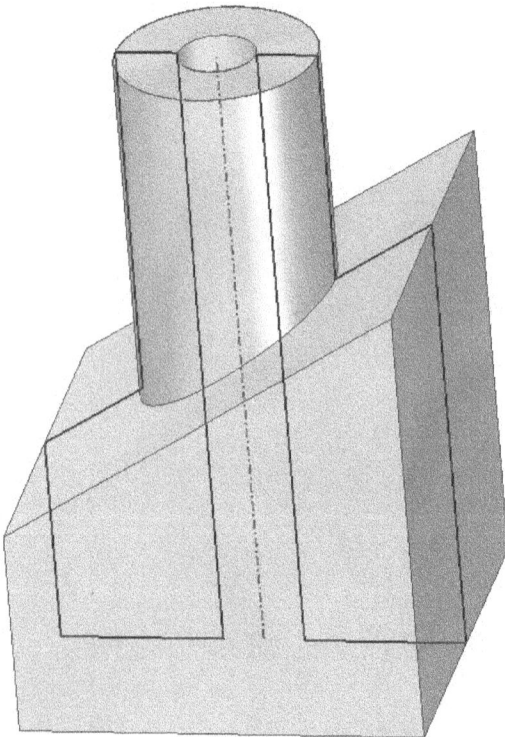

Click the edge of the live section shown below and then choose the red highlighted arrow.

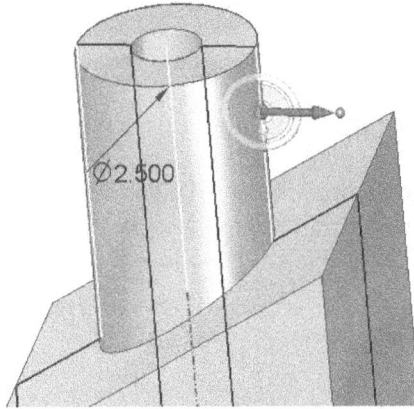

Input a value of 1in and you should get the model shown below. Notice how the live section updated to compensate for your modification. *Note: If you have trouble completing this operation, click the box indicated below by the red arrow. We will go over these in more detail in just a moment.*

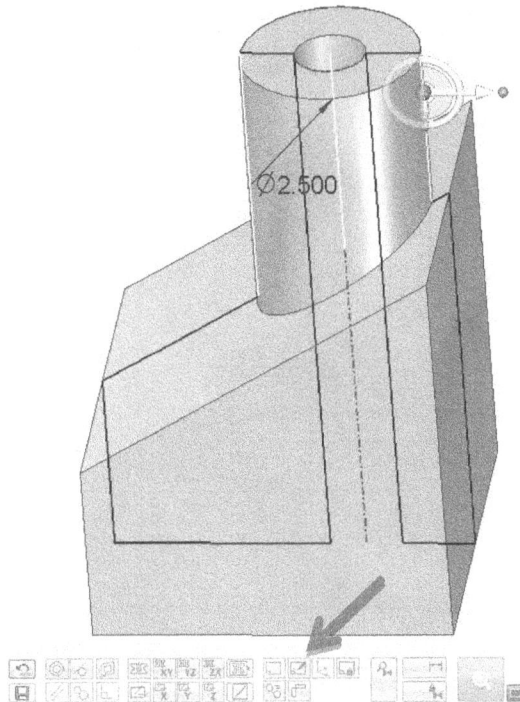

Also, notice how the hole moved with the cylinder even though you only selected the outside of the cylinder. This is due to **Live Rules**. The live rules active for any give feature can be found at the bottom of the screen when you click something to edit it synchronously. They look like the ones shown below.

Live rules are basically a set of rules and relationships that must be adhered to when synchronously editing a part. To demonstrate how they work, we will edit them and then move the cylinder using the live section just as we did previously. Undo what you just did so that the cylinder is in its original location. Click the same edge you did before, but this time deselect the "Maintain Concentric Faces" live rule shown below.

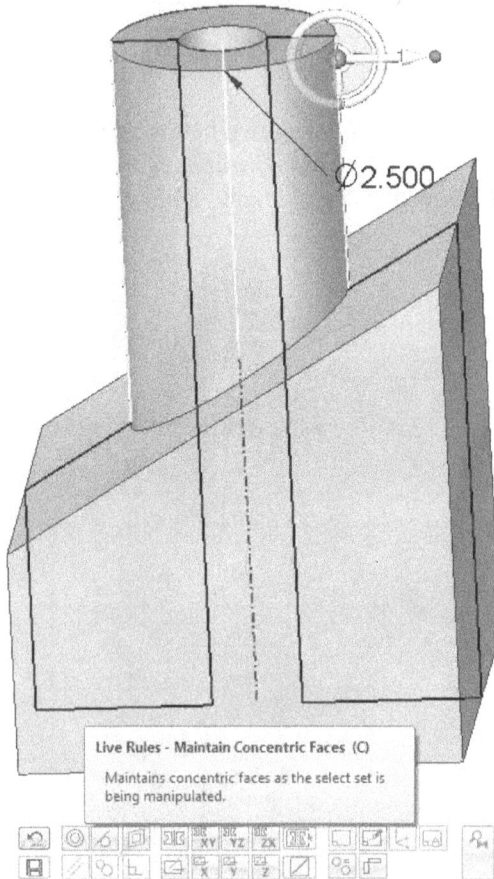

Ø2.500

Live Rules - Maintain Concentric Faces (C)

Maintains concentric faces as the select set is being manipulated.

Now move the cylinder 1in just as you did before. This time, however, your hole should not have updated accordingly.

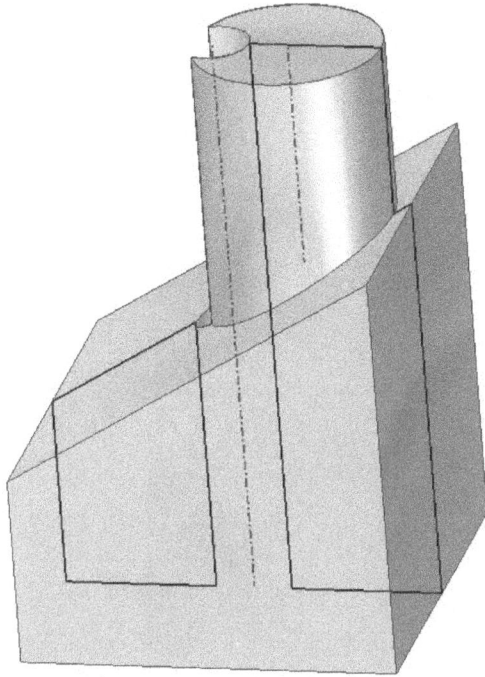

Now click the **Include** tool.

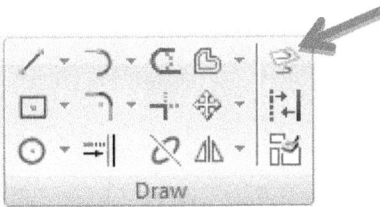

Now select the entire right portion of the live section. (you will have to click the line segments individually)

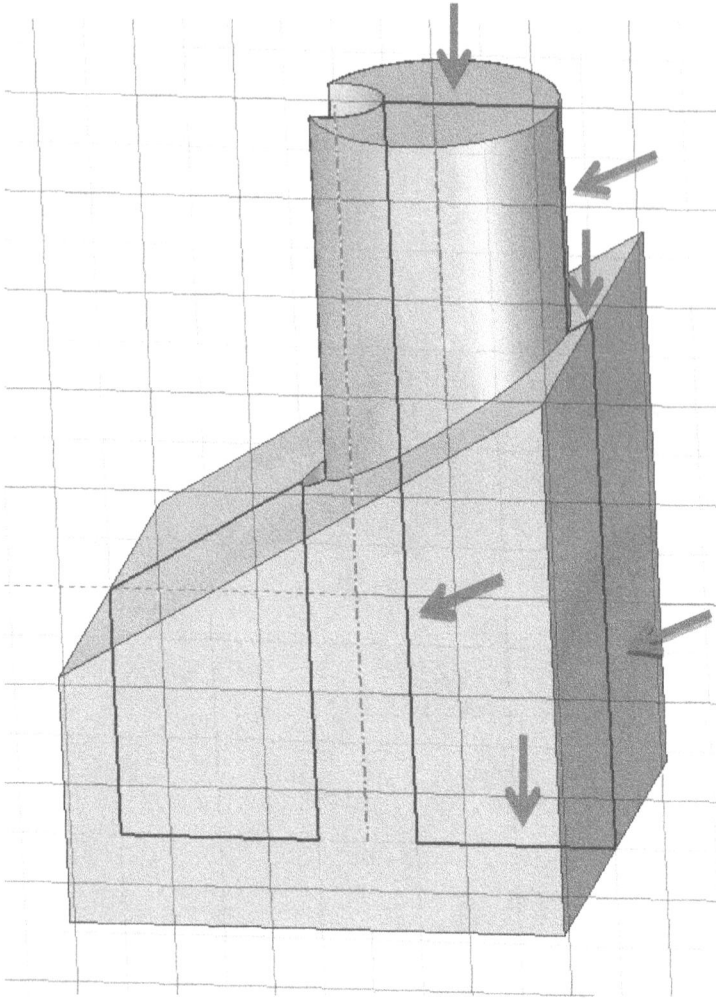

Click **Extrude** and select the curves you just created using the include tool. Extrude them symmetrically by 5in. You should now have the following model.

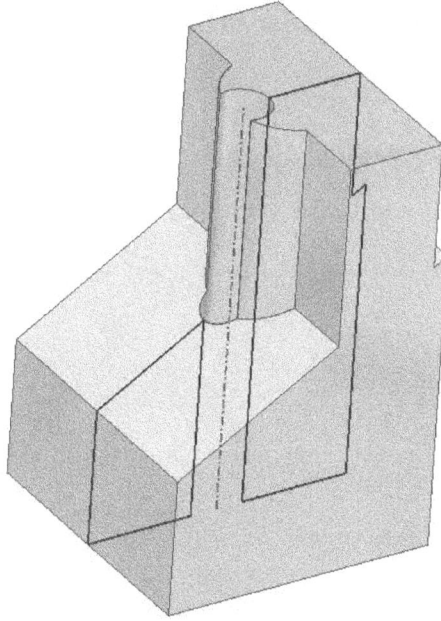

As a final step, use the top edge of the live section to rotate the top face so that it is parallel to the face you slanted earlier.

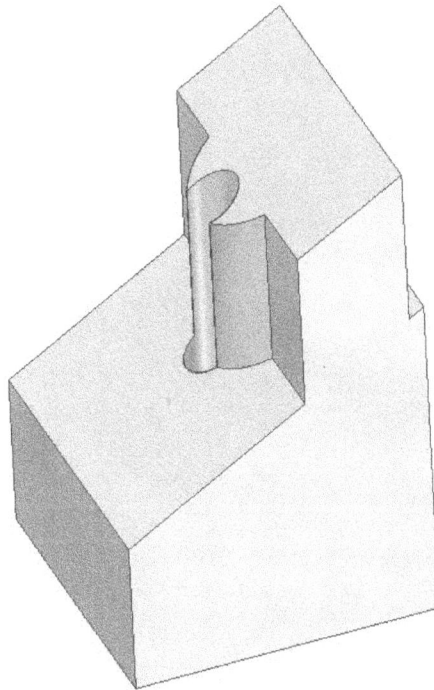

As you can see, **Live Sections** and **Live Rules** are very powerful when used to aid in the synchronous modeling process. Try playing around with them a little more to get more familiar with using them.

Save this as **Synchronous_live_sections.par** so that we can use it in the next lesson.

Exercise complete

Exercise 44: Adding Ordered Features to Synchronous Models

Open **Synchronous_live_sections.par**. Right click anywhere in the design window and click "Transition to Ordered".

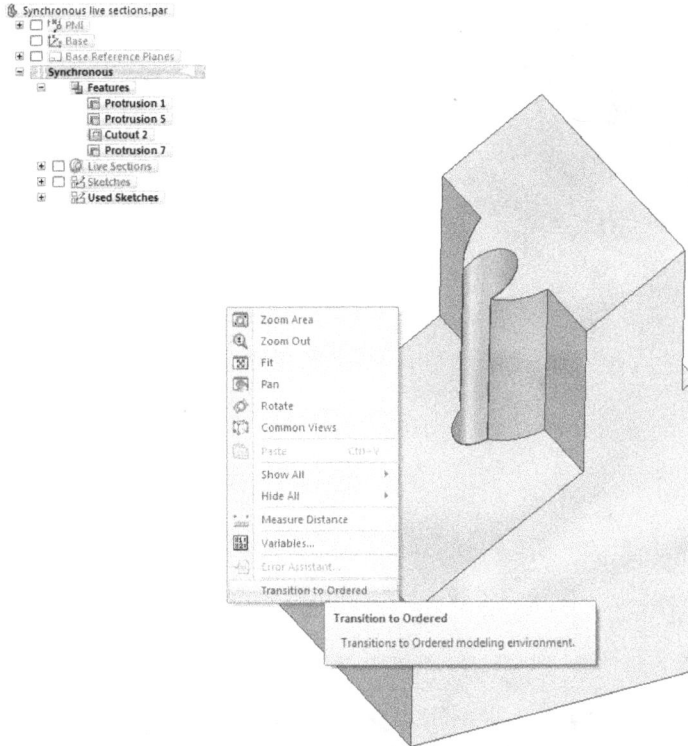

You'll now notice that your PathFinder indicates that you are in Ordered mode.

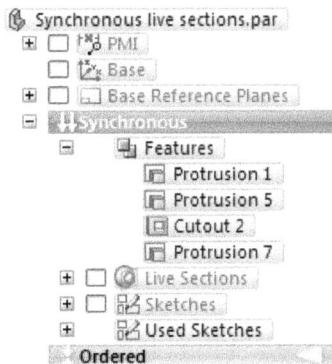

From Ordered mode, you can still synchronously edit the features you created in Synchronous mode, but in addition to this you can add ordered features. This makes for a unique mix between the two modeling environments. The only problem is, when you switch back to

Synchronous mode, your features created in Ordered mode are effectively suppressed until you return to Ordered mode. To demonstrate this, we will start by adding a few ordered features to the model.

Create the following 0.25in x 0.25in square on the slanted front face of the model.

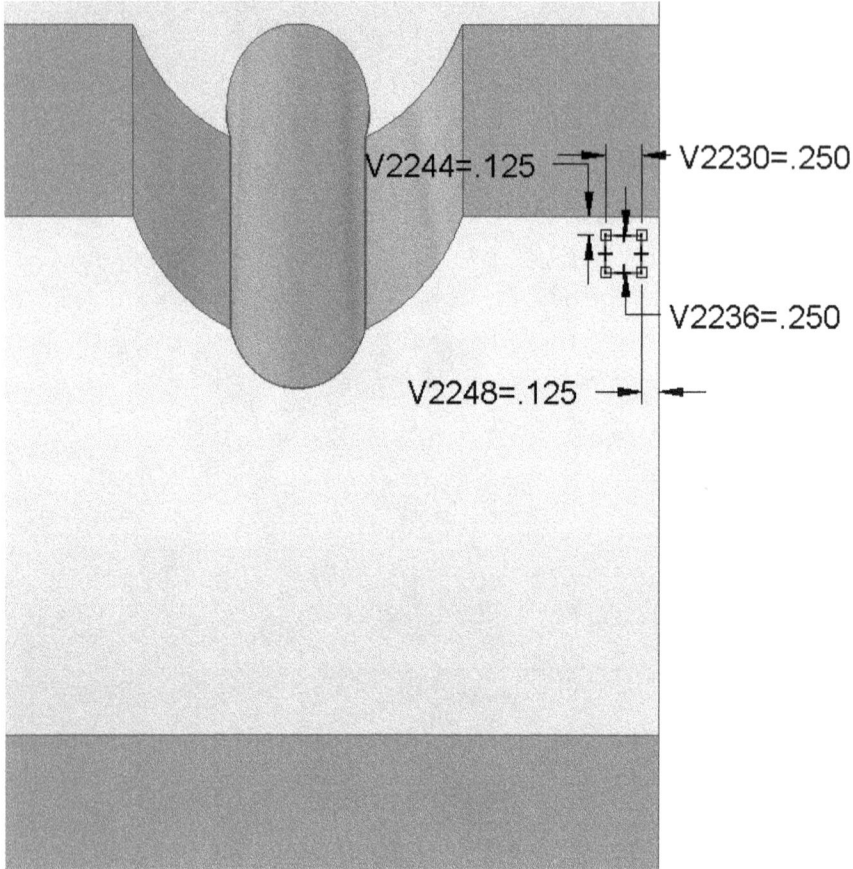

Extrude the square by 0.125in. You should now have the following model.

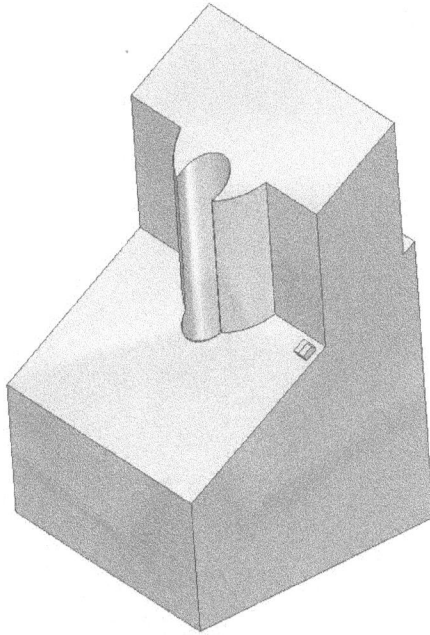

Apply a **Thin Wall** command with a thickness of 0.25in to the model. Choose the two sides of the model as open faces. You should get the model shown below.

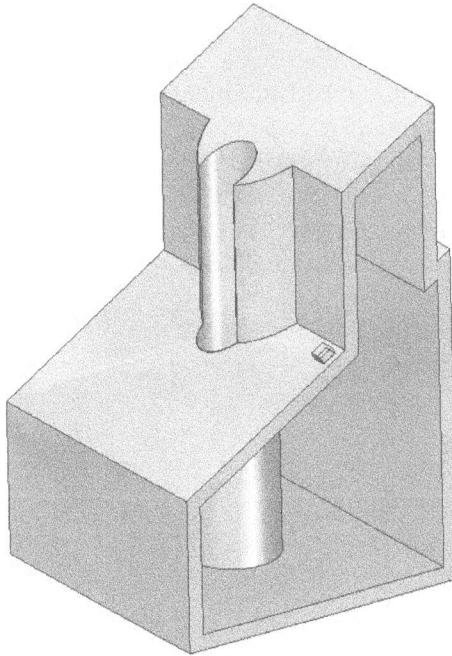

Now let's say we want to **Fill Pattern** that small block on that slanted face. (a feature only available in Synchronous mode) Right click and select "Transition to Synchronous". You'll notice that the changes you've made have disappeared, but are still in the Ordered PathFinder.

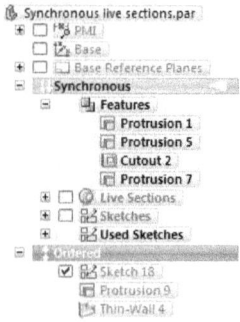

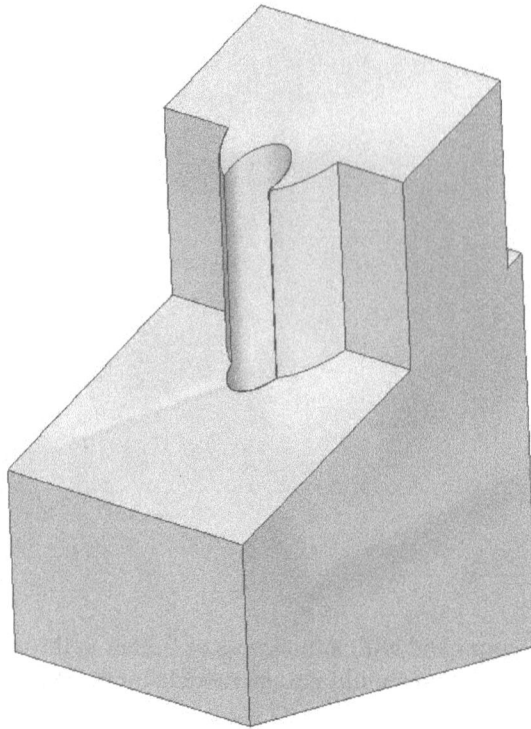

To overcome this obstacle, first transition back to Ordered, then right click the protrusion that designates the small block in the Ordered PathFinder and click "Move to Synchronous". *Note: Be cautious when doing this as it only goes one way. You can't transfer features back to Ordered unless you click "Undo". Also, any feature you created prior to the one you are converting to Synchronous will be converted as well. You'll notice that, since the sketch used to create the block was created prior to the block itself, that sketch will also be moved to the Synchronous PathFinder when you move the block.*

Now convert back to Synchronous. You should have the following model.

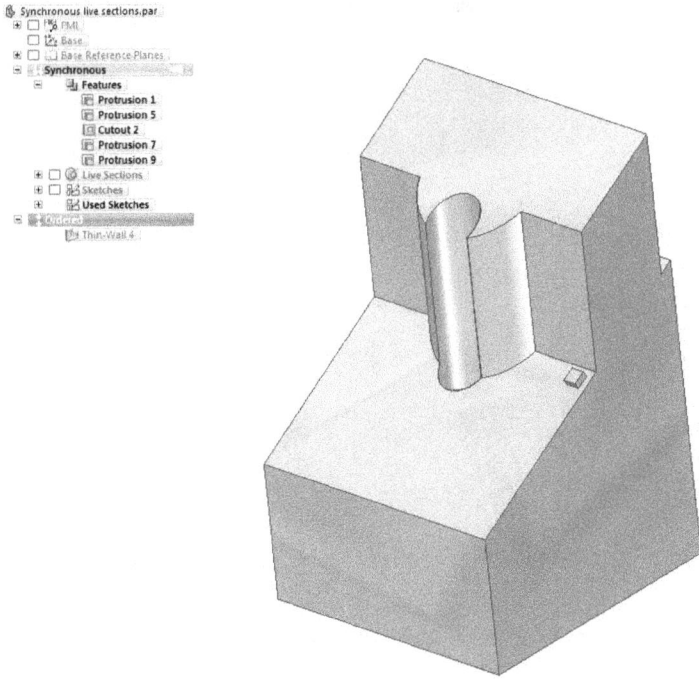

Now select the protrusion in the PathFinder that designates the small block again and then choose **Fill Pattern**.

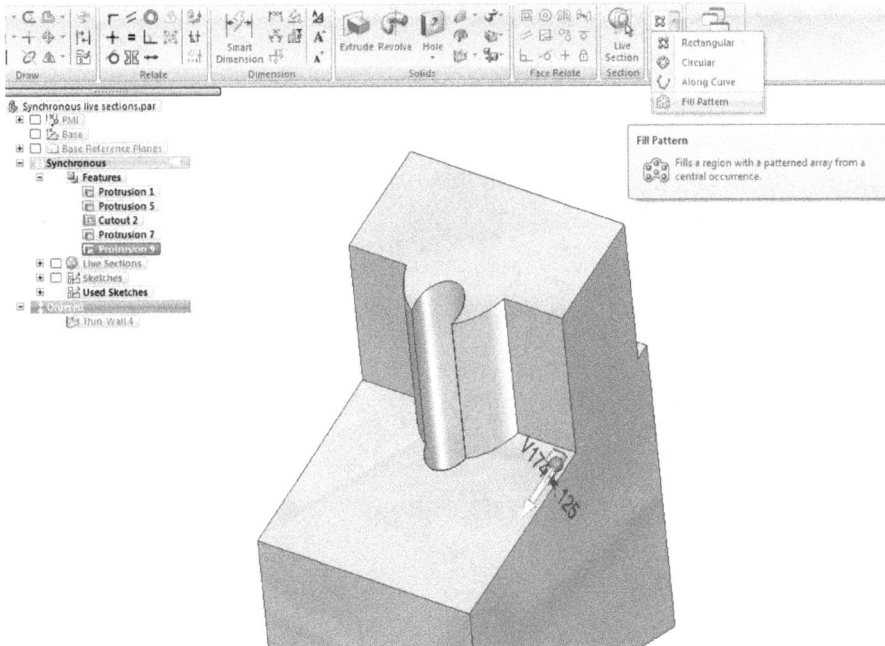

First choose the slanted face that the block is located on as the face to fill. Then input a spacing of 0.5in in both directions as shown below.

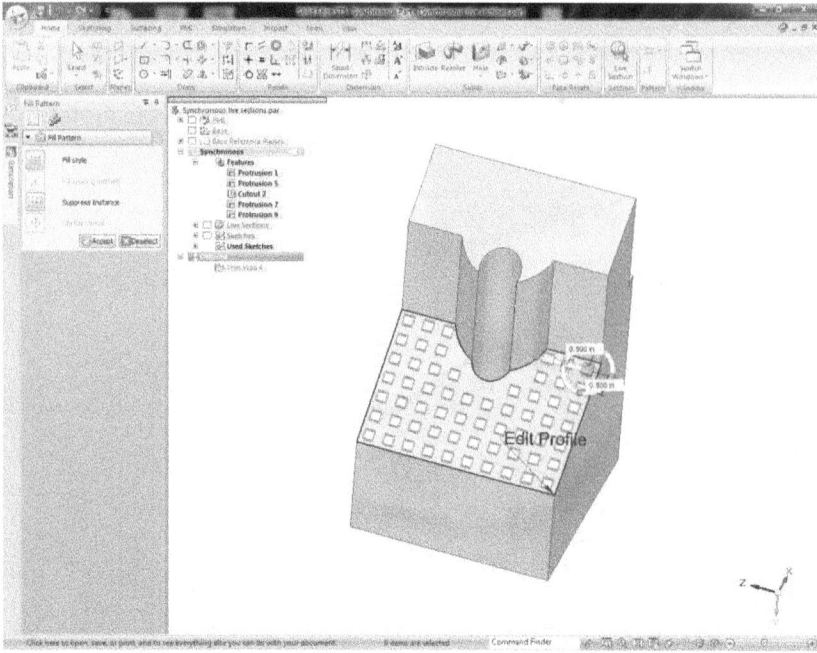

Click Accept. You should now have the following model.

Now transfer back to Ordered mode and you should have the following part.

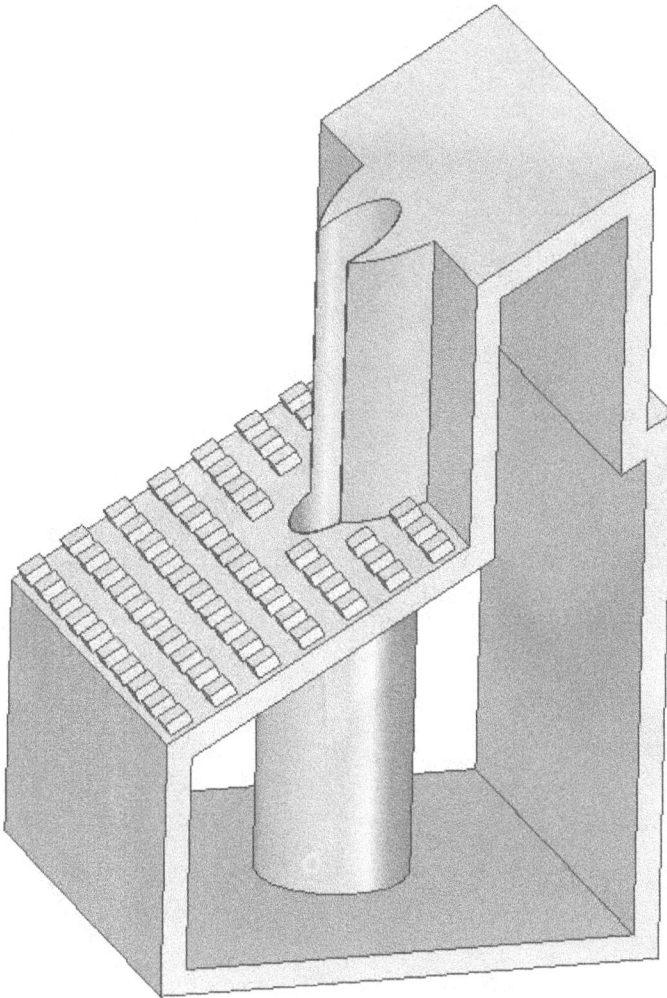

Now, while in Ordered mode, click the back face of the model. (a feature created in Synchronous mode and therefore capable of being synchronously edited even while in Ordered mode) Extend it backwards by 2in and you should have the model shown below.

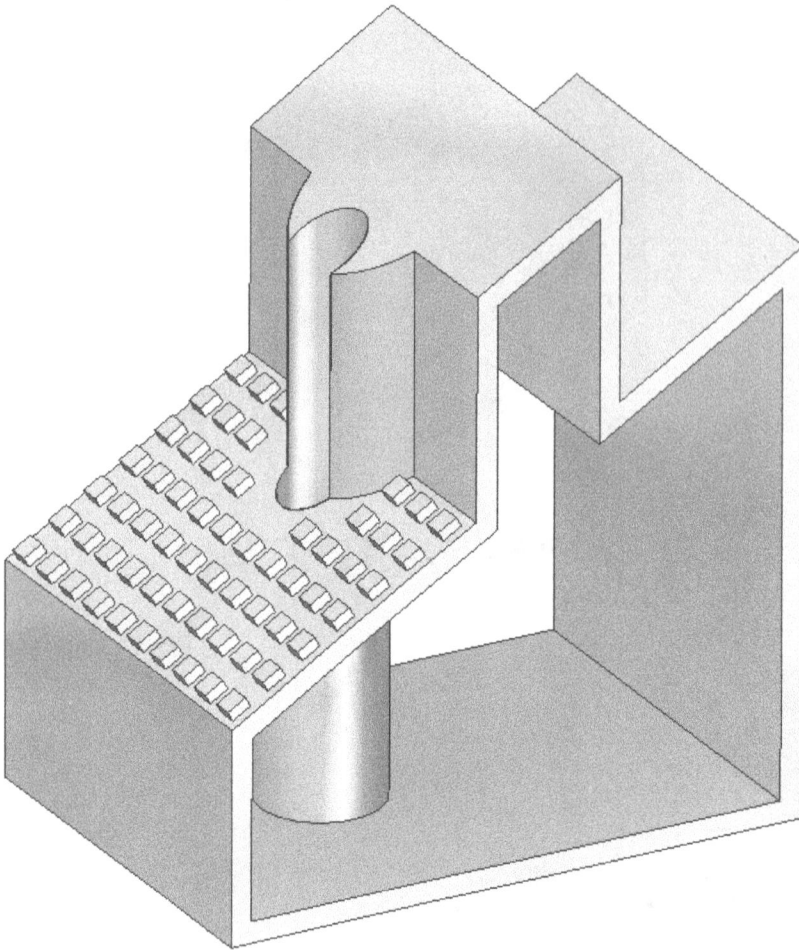

Exercise complete

Exercise 45: Editing Imported Parts using Synchronous Technology

Before we begin, go to designviz.com and download the Sold Edge ST5 work parts. You'll notice that this is not a SolidEdge part file. This part was created in NX 8.5, but that won't stop us from Synchronously editing it. The wonderful thing about Synchronous mode is that the individual steps taken to create a part are insignificant, and therefore we are able to edit parts without having any history about how they were created.

Open Lesson 45 Part File. When prompted to do so, choose the **ANSI part** template. You should now have the following part. Notice how the part has no design features and is simply a "Body Feature".

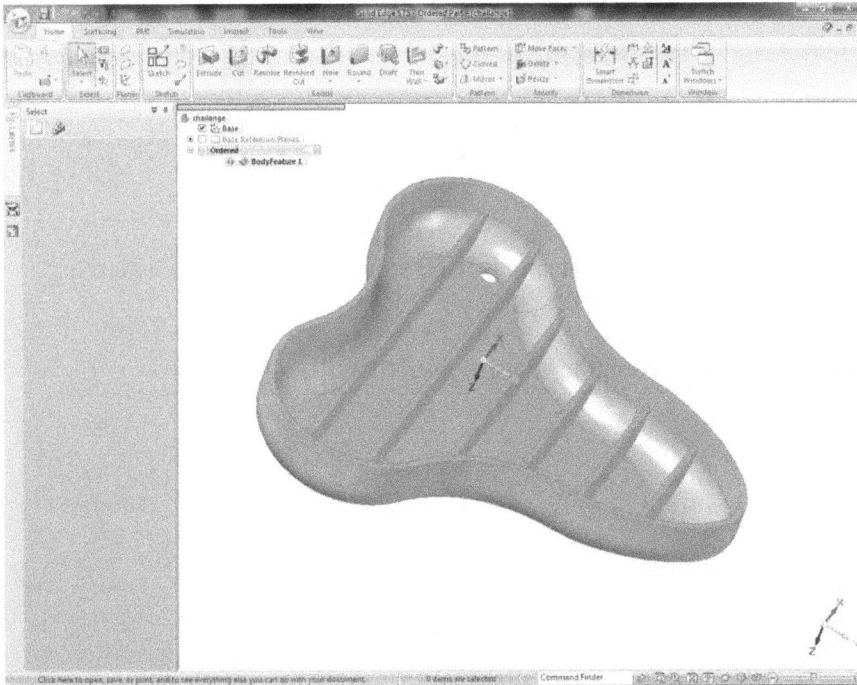

Start by right clicking the "Body Feature" in the PathFinder and clicking "Move to Synchronous". This will allow us to Synchronously edit the part.

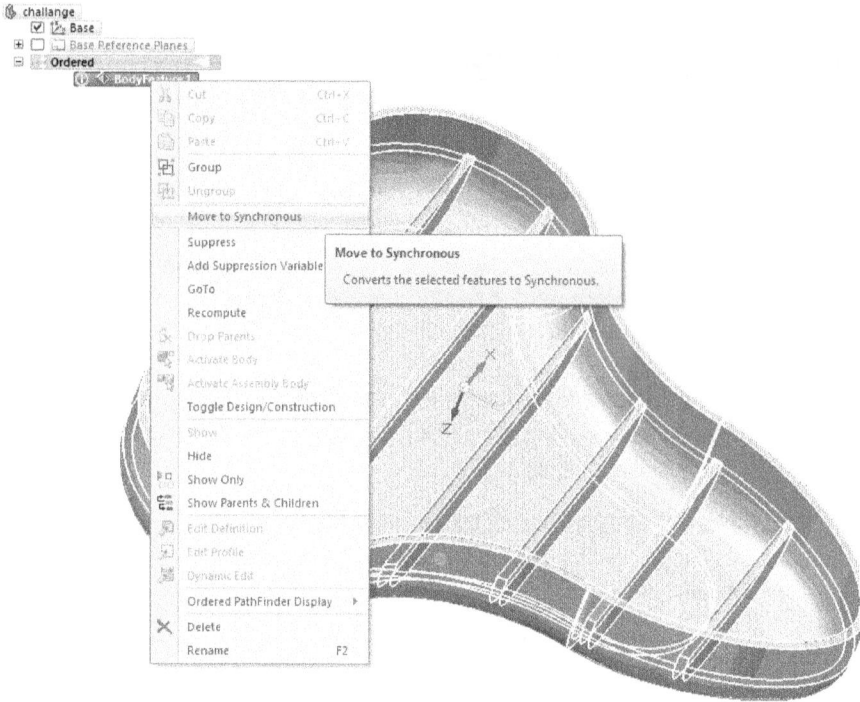

Once you have moved the part to Synchronous mode, right click and select "Transition to Synchronous" so that you can begin editing the part.

Now click the **Coplanar** button located in the "Face Relate" box.

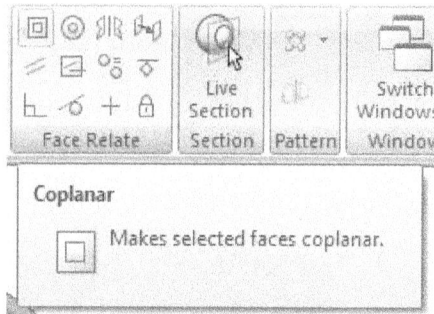

Select one of the ribs as shown below, and then click Accept.

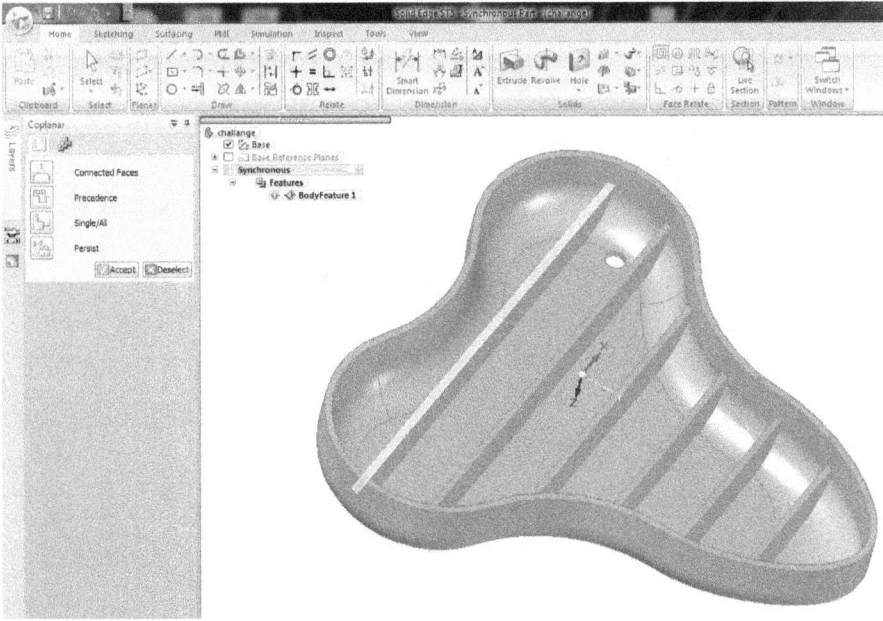

Now select the top edge of the part as shown below. Click Accept.

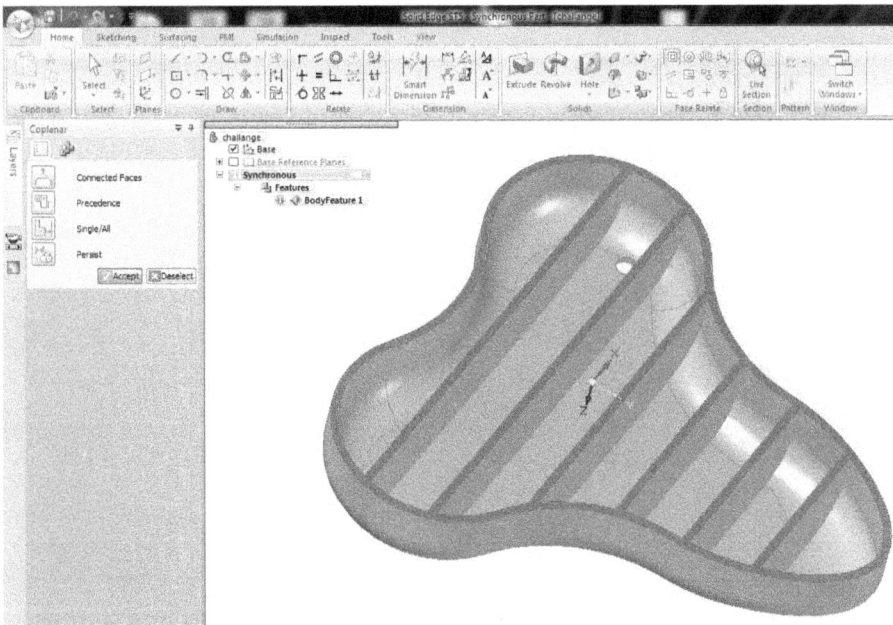

You'll notice that, since the ribs were already coplanar with each other, all of the ribs will extend to become coplanar with the upper edge of the part.

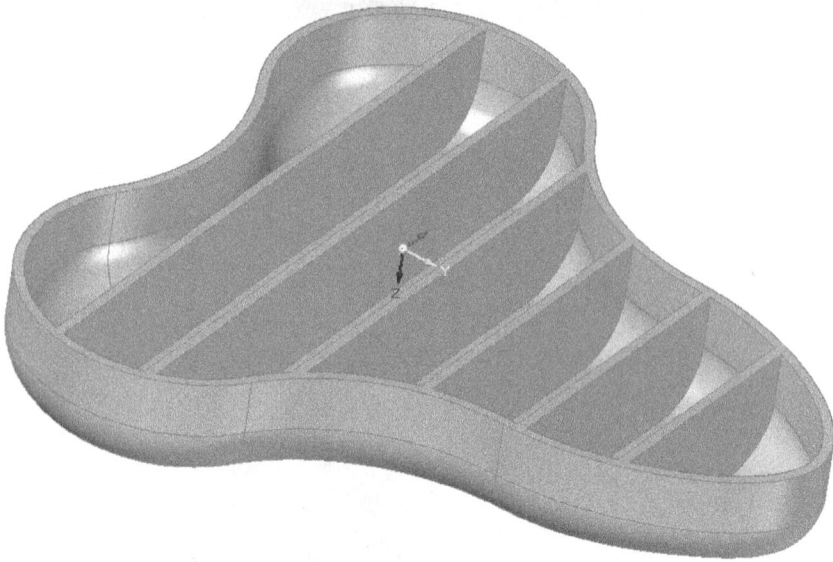

Now select the top face of the ribs/upper edge.

Drag the small blue ball below the arrow to the X-axis arrow so that it is oriented as shown below.

Using the wheel, tilt the top face of the part by 10 degrees. (or -10 degrees depending on how your part is oriented) You should now have the model shown below. *Notice how the ribs updated to remain coplanar.*

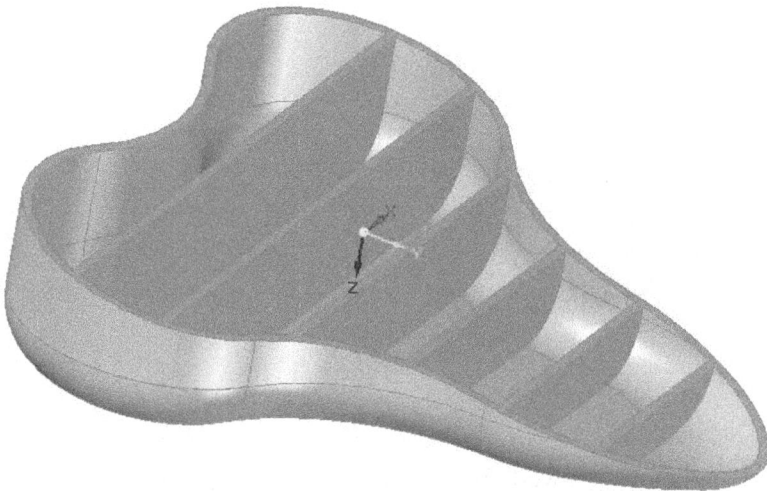

Now click the **Symmetry** button located in the "Face Relate" box.

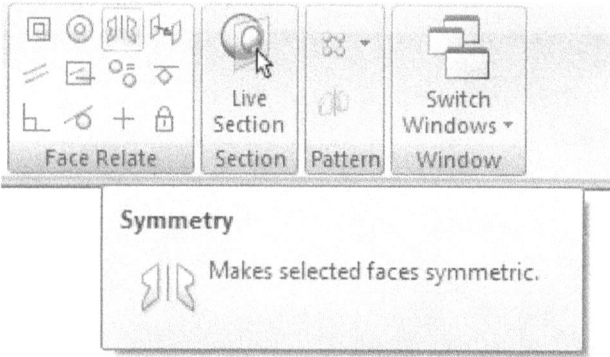

Symmetry

Makes selected faces symmetric.

Select one of the holes, click accept, and then select the other hole and click accept. You'll then be prompted to select a plane about which they are symmetrical. Choose the YZ-plane as shown below.

Now click one of the holes and move it inward by 1in. You'll notice that, since they are symmetrical, both holes move.

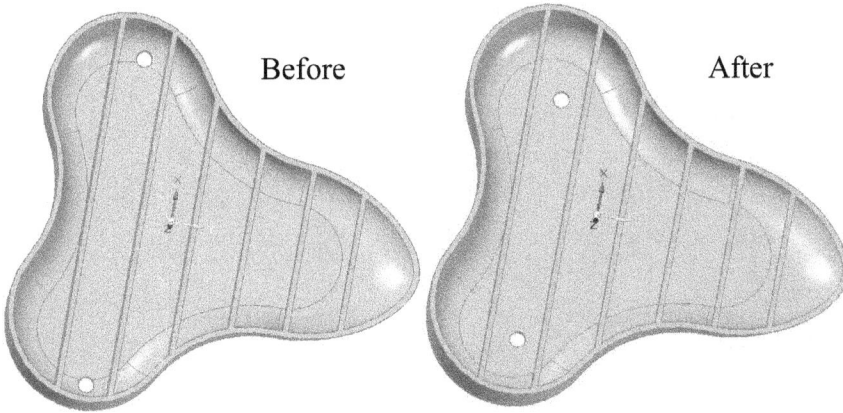

Before   After

Now select one of the holes and edit it's diameter to be 0.75in. Just as before, you'll notice both holes change.

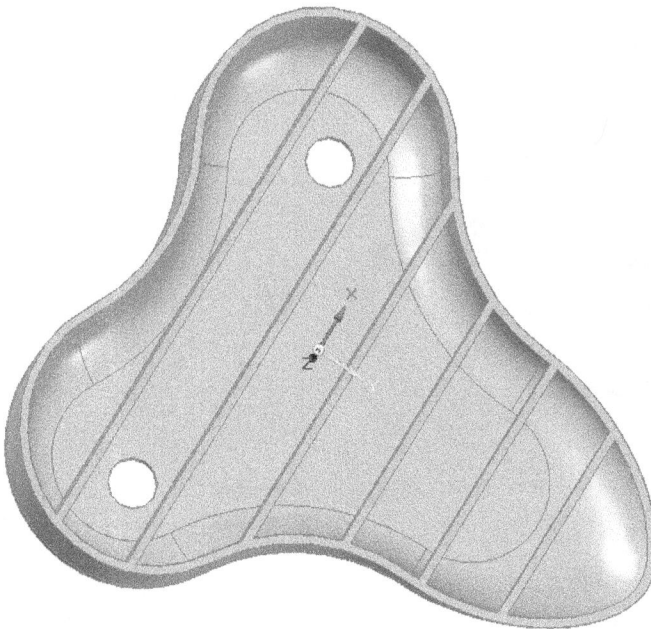

Below is a comparison between the original and the new parts.

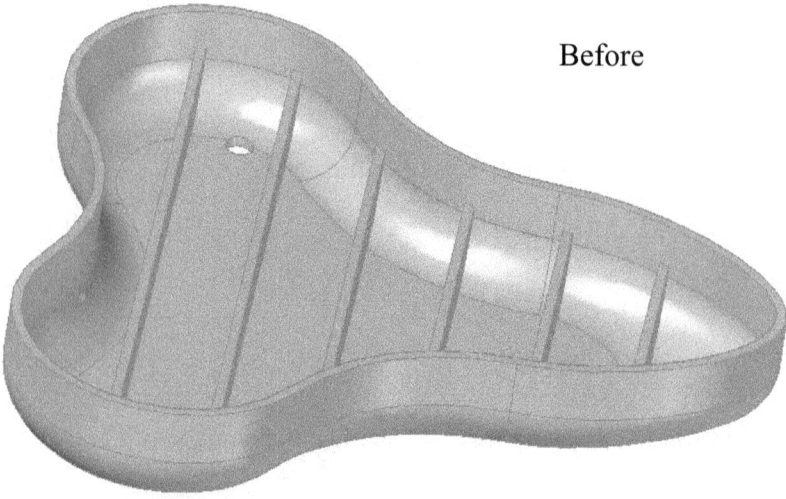

Before

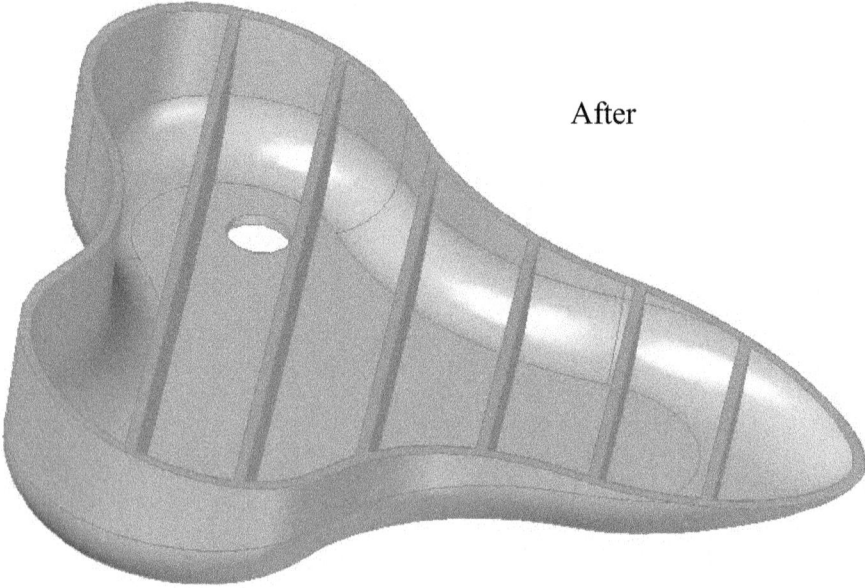

After

Exercise complete

Using Synchronous Modeling in an Ordered Part File

The following figure is a traditional modeling part file. It consists of a protrusion, a cut a round and several ribs and bosses. In this exercise we are going to edit some of the surfaces on this model using the 'synchronous-like' modeling tools found in the traditional modeling environment. *You can open this part from your digital download.*

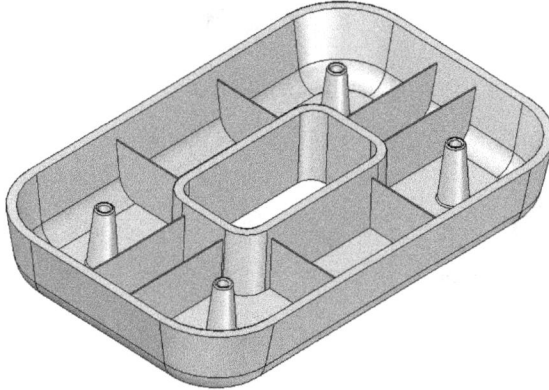

The tools we are going to use are found in the Home tab under the **modify** tab. Select **move face**.

The first step is to select the faces to move. Select all the top faces of the ribs, this includes the top face surrounding the cutout.

Select all top rib faces and cutout face (9 total)

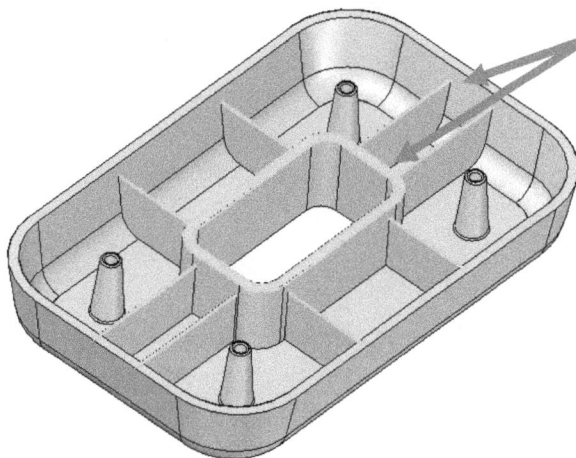

Next select the movement step. Select **'along face normal'** from the selection intent and pick the top face as shown in the following figure. Pick a start location for the move face - select one of the points along that same top face.

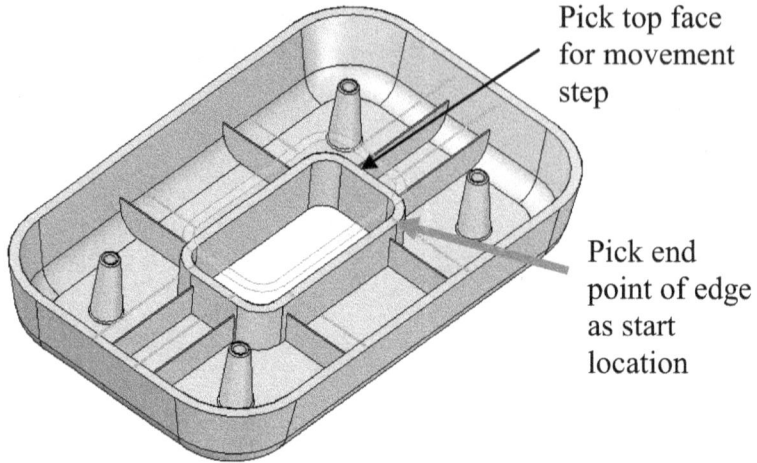

Pick top face
for movement
step

Pick end
point of edge
as start
location

Moving the cursor in the graphics display will automatically move the face. A more accurate way to achieve the desired distance for the move is to use the 'To Point' distance box in the move face dialogue box. Change it to -3mm. The faces should now move the specified distance.

Now let's try to change the thickness of some of the ribs. Choose **Offset faces** found in the same menu as move face. The first step is to pick the face, chain, feature or body to offset. Pick face and then choose one side face of a rib. Choose a direction using the cursor (pick face away from the center).

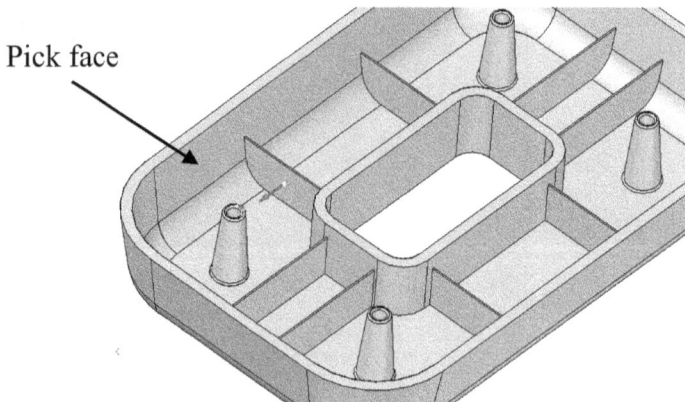

Pick face

Finally choose an offset value of 1mm. Repeat this for all the other ribs.

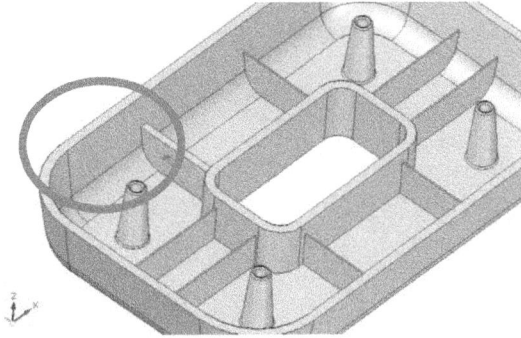

We are going to change the angle of the ribs now. Select **rotate faces** from the same menu. Select the top face of one of the ribs as shown in the following figure.

Select Top
Face

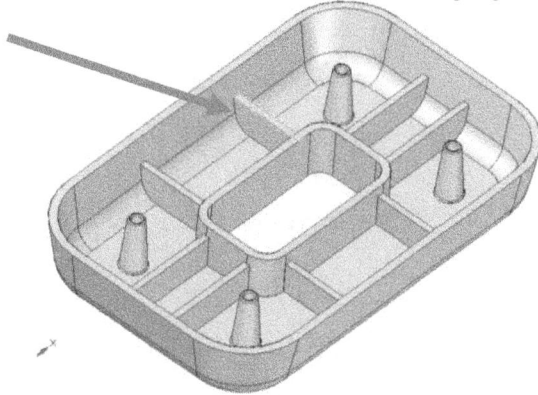

Select an edge of the geometry to rotate around.

Select Edge
to rotate
around

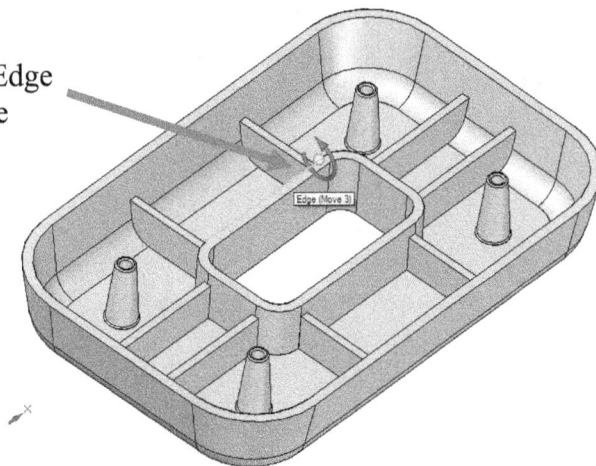

Finally choose the angle to rotate the face. Be Creative. Repeat this for all the other rib faces.

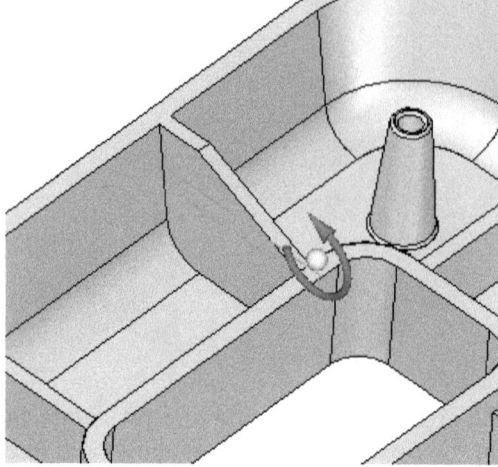

The model should now look something like the following figure.

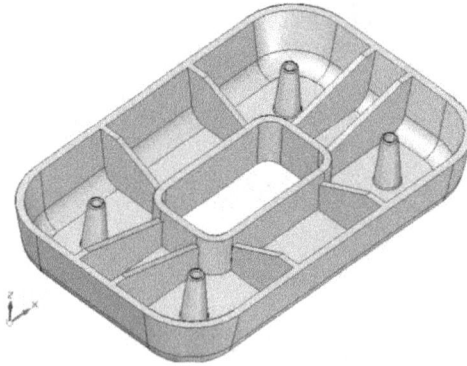

The screws that we were going to use for this part have changed and we need to make a quick change to the hole size in the bosses. Use the **Resize Hole** feature, select the hole for each boss and change the size to 1.5mm.

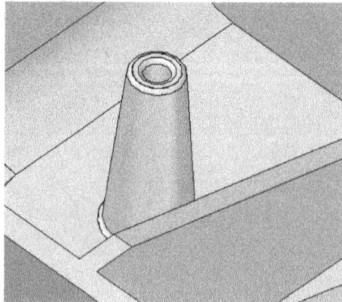

Now change the size of the rounds at the foot of the bosses using **Resize Rounds.** The same steps are required as the Resize Hole except pick the rounds instead. Change rounds to 1mm.

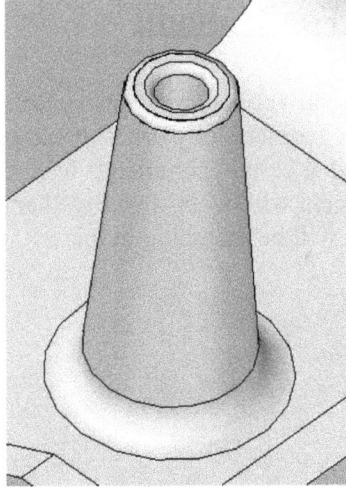

Try using the **delete command s** to delete holes, rounds, features and faces. They are simple to use - just select the faces and delete them!

Exercise Complete

Exercise 46: Fish Food For Thought

Now that you've accumulated a relatively comprehensive Solid Edge skill set, it's time to put your knowledge to the test with a geometrically difficult and organic model. This type of model forces you to go beyond what certain commands were designed to do and innovate new ways of modeling as you experiment with what works well and what doesn't. Below is an example of Largemouth Bass we will be modeling in this

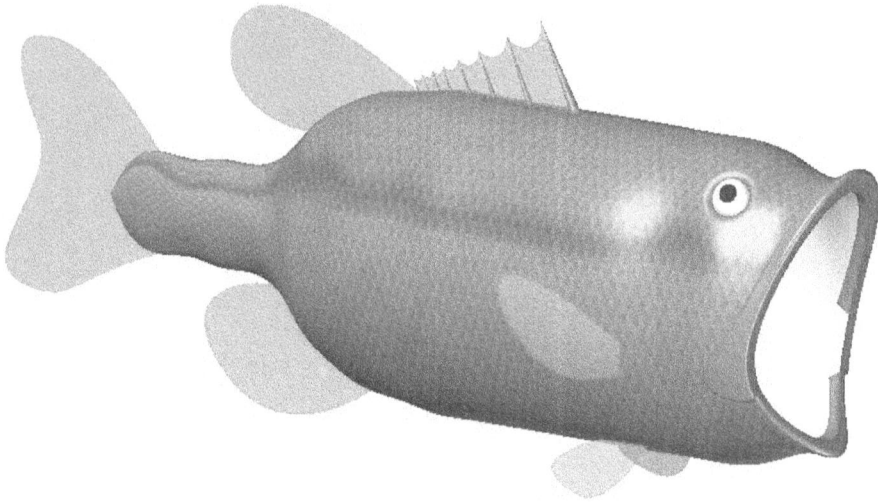

exercise.

Before you jump right into modeling, it's always a good idea to size up the task at hand and brainstorm the best way of tackling the modeling process for said task. The easiest way to create the curved geometry that will become the fish's body is to use a **Blue Surf**. A Blue Surf will ensure that the fish is comprised of flowing curves rather than sharp edges. For this to work, we need to create many cross sections of the fish. To accomplish this, we will begin by creating a series of planes parallel to the Front Plane.

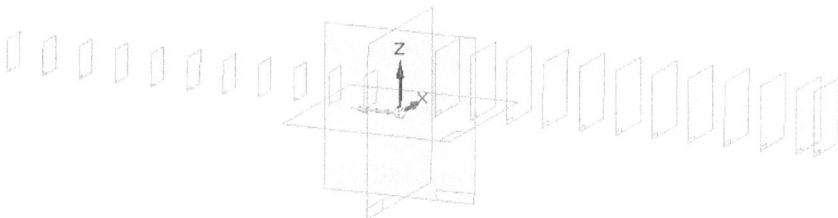

From here, we can begin sketching the various cross sections of the fish on the parallel planes we just created. The body of the fish is composed primarily of ellipses with the sole exception being the point where the tail tapers down. An example is shown below from two different viewpoints. *Note: There are no set dimensions, but just to give you a frame of reference, the largest ellipse is about 4in tall.*

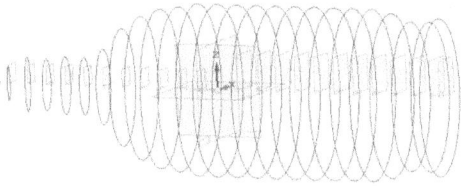

Now you can create a Blue Surface across all of the sections as mentioned earlier. To do this, select each cross section in order starting from the front, and ending with the point at the end of the tail. Your model should now resemble the following image. *Note: Ensure "Close Ends" is selected in the Blue Surf options under "End Capping".*

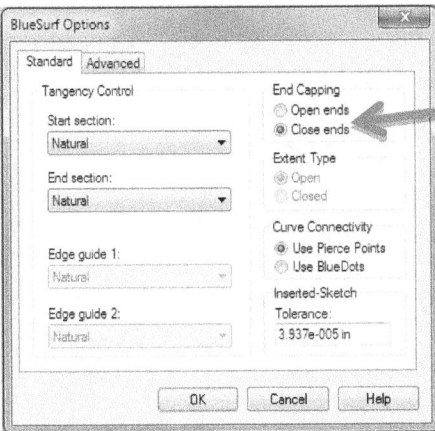

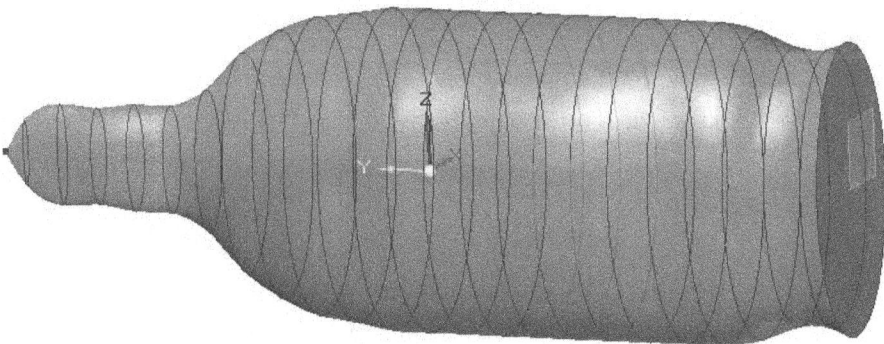

Convert the model to a design body if it is currently a construction body. This is done by right clicking it and then selecting "Toggle Design/Construction".

Now for the fins. To create the tail fin, create a sketch on the Right Plane. (the sketch should be perpendicular to sketches used to make the body) use the **Curves** tool in the sketcher to sketch something similar to the tail shown below. **Extrude** it symmetrically by 0.005 in.

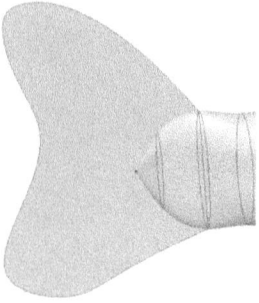

You should now have the following model.

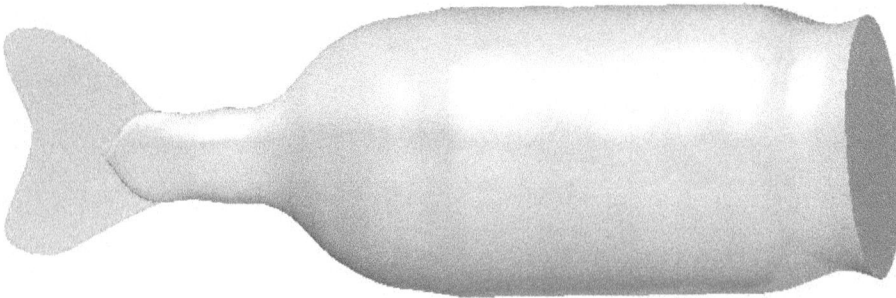

For the dorsal fin and bottom fin, create a sketch on the Right Plane again, and sketch the following curves.

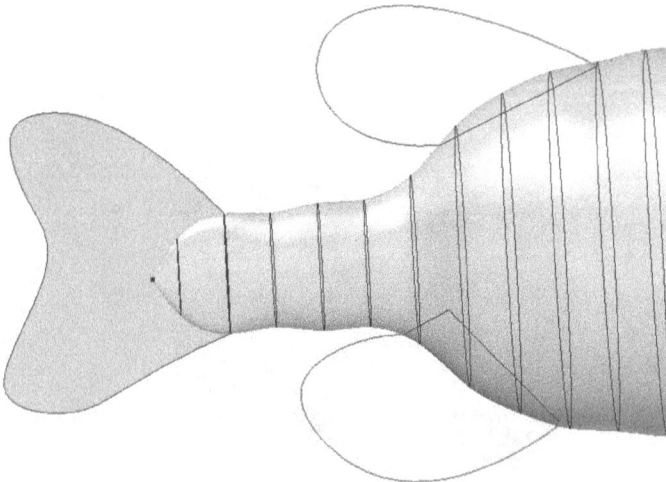

Extrude these fins symmetrically to 0.005 in as well. You should now have the following model.

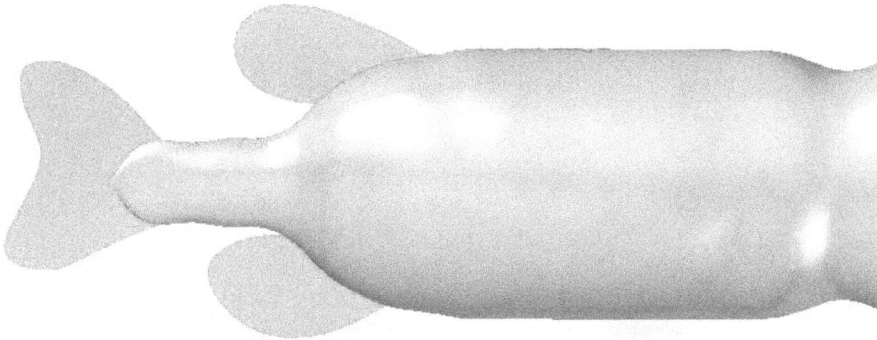

The back spines aren't quite as easily modeled as the previous fins. We're going to start with the spines themselves and then add the webbing later. The spines are created using the **Revolve** tool. Just draw small triangles similar to the ones below and revolve them about one of their longer edges. *Note: To ensure that the angles of all of the spines were the same, I first sketched a large triangle to house the spine sketches and then set the left side of every spine triangle perpendicular to the large triangle.*

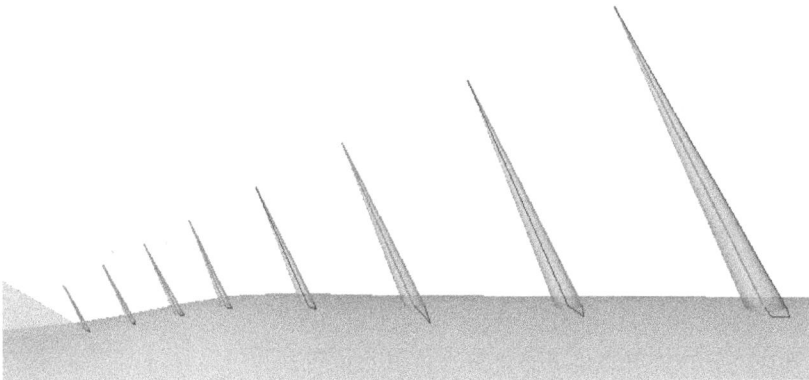

To create the webbing, I sketched the following sketch onto the same plane I used when sketching the previous 3 fins.

As with the other fins, **Extrude** the sketch symmetrically by 0.005in. You should now have the following model.

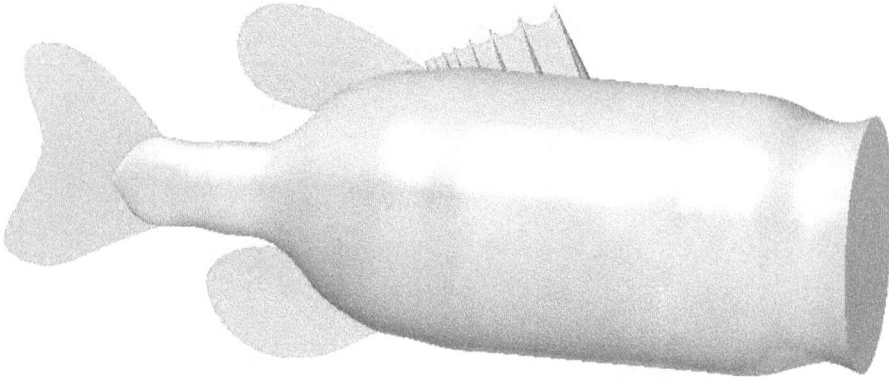

For the side fins, begin by creating the following sketch on a plane that is parallel to the right plane.

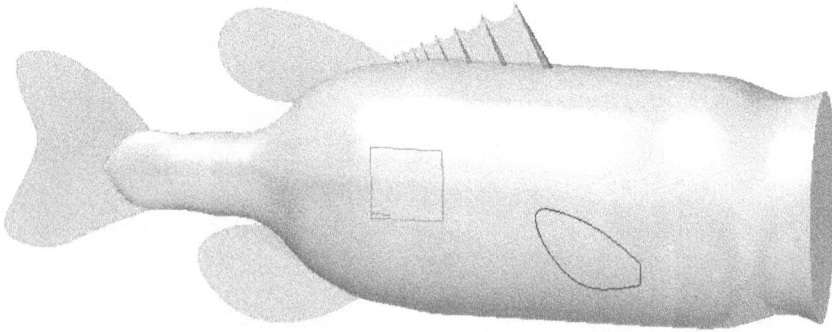

Use the **Project** command located in the "Surfacing" tab to project the sketch of the fin onto the side of the fish. You should get a result similar to the one shown below.

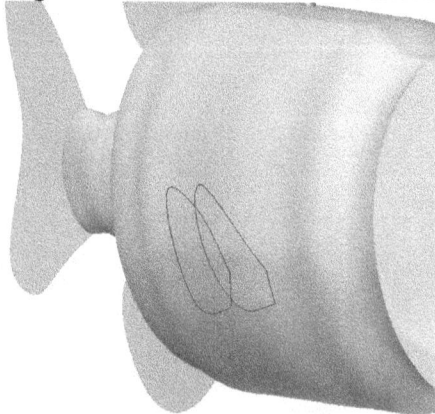

Now use the **Bounded** command to create a surface out of the projected fin. Once you have a bounded fin, use a combination of the **Move Face** and **Rotate Face** commands to make the bounded fin jut out of the fish's body at a slight angle as shown below.

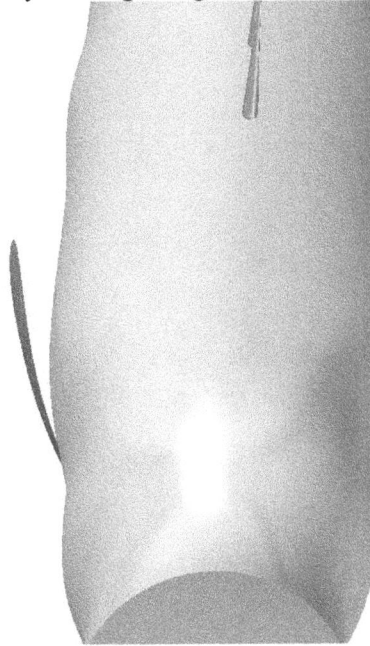

Once you have the fin in position, use the **Thicken** command to make it a solid body. Thicken it by 0.005 in. You should now have the following model.

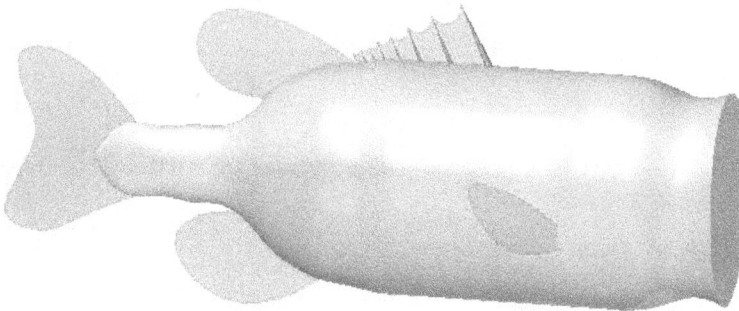

Now use the **Mirror** command to create an identical fin on the other side of the fish.

Once that is done, you can begin work on the mouth of the fish. Create a sketch similar to the one shown below on a plane parallel to the Right Plane.

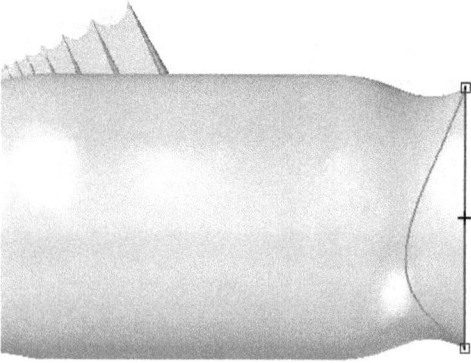

Use the **Cut** command to remove this section from the body of the fish. (this will be hollowed out later to create the fish's mouth) Now create the following sketch on the same plane as your previous sketch.

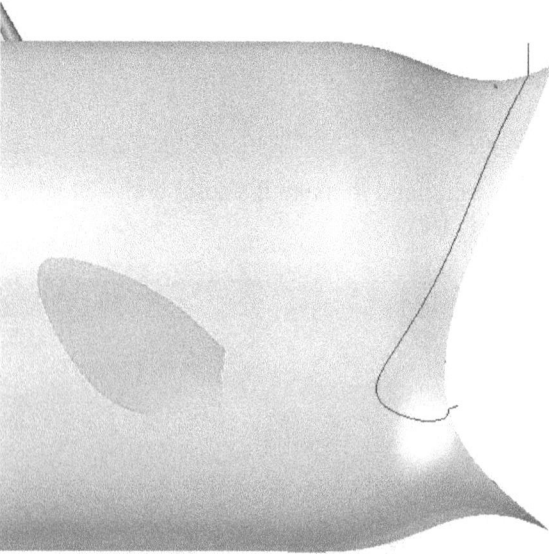

Use the **Surface Extrude** command to create the following plane. We will be using this plane to split the fish so that we can work on just the part of the mouth we want to add detail to.

Use the **Split** tool (located in the "Add Body" menu) to split the fish. You should now see the following model. If you see the rest of the fish instead of just the lip, just right click the lip and select "Activate Body".

Use the **Thicken** tool to thicken the selection shown below by 0.125in.

Now use the **Union** command to unite the lip back with the body. You should get the following result.

Newly created lip

Now create a plane parallel to the front plane near the fish's mid-section. Use this plane to **Split** the fish. *Note: If you split one of the fins as well, you will have to unite it again after we hollow out the mouth.*

Now use **Thin Wall** to hollow out the mouth to a thickness of 0.25in.

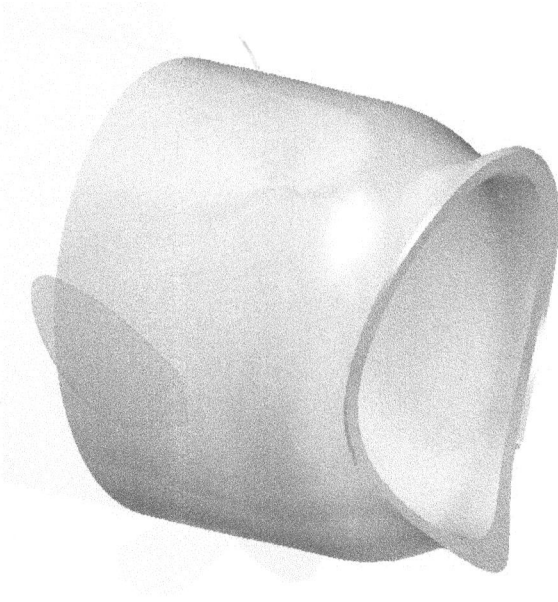

Unite the two halves of the fish again.

Now for the final fins. Create the following sketch on a plane parallel to the right plane.

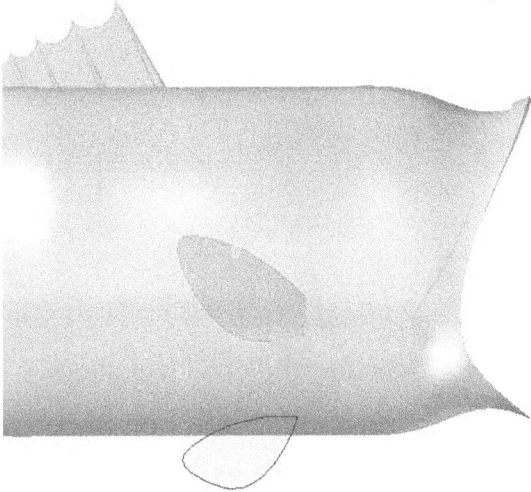

Use the same method as before to create and position these fins. (bound it, use move face and rotate face to position it, thicken it, and finally mirror it to create the other fin) They should look like the following picture upon completion.

To create the eyes, create an angled plane and then draw the following sketch.

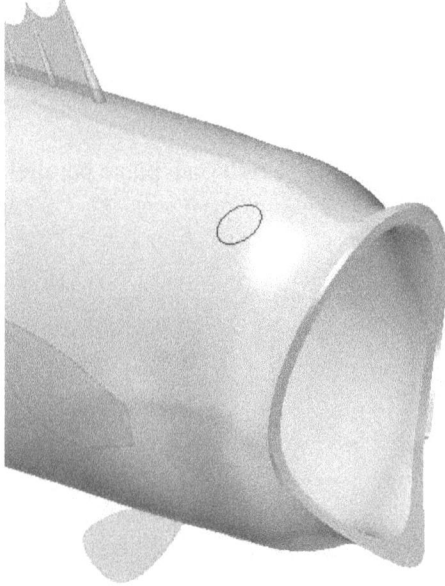

Use the **Project** command to project the sketch onto the fish and then mirror it to create a second eye.

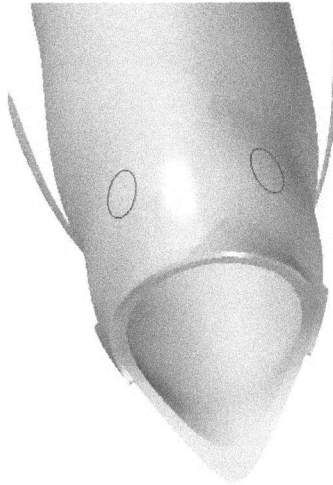

Use the **Bounded** command to turn your projected eyes into surfaces. **Offset** both of these surfaces by 0.125in toward the inside of the fish. Then use **Lofted Cut** to cut from the bounded surfaces to the offset ones. You should end up with eye sockets.

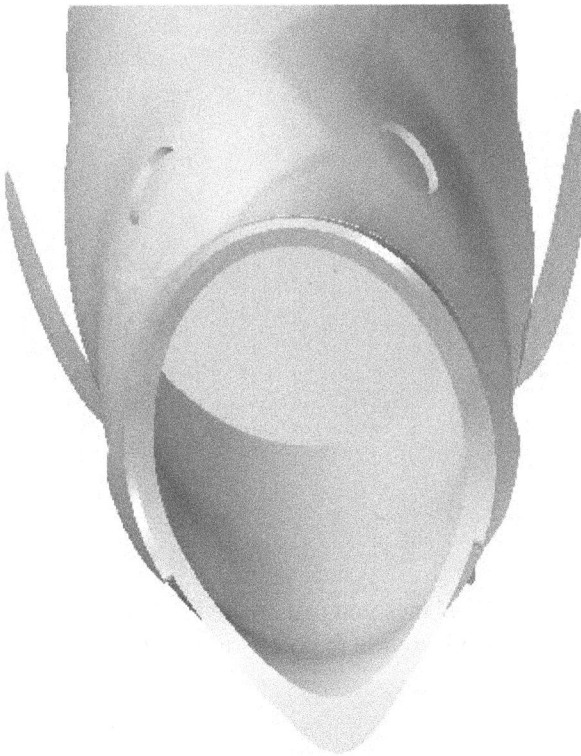

Create a plane on the inside of each of the eyes using "By 3 points". We will use these planes in just a moment.

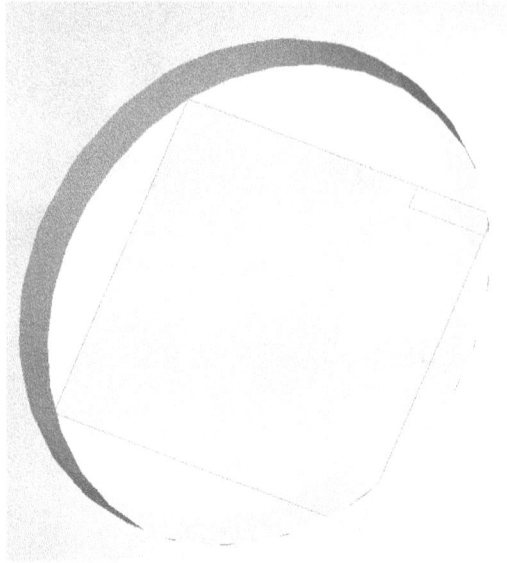

Now apply a **Round** to the entire eye socket of both eyes.

Then create sketches in the eye sockets on the planes you just created. Offset the circle around the eye socket using the **Include** tool and then **Extrude** the offset circle up 0.25in. You should now have the following model.

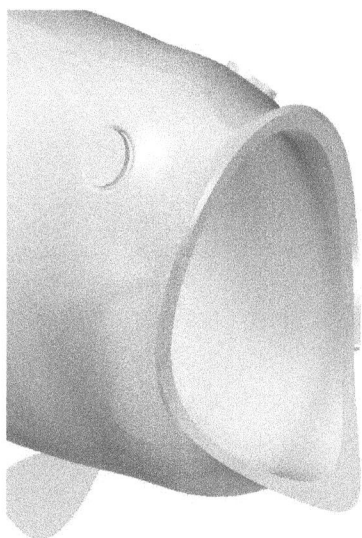

As a final touch on the eyes, add a **Round** of 0.25 in to the top edge of the eyes.

Largemouth bass don't have teeth in the traditional sense, but hard, sandpaper like plates on the inside of their mouths used to grab ahold of smaller fish. To add these, use the same process as you did earlier to add the lip to the fish's mouth. (create a sketch, extrude the sketch so that it intersects the fish, split the fish using the extruded surface, use thicken to create the teeth/plates, and then unite it back to the original body) After doing all of that, add slight rounds to the teeth plates. Here are some finished teeth plates. (I have colored them orange and added edge lines for clarity purposes)

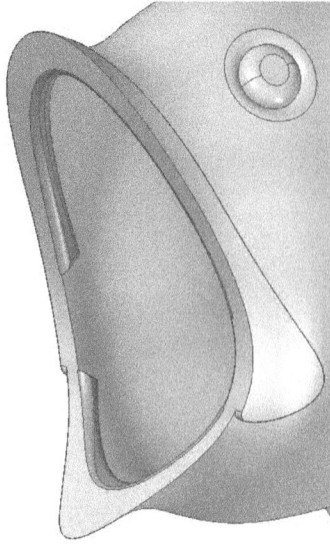

Now use the **Round** tool to add rounds to the edge of the fish's lip.

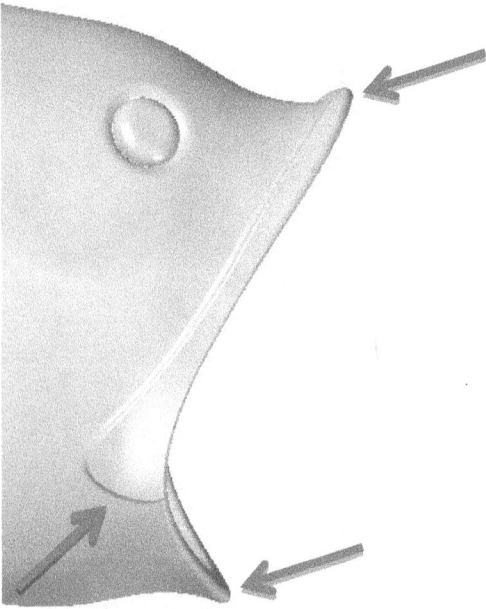

Now select "Blend" from within the **Round** tool.

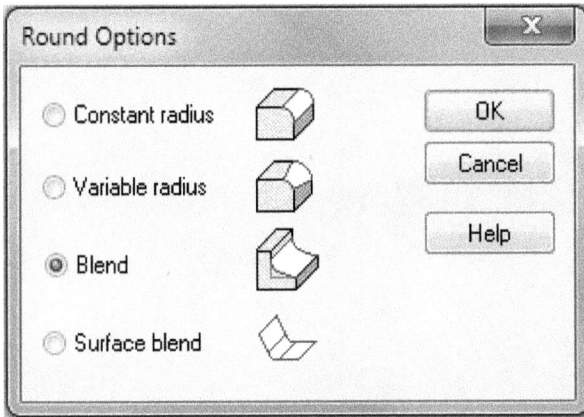

Use this feature to blend the bottom edge of the lip to the fish's head.

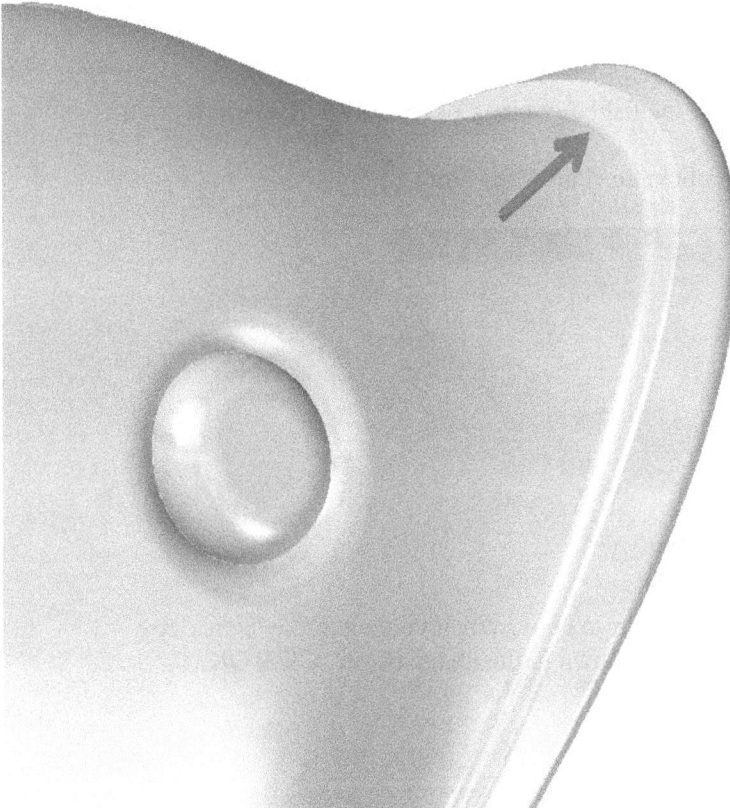

As a final touch, go to the View tab and under "Style" click **Part Painter**. Use this to give your fish a realistic coloration.

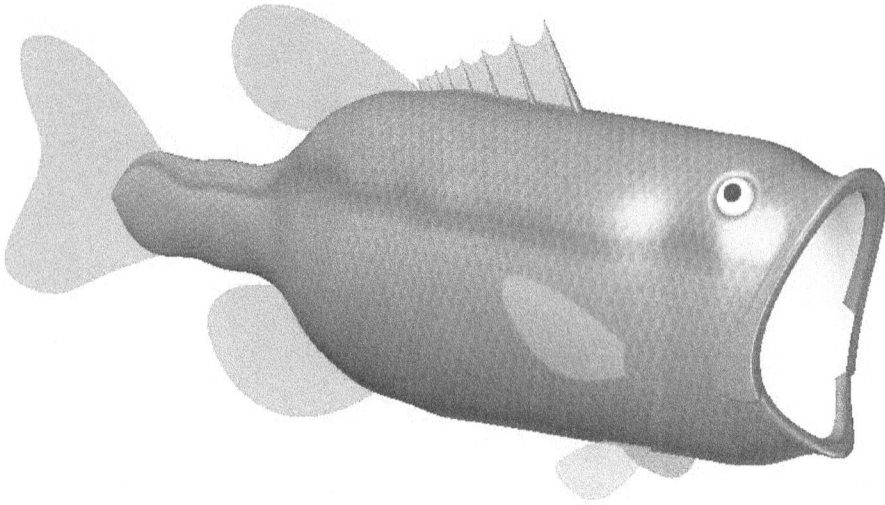

Congratulations, you're done with the fish! Save it as **Largemouth_Bass.par**.

Open a new **ANSI part file** and click the "Part Copy" tool.

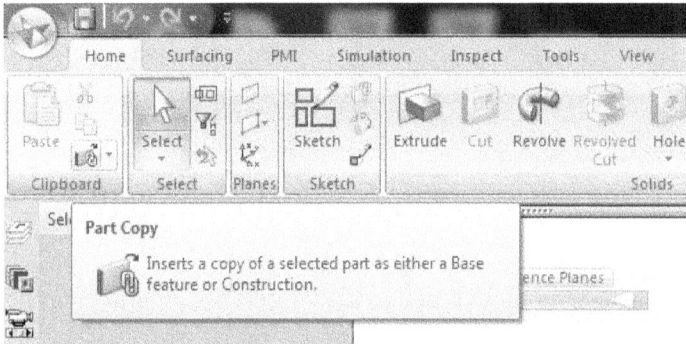

Select the fish you just created as the part to copy. When you get to the "Part Copy Parameters", change the scale to 0.2 as shown in the image below. Click OK.

Part Copy Parameters

☑ Link to file   [icon]   ☑ Copy colors

Family of Parts member: [_____ ▾]

Coordinate system: [_____ ▾]

◉ Copy as Design body
   ☐ Merge Solid bodies

○ Copy as Construction body

☐ 📦 \\MYBOOKLIVEDUO\Public\DV BOOKS\SOLIDEDGE B
   ☑ ◈ Design Bodies
     ☐ ◈ **Design Body_15**
   ☐ ◈ Construction Bodies
     ☐ 🔲 Solid Bodies

☐ Mirror body   ☐ Flatten part
   ○ about Top (xy) plane
   ○ about Right (yz) plane
   ○ about Front (xz) plane

	X	Y	Z
◉ Scale:	0.2	0.2	0.2

   ☑ Use uniform scale

○ Shrink factor: [____]

☑ Show this dialog after the Part Selection Step*

*This dialog can be shown by clicking on the Part Copy Parameters Step.

[OK] [Apply] [Cancel] [Help]

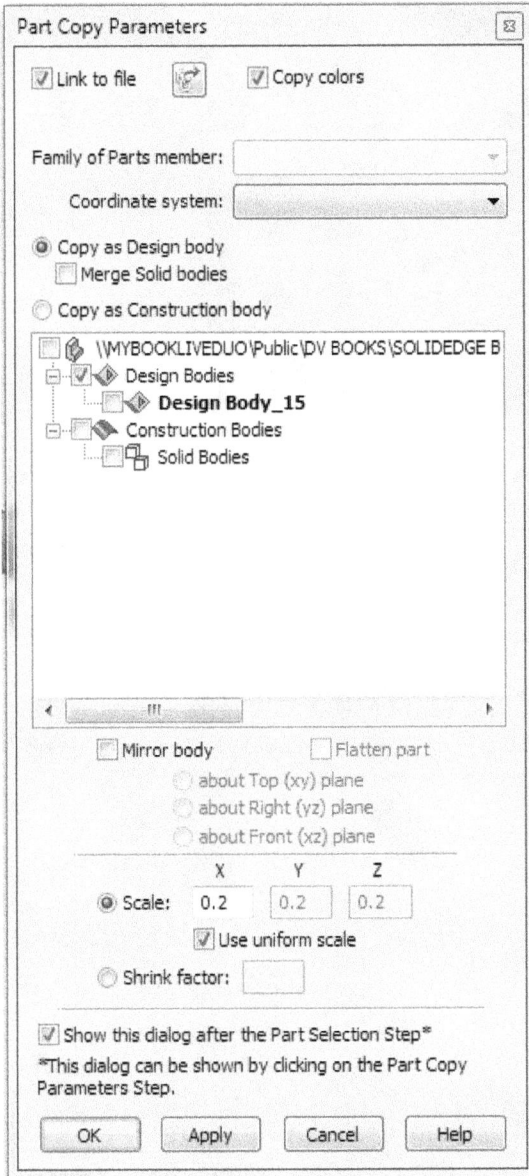

Now that we've scaled down the fish, give it a new paint job. Have fun and be creative with this part!

Save this fish as **Small_Fish.par**.

Open a new assembly file and import both of your fish.

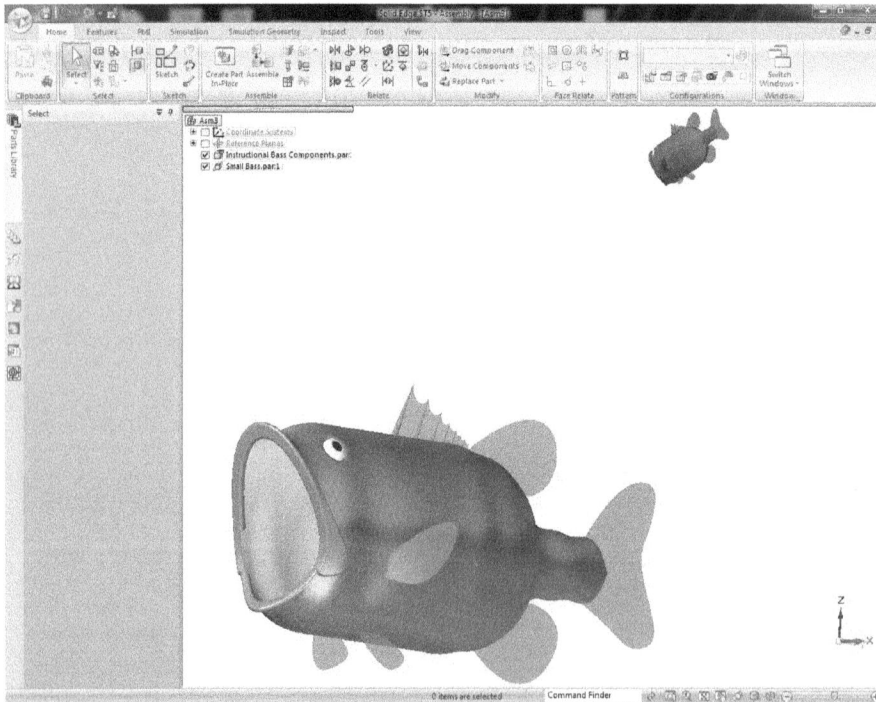

Orient your fish using the **Move Component** and **Drag Component** commands as I have done below.

Save the assembly as **Fish_Food.asm**.

Exercise complete

Exercise 47: Assembly 1

Assemblies in Solid Edge are very straight forward. The first step is to create an assembly file. An assembly file is very similar to a normal part file. However instead of creating geometry in the assembly file the designer adds existing part files into the assembly file. The first part added to the assembly will be 'Grounded' i.e. fixed in space. *Note: it is possible to remove the 'grounded' restriction.* All other parts added to the assembly have to be positioned using the **'Relate'** tools. The relate tools instruct the parts how to be positioned in relation to the other parts in the assembly. For Example, the designer can tell a face from Part A to be attached to the face of Part B. Once this relationship is created the two faces will always be attached unless that relationship is removed.

Open a new **ANSI Assembly** file located in the new tab under the Solid Edge button in the top left corner of the screen.

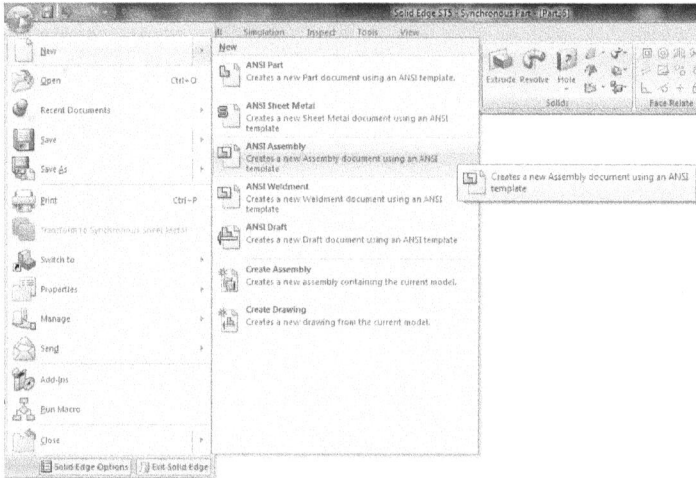

Note: Confirm that the units of the new assembly file correspond with the units of the parts to assemble. It is possible to can change the units in the same way a basic part file, shown at the beginning of this book.

Once the assembly file is open parts can be added to the assembly. The first part imported into the assembly will automatically be positioned on the coordinate system as the Base part. To import a file, go to the Parts Library located next to the Pathfinder in the left menu, as shown in the following figure.

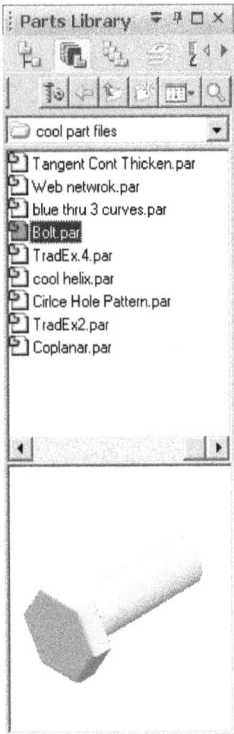

*Note: if the Pathfinder/Part library box is not open or if you delete the box, you can reopen it in the **view tab** by clicking on the ⬜ symbol in the **Show** options box and selecting **Part library** from the dropdown menu.*

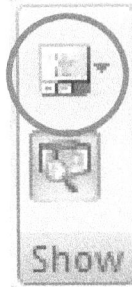

In the **Part Finder** browse the files for the **Bolt.par** file. You may have created this in an earlier exercise. Click and drag it into the screen. Drag the part anywhere on the screen and it will snap to the center of the coordinate planes. *Note: all parts are available from designviz.com as a digital download.*

Place the **Nut.par** file in the screen as well.

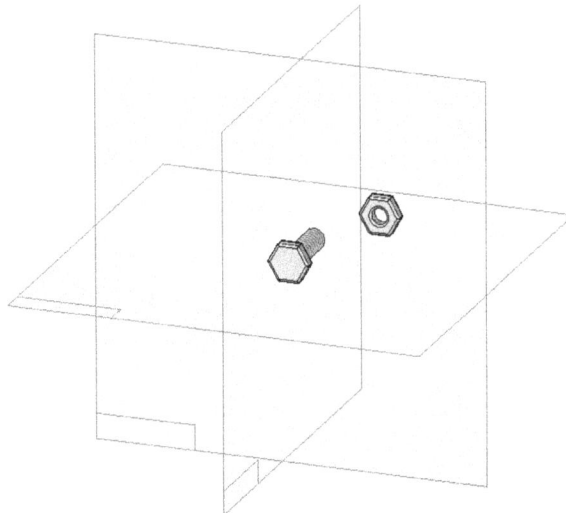

It is possible to move a part once it is in the window, in the Home tab click on the **Drag Component tool** in the Modify options box. *Note: It is often an easier method to move the*

component to roughly the right orientation and closer to the position that the part is going to be assembled. This will cause less uncertainty in the solution SE will provide.

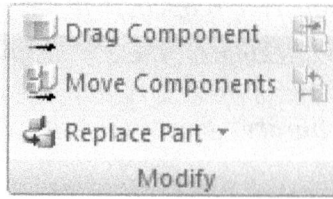

Click **OK** on the box that pops up and click and drag the part to the position desired.

Begin to assemble the parts, select the **Axial Align** constraint.

Click on the hole of the nut, and the cylindrical face of the bolt.

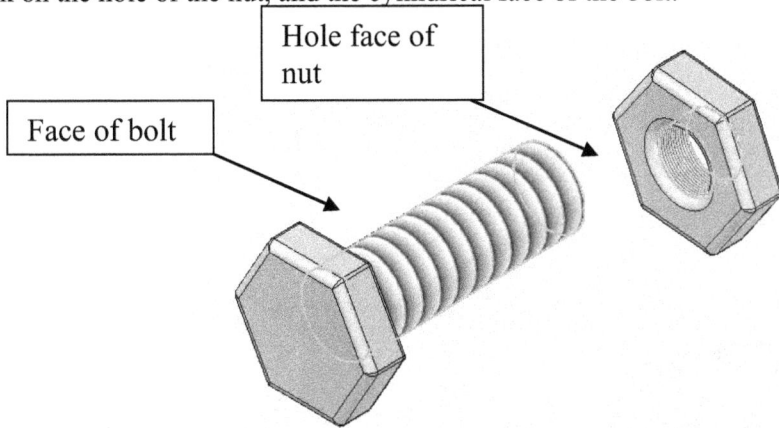

Hole face of nut

Face of bolt

Note: It is possible to move the nut along the center axis of the bolt. Now click on the **Mate tool** in the Relate options box.

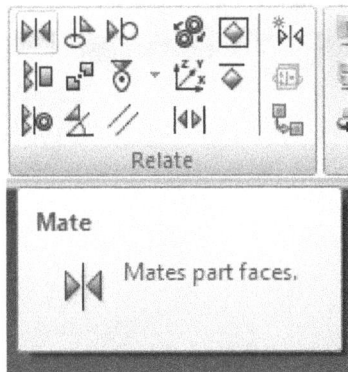

Relate

Mate

▷|◁   Mates part faces.

Click on the front surface of the Nut, and the corresponding surface of the bolt, as shown below.

Face of nut

Bottom face of bolt

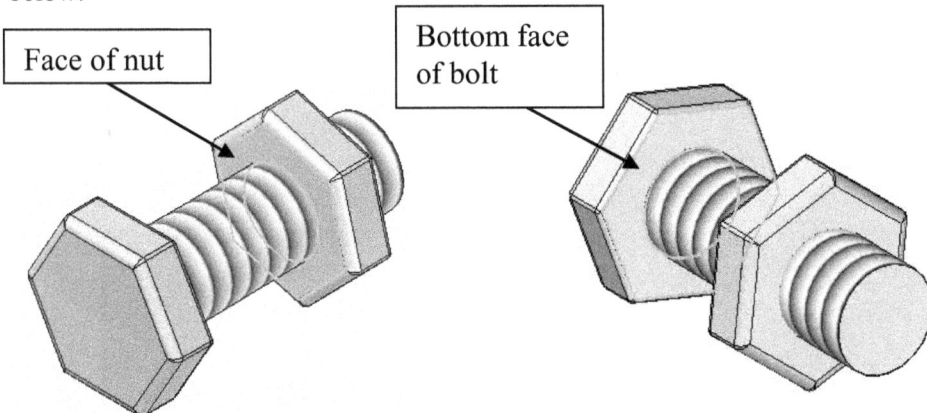

The finished assembly should look like this:

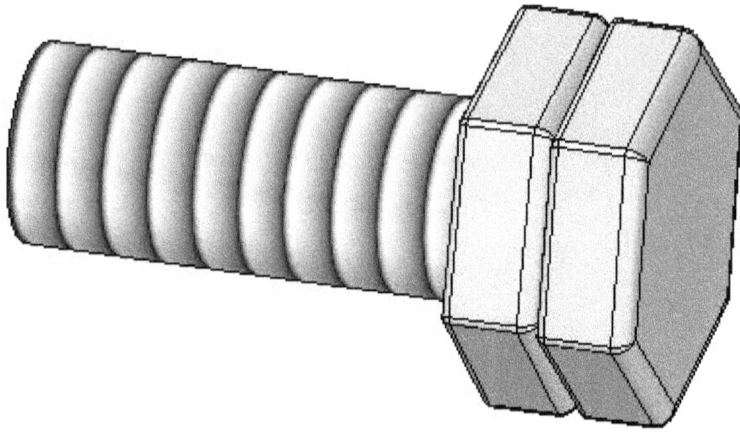

Exercise Complete

Exercise 48: Assembly 2

Create a solid model that looks like the part shown below.

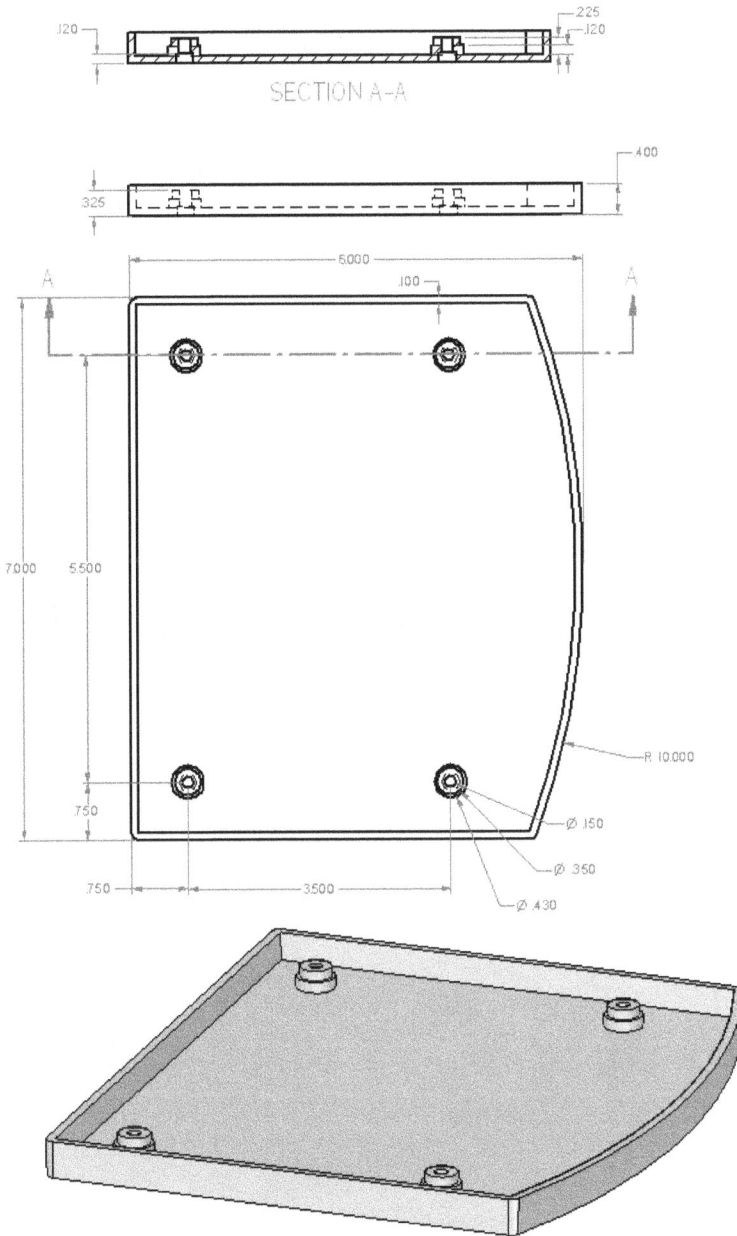

Save this file as **Bottom.par**

Open a new part file and create a part that looks like the following figure.

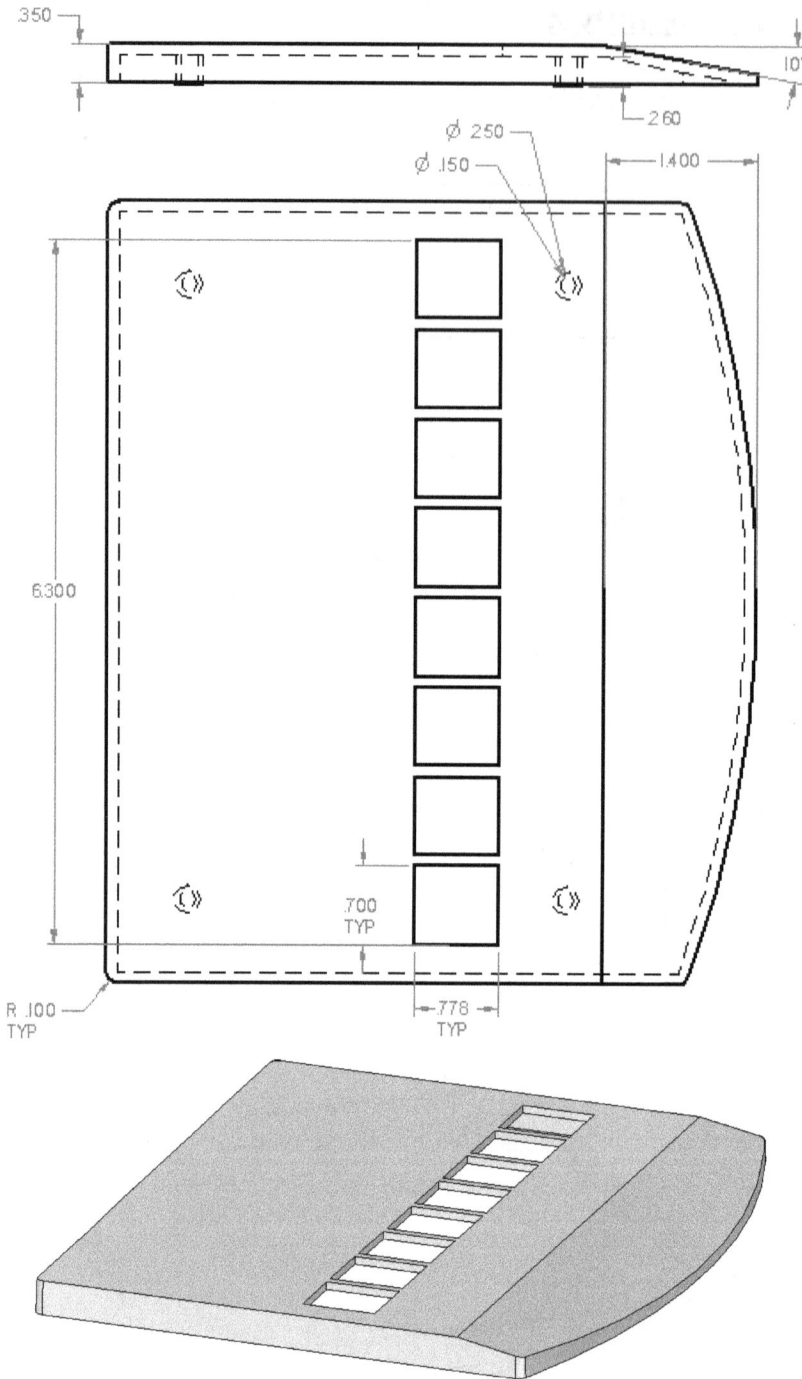

Save file as **Top.par**

Open a new part file and create a part that looks like the figure shown below.

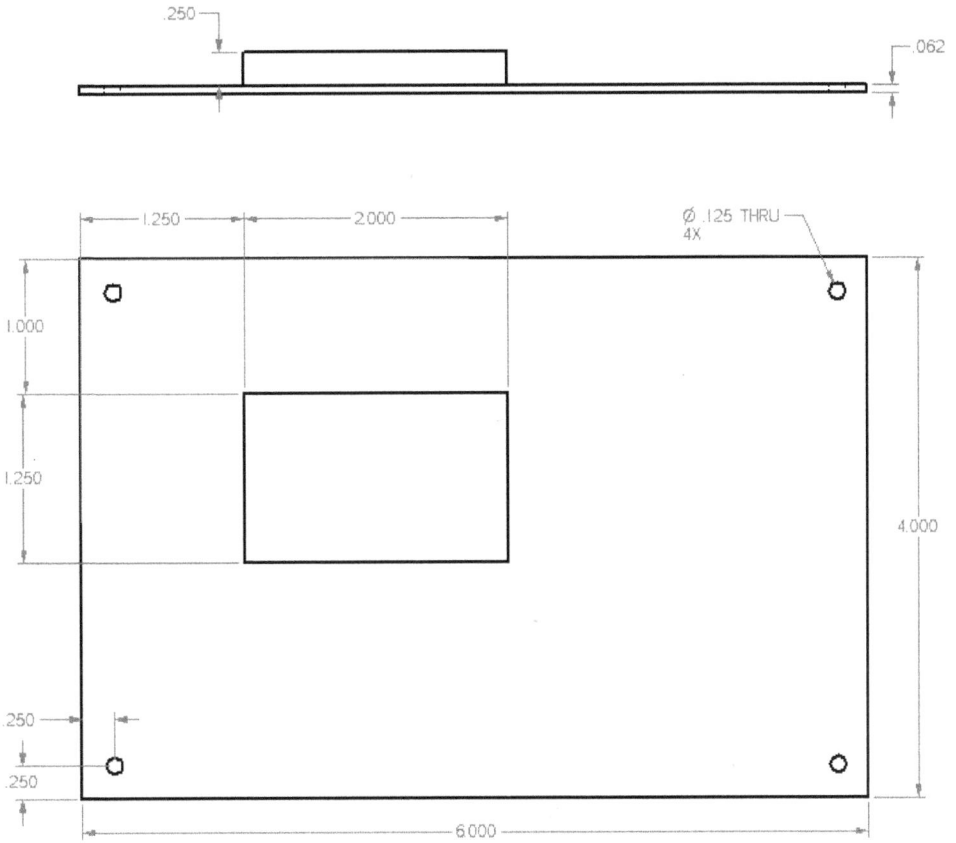

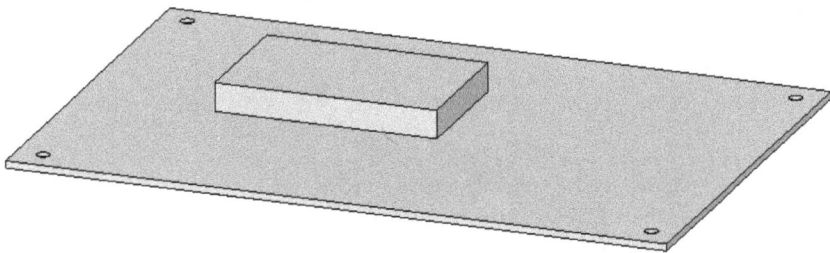

Save file as **PCB.par**

Open one more new part file, and create a part that looks like the following figure.

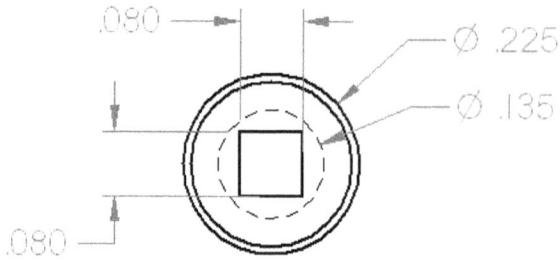

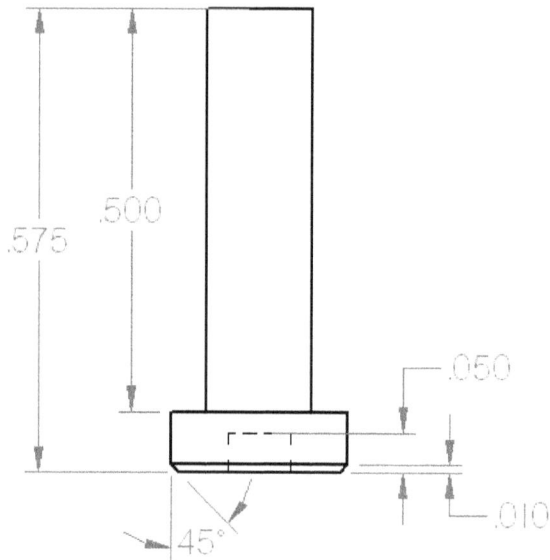

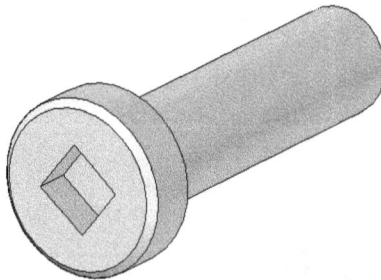

Save the file as **Bolt.par**

Create a new ANSI Assembly

Note: The units of the new assembly file must correspond with the units of the parts that are being assembled. It is possible to change the units as a basic part file shown at the beginning of this book.

Once inside the assembly, import the parts. The first part imported will automatically be positioned on the coordinate system as the Base part. To import a file, go to the **Parts Library** located next to the **Pathfinder** in the left menu.

Note: if the Pathfinder/Part library box is not open or if the box is deleted, reopen it in the view tab by clicking on the ▣ *symbol in the Show options box and selecting Part library from the dropdown menu.*

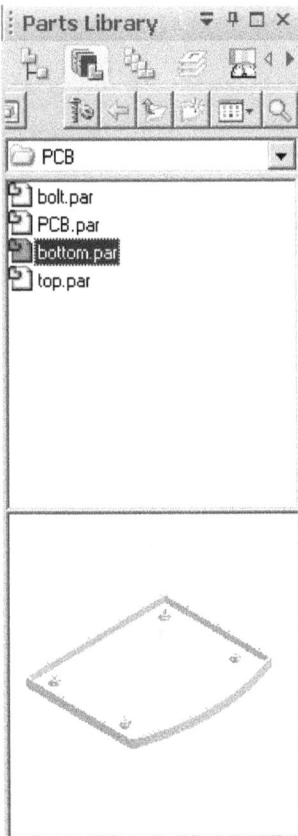

Drag the **Bottom part** in the window and it should snap to the center of the coordinate system.

Next drag in the **Bolt**, repeat this process until 4 bolts are in the screen.

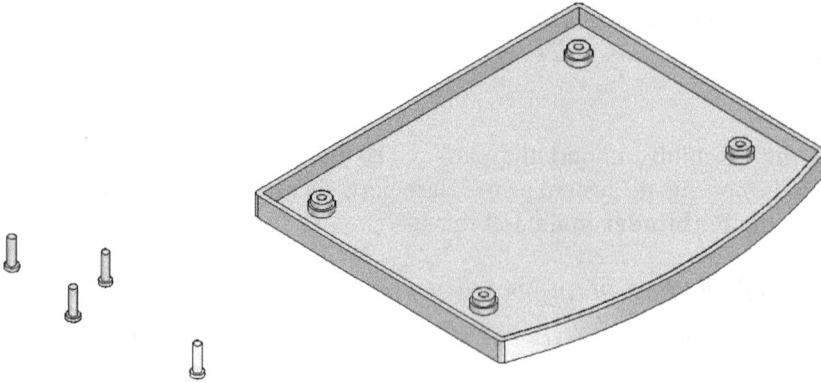

Rotate the assembly to see the base of the **bottom tray**.

To begin to assemble the parts, select the **Axial Align** constaint.

One by one click on the shaft of a bolt, then the cylindrical face of one of the holes (repeat until all four bolts are aligned with a different hole). *Note: You can reverse the axial align if the bolt is facing the wrong direction.*

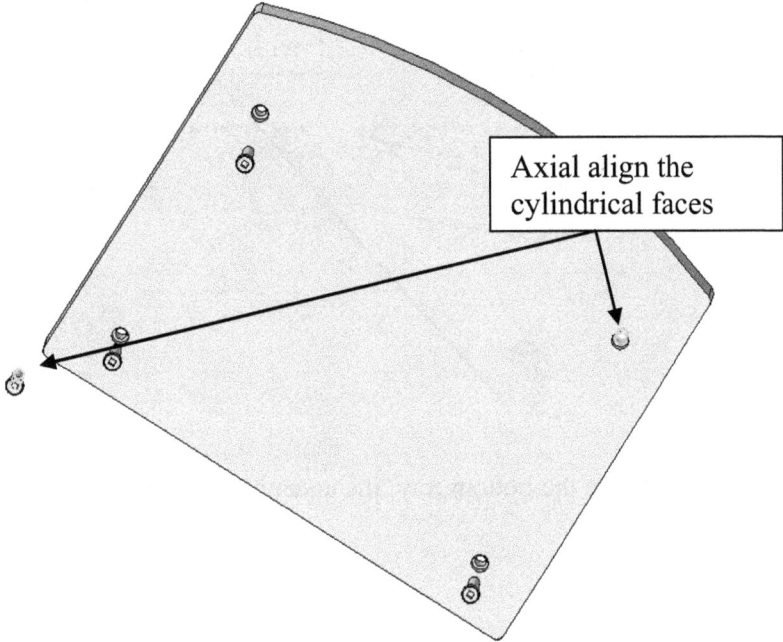

Axial align the cylindrical faces

Now click on the **Mate tool** in the Relate options box.

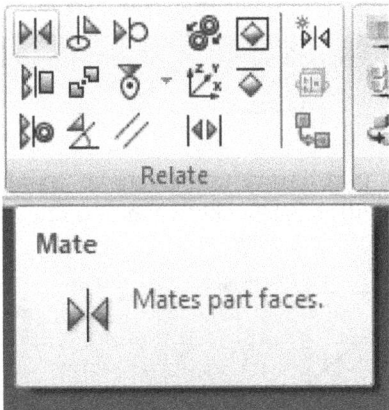

Relate

Mate

Mates part faces.

Next mate each bolt surface with each hole surface as shown below.

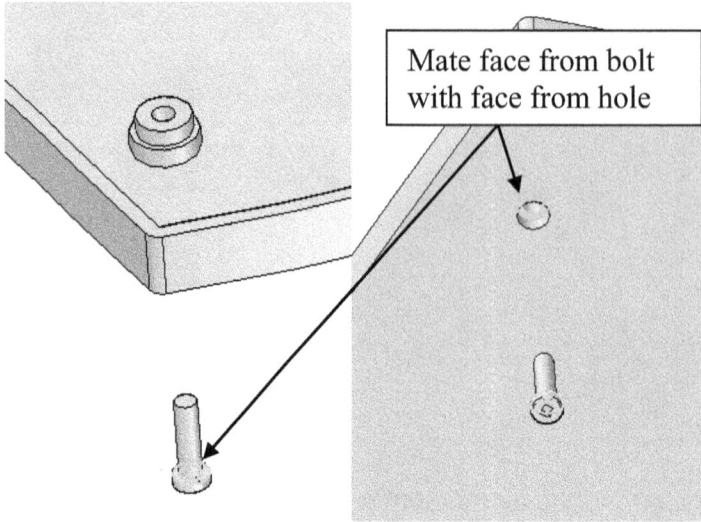

Mate face from bolt
with face from hole

With all the bolts mated to the bottom tray, the assembly should now look like the following figure.

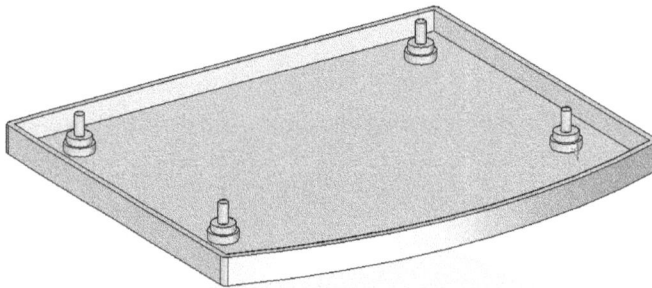

Now drag in the PCB from the part library and create an **Axial Align** between the hole in the PCB and the Bolts (Only create align relates between two sets of the bolts and holes).

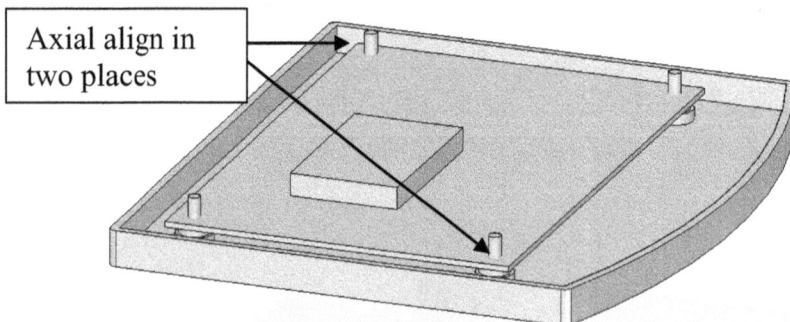

Axial align in
two places

Drag in the Top cover and create an **Axial Align** between the Bolts and the holes in the Top cover (only create 2 relationships).

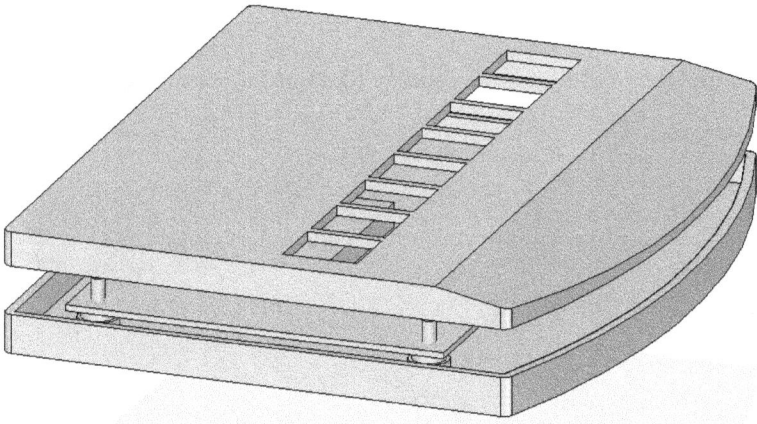

Now create a **Mate** between the surface of one of the Top cover's holes and the top of the PCB.

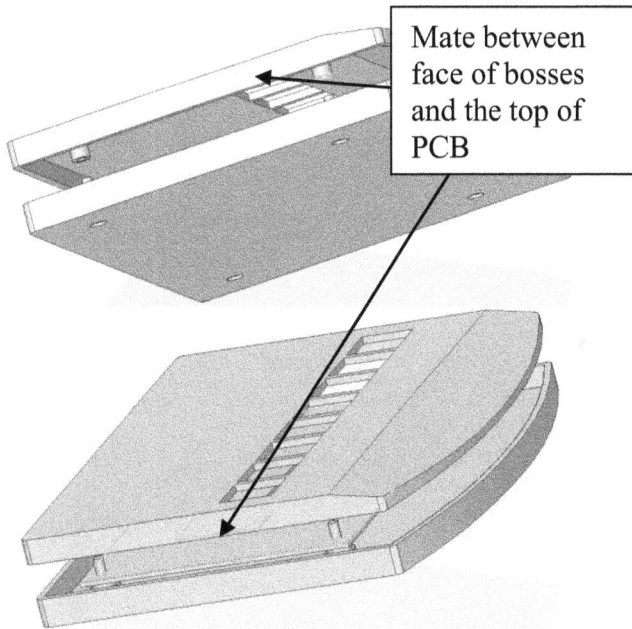

Mate between face of bosses and the top of PCB

The finished assembly should fit together nicely like the following figure.

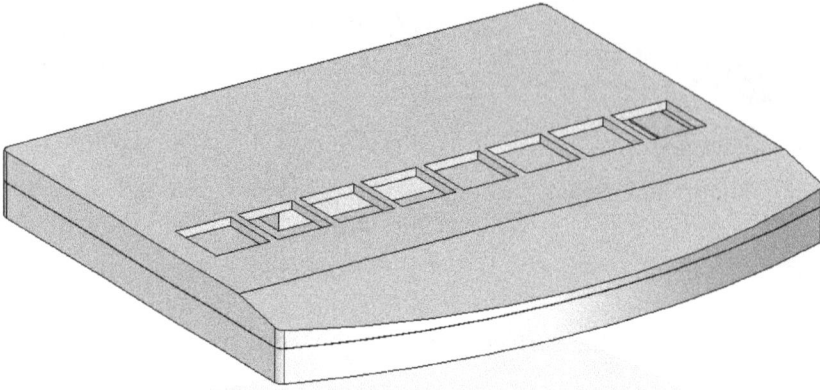

Exercise Complete

Exercise 49: Assembly 3

Create a solid model that looks like the part shown below and **Save** it as **Pipe.par**

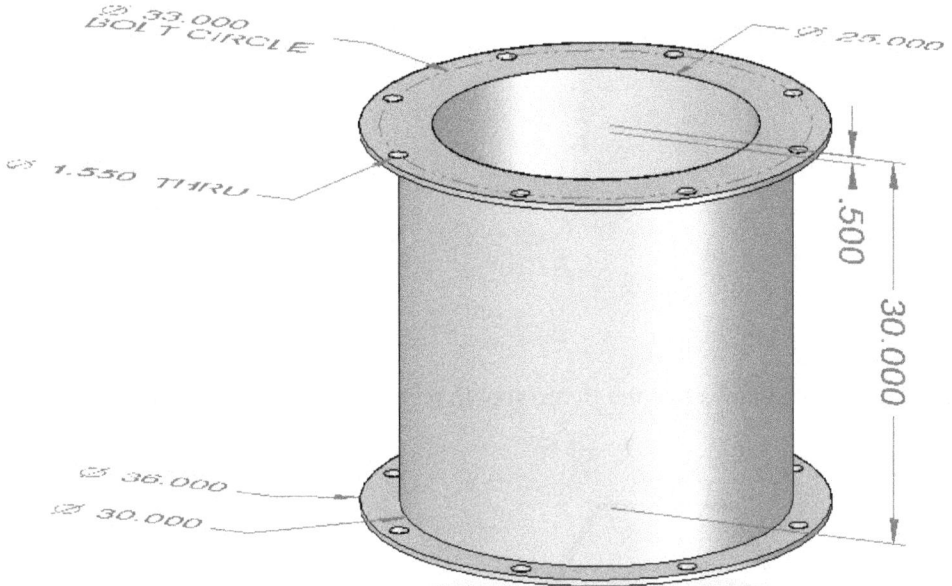

Create another part using the revolve tool that looks like the figure shown below and Save this one as **Bottom.par.**

Note: This part will mate with pipe.par so the number of holes, and their distance apart are the same.

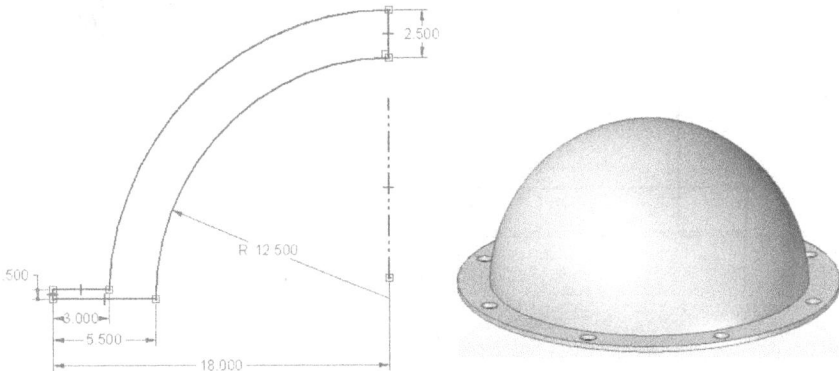

Revolve another part that looks like the figure shown below and Save it as **Top.par.** Create the same pattern as the last two parts as shown below.

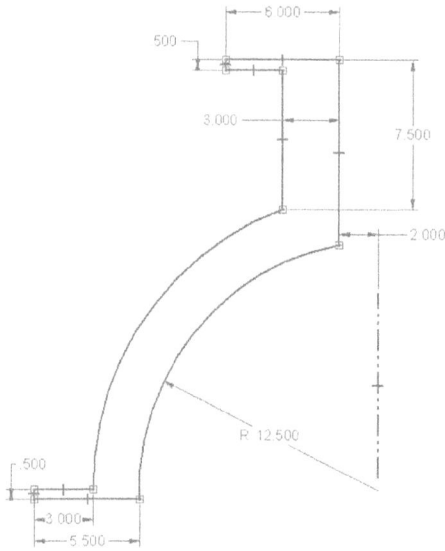

Create a new part that looks like the figure shown below and Save it as **Bolt.par**

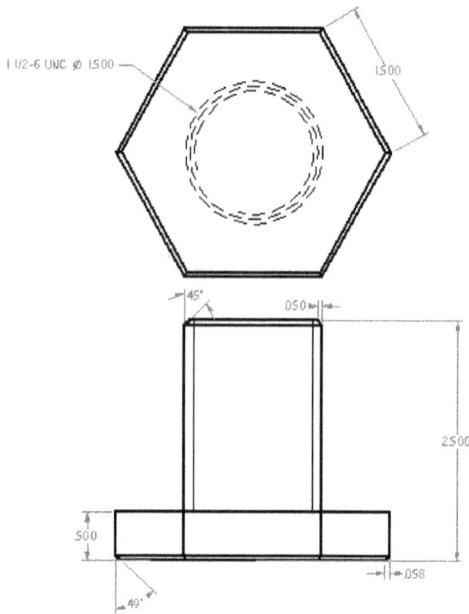

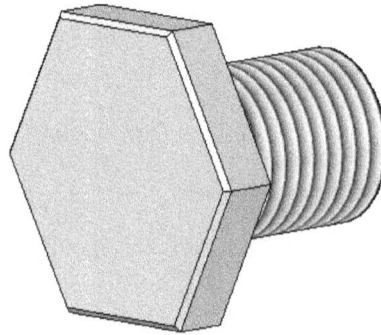

Create a **Nut.par** and **Washer.par** as separate part files to go with the Bolt. Furthermore, as the dimensions of the nut are different than the set dimensions in the Solid Edge program, in order to create the thread within the Nut, use the Hole tool and change the hole to **'Threaded'** by using the options button.

Note: the Nut has the same dimensions as the head of the bolt...

Ø 1.55

Ø 2.60

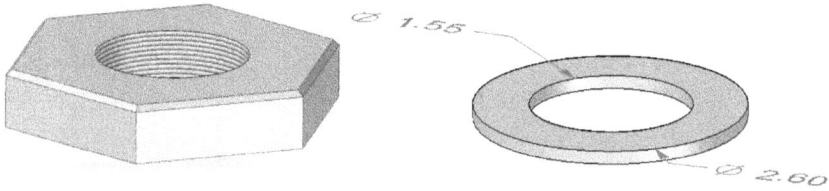

Create **a new ANSI Assembly** and drag the **Pipe.par, Bottom.par** and **Top.par** into the screen. Instead of using the **Relate tool** to assemble the parts, let's use the **Assemble tool** located in the Assemble options box. A menu should appear on the left that looks like the figure below. *Note: these are the same tools as the relate tools.* Click on the button and select **Axial Align** and align a hole from the top part with a hole on the pipe. Align a second top hole with the corresponding pipe hole.

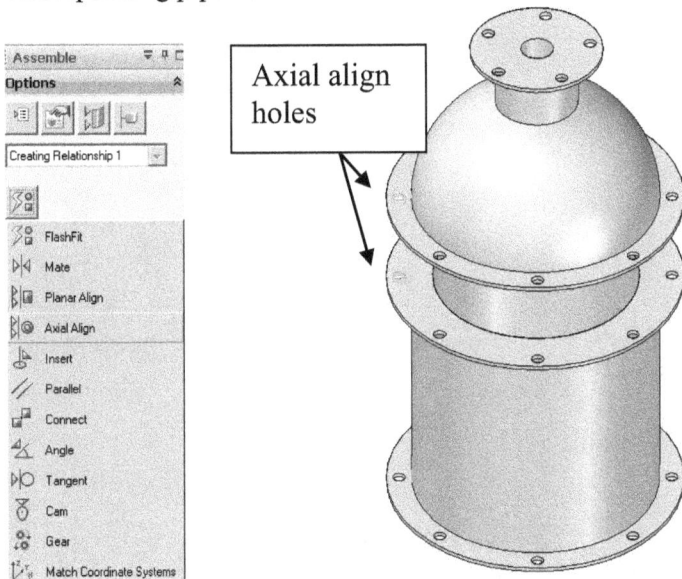

Axial align holes

Note: When in the Assemble tool select multiple sets of alignments without having to re- select the tool in between sets.

Hit Esc to exit the **Assemble tool.**

Use the same procedure to mate the **Bottom part** to the other side of the **Pipe.**

Click on the ⬚ button again, this time select **Mate** and click on the bottom surface of the Top part, and the corresponding surface on the Pipe.

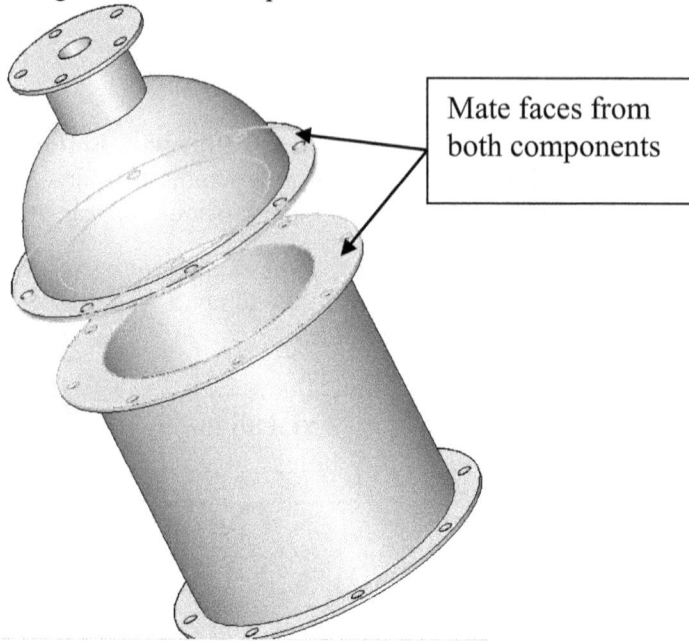

Mate faces from both components

Repeat this process for the Bottom part to the other end of the Pipe and the result should look like the figure below.

Hit Esc to exit the Assemble tool.

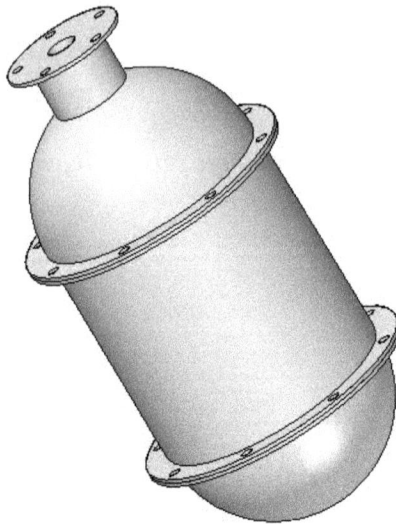

Drag two Bolts, two Washers, and two Nuts into the screen.

Click on the **Insert tool** ⚲ and align/mate the Bolt with one of the holes on the Top piece.

First, click on the Bolt and the Hole to align them then in the same step mate the surface of the head of the Bolt to the Top piece. *Note: The Insert tool works exactly like using the Axial Align and Mate tool all in one.*

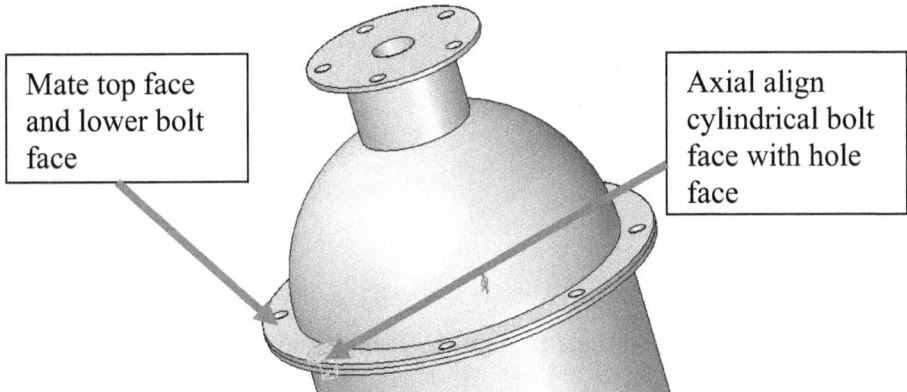

Mate top face and lower bolt face

Axial align cylindrical bolt face with hole face

Axial Align the Washer and the Bolt and Mate it to the surface of the Pipe.

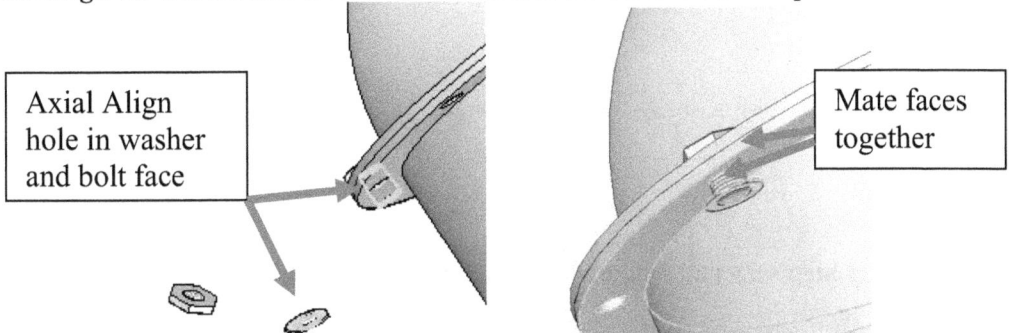

Axial Align hole in washer and bolt face

Mate faces together

Use the same procedure to **Align** and **Mate** the Nut to the Washer, as shown below.

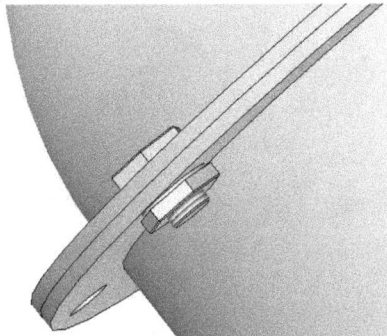

Place the second Bolt, Washer, and Nut into the hole of the Bottom directly below the Top hole.

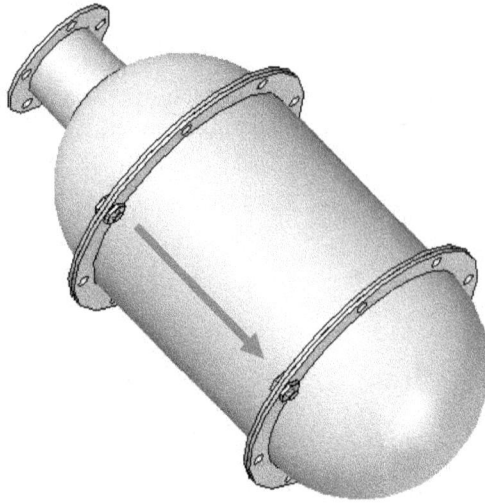

Instead of placing all 16 bolts, washers and nuts by mating and aligning, let's use a pattern to do it all in one step!

Click on the **Pattern tool** in the **Pattern** options box.

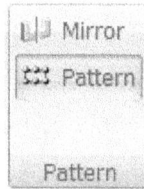

In the **Select Part Step** select all 6 parts shown below and, **Accept.**

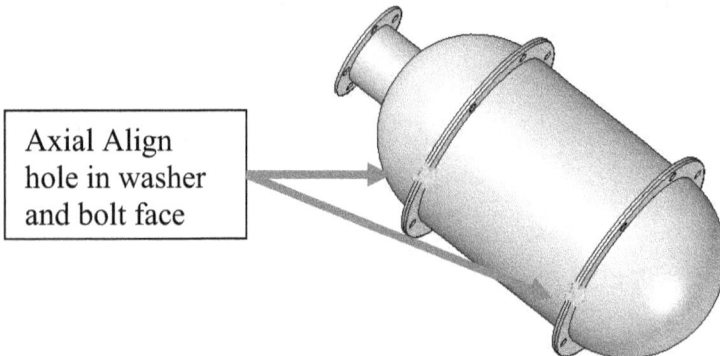

Axial Align
hole in washer
and bolt face

In the **Define Patten** Step when the Select Part button is highlighted, click on the Pipe.

Then when the **Select Pattern Feature** is highlighted select the pattern of holes as shown below.

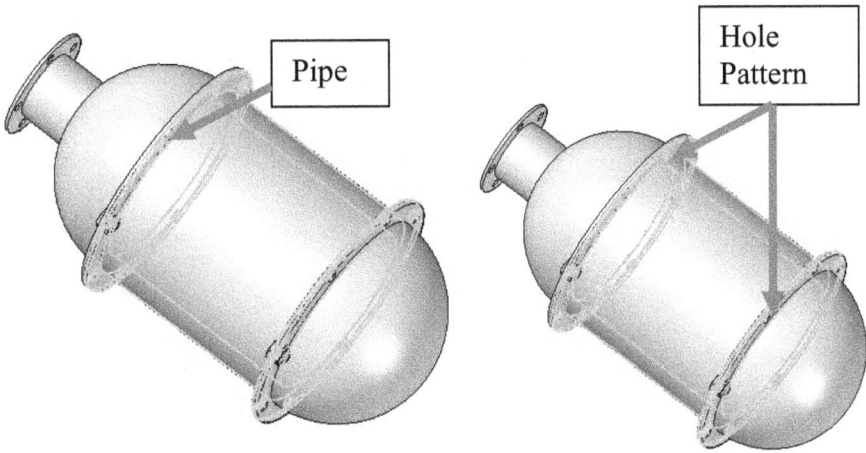

Select the original hole as the **Reference Position** and click **Finish.**

A Bolt, Washer and Nut should appear in each of the remaining 14 holes. The finished assembly should look like the figure below.

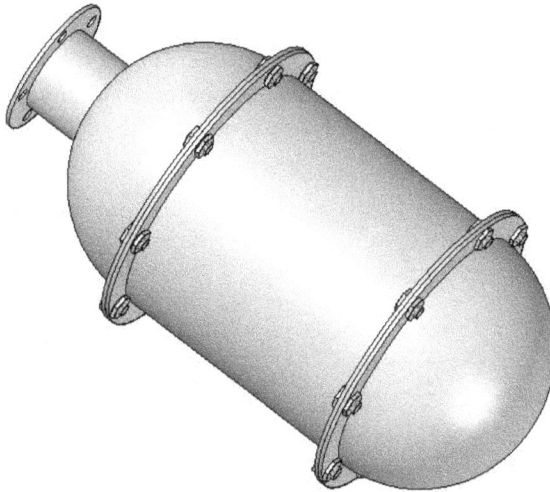

Exercise Complete

Exercise 50: The Gear Relationship

In a new part file click on **Engineering References** located under the Tools tab in the Environs options box.

In the dropdown menu select **Spur Gear Designer.**

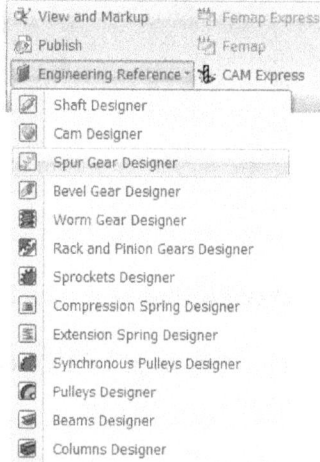

A window should appear that provides different parameters to change to create the exact set of spur gears desired. For our purposes click **Create,** because the default parameters will work fine.

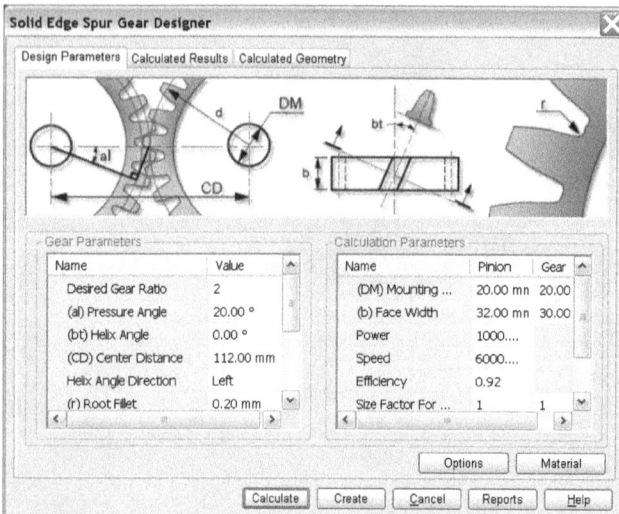

*Notice: You can also calculate the strength of the gear and see a bunch of other cool results in the **Calculated Results tab***

Click **Create** – SE will ask to Save each gear as a separate part file, Save the gears as **SPUR GEAR1**.par and **SPUR GEAR2.par.**

Note: The Engineering References tool creates parts using the units specified in the original part in this case we are using millimeters.

Back in the original part file create a part that looks like the figure below. This is going to act as a base to position the gears.

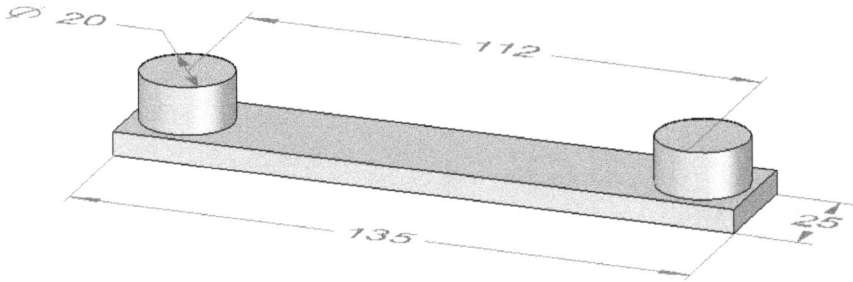

Save this file as **SPUR GEAR BASE.par.**

Open **SPUR GEAR1** and add a small handle as shown below.

Note: Be creative with the dimensions of the handle.

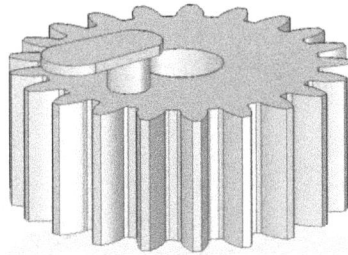

Open **SPUR GEAR2** and add a hole as shown below.

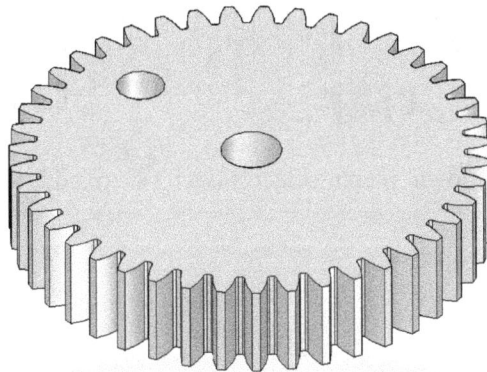

Create a new Assembly file and import the **SPUR GEAR BASE** as the grounded part, then import the 2 gears from the **Part Library.**

If the gears look like the ones on the left below, they are in their *simplified part form* change them into their *Designed part form* by right clicking on the part and selecting Use Designed Part (Do this for both gears). *Note: the simplified part form is intended to save computer memory which is great for larger assemblies.*

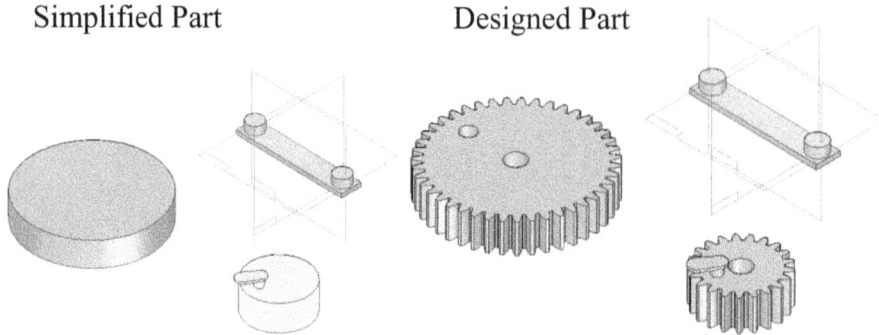

Simplified Part   Designed Part

Use the **Insert tool** located in the Relate dropdown menu to **align** and **mate** each gear hole to one of the pegs on the Base.

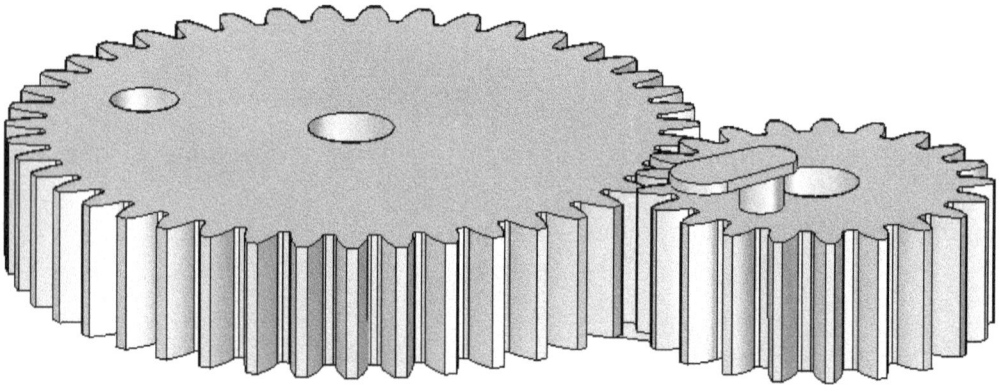

Back in the Relate drop down menu select **Gear** to add a gear relationship to the assembly.

In the menu on the left, make sure the **Rotation-Rotation** 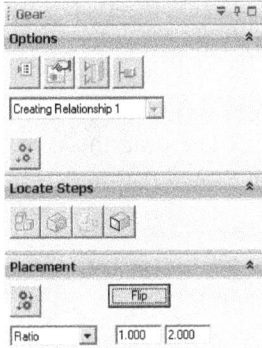 button is selected, and **click** on the center circle of the large gear, then the center circle of the small gear.

If both arrows are rotating the same way (as shown below) click **'Flip'** on one of them
Type in 2.00 when the small gear is selected to specify that the gear ratio is 1:2.

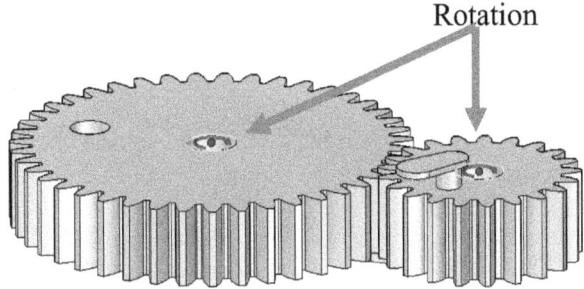

Rotation

The rotational arrows should look like the ones shown below to ensure the gears spin in the right direction.

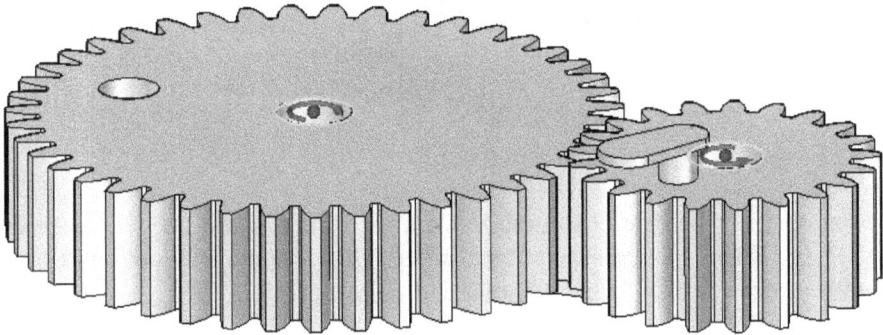

Right Click to exit the Gear menu.

Click on the **Move Part** tool in the Modify options box and click on the **Rotate** button from the menu on the left.

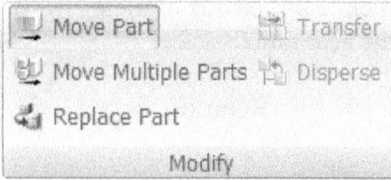

Click **OK** on the pop up window, the default settings will work for this exercise

Click on the small gear to place the coordinate system then click and drag the gear. The gears will now be spinning in opposite directions.

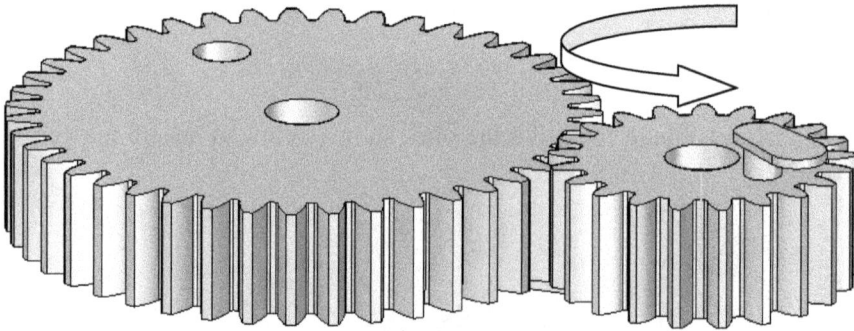

Hint: Rotate the gear slowly to see all the action.

If it is not possible to place the coordinate system then the rotation may be locked. Click on the **SPUR GEAR BASE** located in the **Pathfinder**. This will show all the constraints, shown below, that have been placed that involve that particular part.

Notice: next to the Axis align constraints (one step of the Insert command) it says [rotation locked]

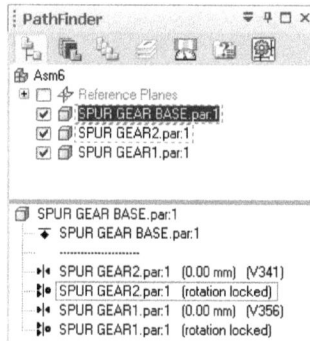

Click that constraint, and select the Rotate button in the Placement Step directly above the **Pathfinder** (Do this for both axial aligns that are locked).

Retry the **Move Part** Step and the gears will spin!

Click on the **Motor tool** in the Motors dropdown menu in the **Assemble** options box.

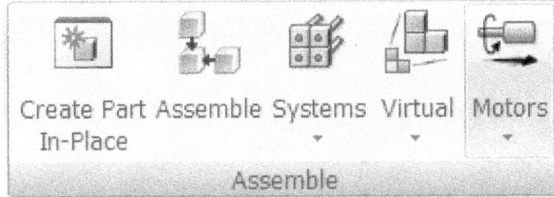

In the **Select Moving Part** Step click on the small gear.

In the **Motor axis Step** type 50.000 deg/s into the appropriate box then click on the center cylinder to place the direction to rotate.

Click **Finish.**

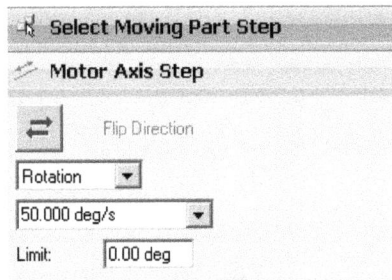

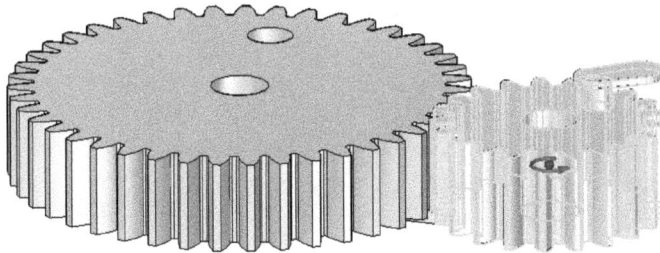

Back in the **Motors** dropdown menu, click on the **Simulate Motor tool** and click **OK** when the window pops up.

Click the **Play** button at the top of the **Animation bar.**

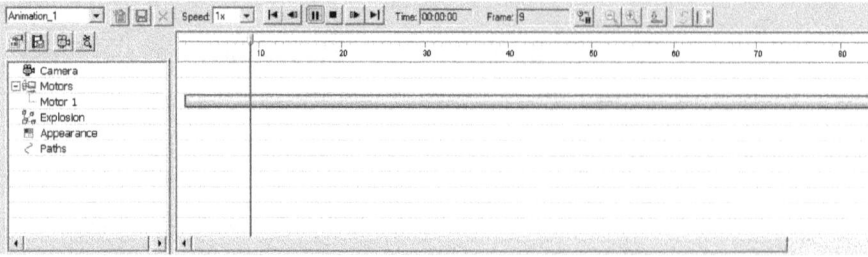

The gears will now spin.

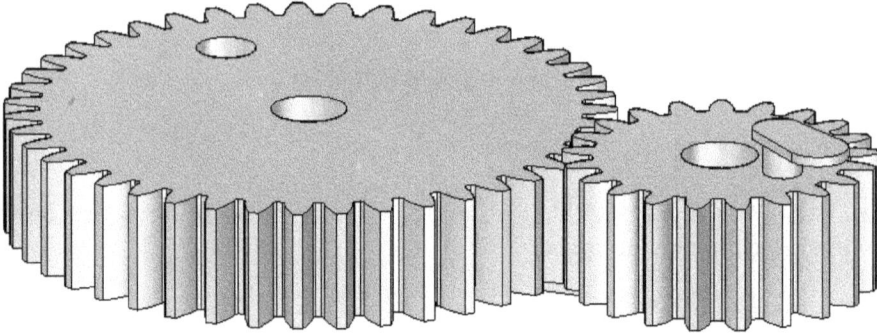

Exercise Complete

Exercise 51: Assembly 4 "Do Nothing"

Create a solid model that looks like the part shown below and Save it as **Base.par**

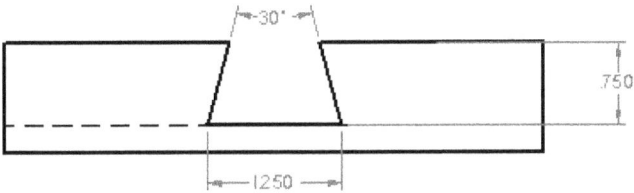

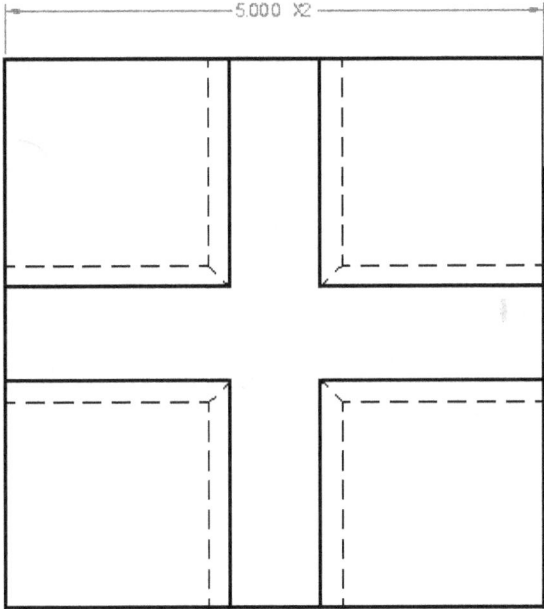

Create another part with a threaded hole in the center that looks like the figure shown below and Save it as **Key.par**

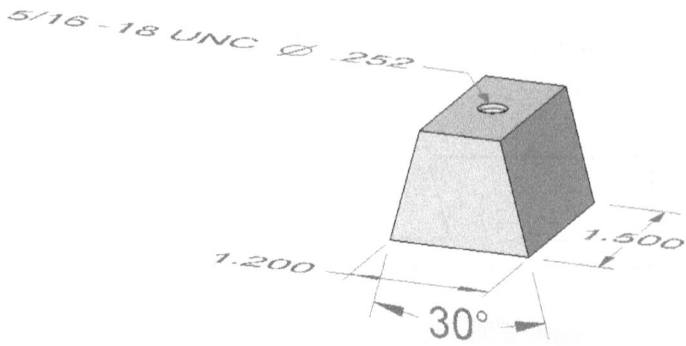

Create another solid model that looks like the figure below and. Save it as **Crank.par**

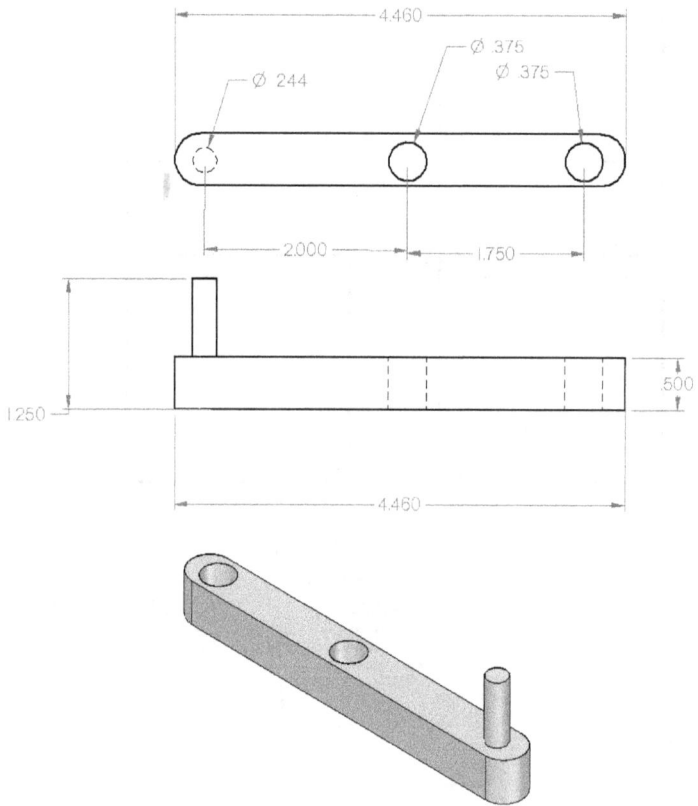

Now, in your internet browser address bar enter the following address.

http://www.mcmaster.com/#90298a619/

This address will show a ready-made shoulder bolt that can be purchased from the McMaster-Carr website. *Note: DO NOT purchase this part for this exercise; this is simply to simulate using vendors components.*

The website will show a bolt drawing that looks like the following figure.

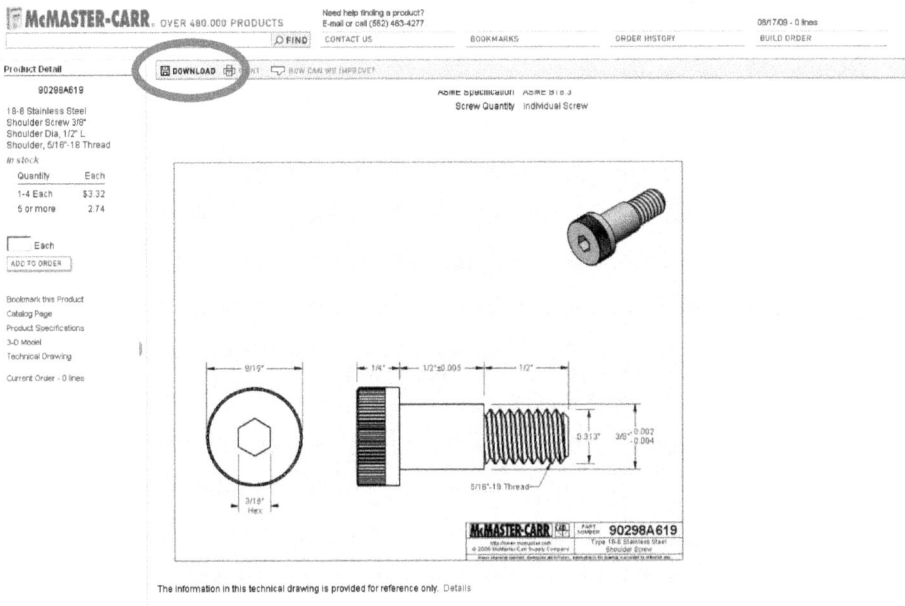

Click on the **Download** button at the top left corner of the page.

Once the **Download** button is clicked, choose what type of file to download. In this case the **3-D STEP** file will work best, so highlight it and click Save.

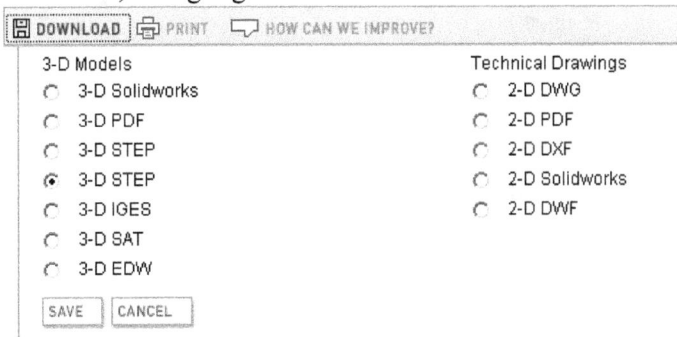

A window that looks like the following figure will appear. Click **Save** and name the part **shoulderbolt.stp** to be used later in the exercise.

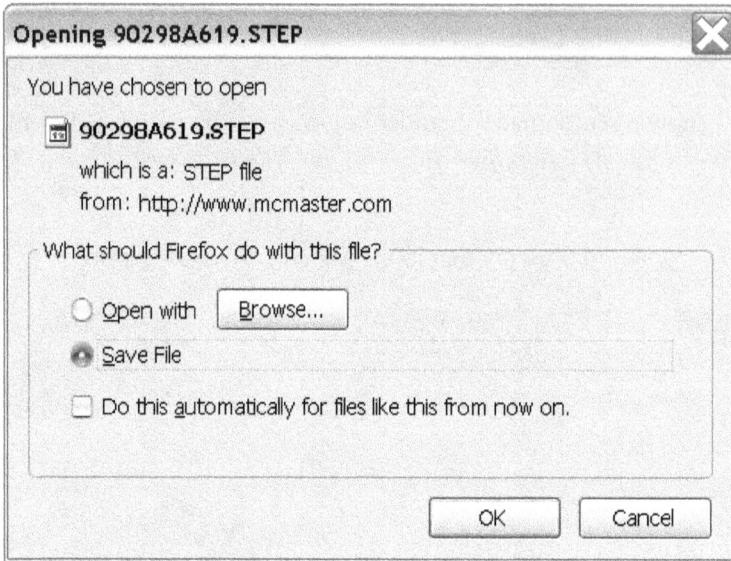

In Solid Edge click **Open** to bring the shoulder bolt into a new window.

To find the file, change the File type to **STEP documents** (*.step,*.stp).

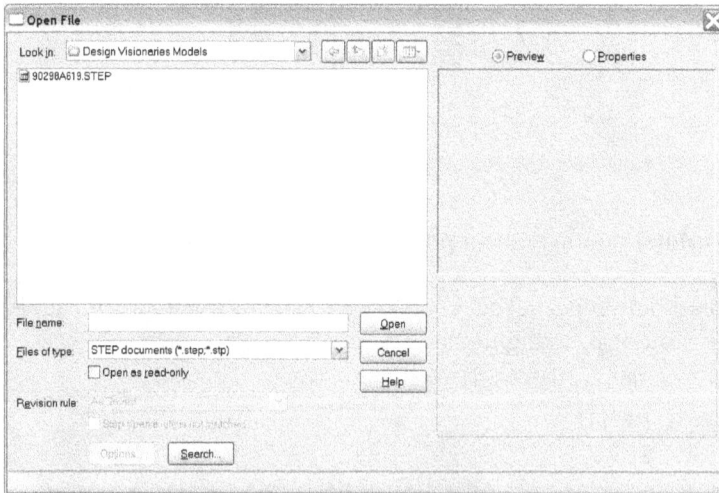

When found, click **Open.** It will ask what type of file to open it as. Click on **ANSI part.par** then hit **OK.**

A box might pop up that looks like the following figure, click **yes.**

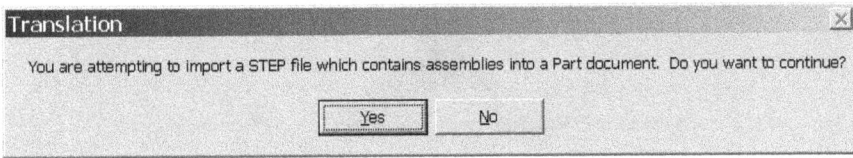

The shoulder bolt should now be in a part file. *Note: your part may have been brought in with mm conversion (if your default for SE is mm).* If this happens, this is a quick way to scale the part:

Save this file as **Shoulderbolt.par** and **open a new part file.**

In the Home tab click the **Part Copy tool** in the Clipboard options box.

In the **Select Step** find Shoulderbolt.par and click **Open.**

In the pop up window type 25.4 into the Scale X box. Make sure **'Use Uniform Scale'** is selected.

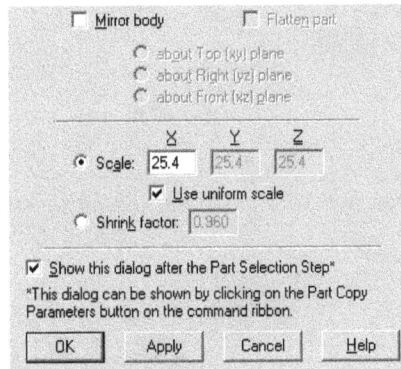

Click **OK** then **Finish** and a larger version of the shoulder bolt should appear in the window.

Save this as **Shoulderbolt1.par.**

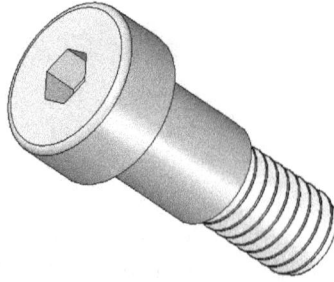

Open a new ANSI Assembly file.

Note: remember to change the units to inches and the style to ANSI if necessary.

Drag the Base from the **Parts Library** into the screen first followed by the Key.

Mate the bottom of the Key with the Base.

In the **Mate** menu on the left type 0.025 in into the **Placement** box and mate one angled side on the Key with the corresponding side on the Base.

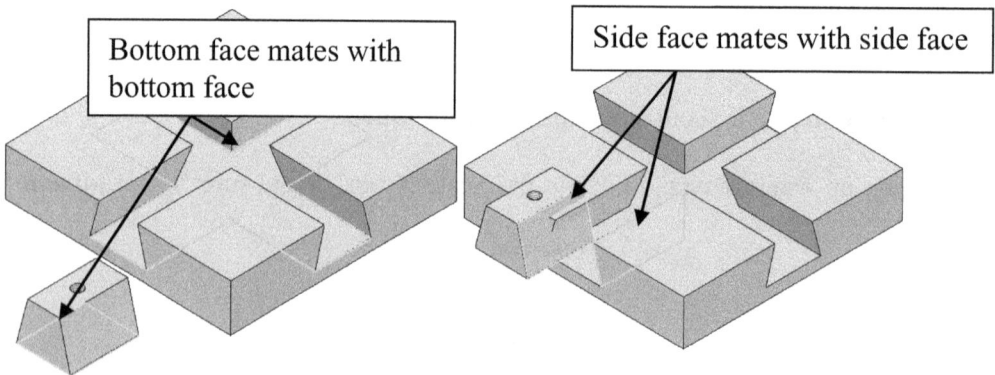

Bottom face mates with bottom face

Side face mates with side face

Drag another Key into the screen and follow the same mating steps to mate it to the groove to the right.

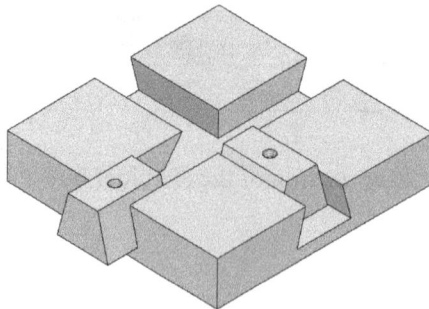

Next, drag the **Crank** into the screen, along with 2 Shoulderbolts.

Axial Align and **Mate** a shoulder bolt into each of the two holes of the Crank.

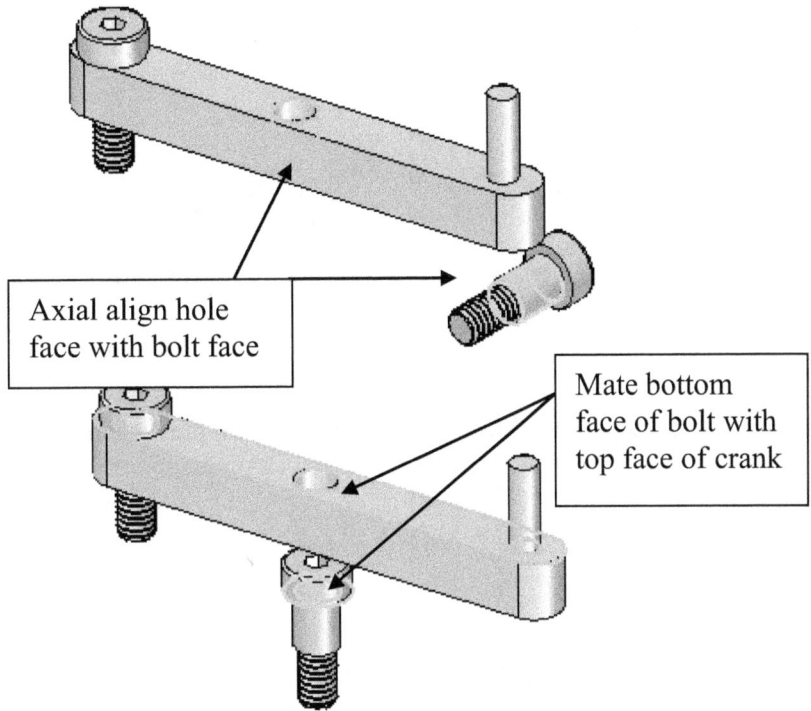

Axial align hole
face with bolt face

Mate bottom
face of bolt with
top face of crank

Axial Align the bolts with each hole on the two keys.

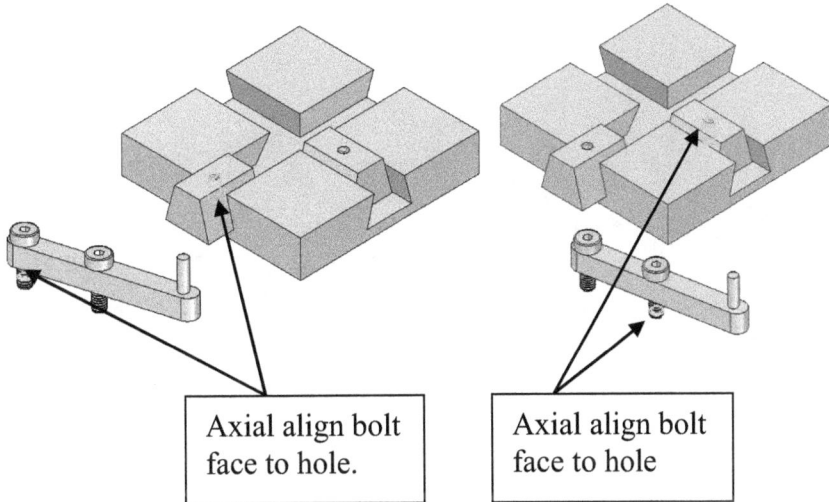

Axial align bolt
face to hole.

Axial align bolt
face to hole

Mate the bottom of the Crank to the top of the one of the Keys.

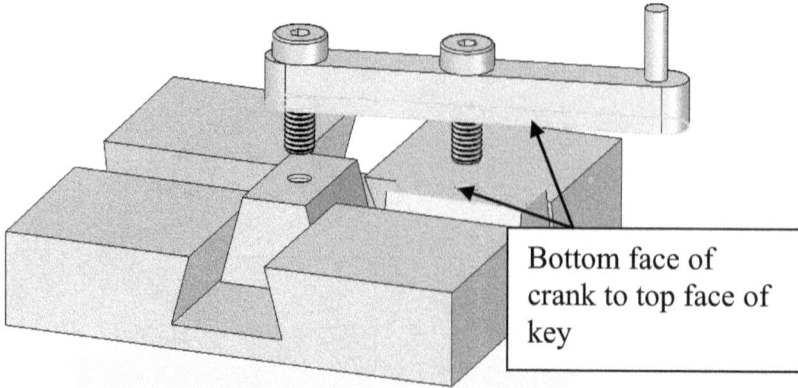

Bottom face of
crank to top face of
key

Click on the Motor tool in the Motors dropdown menu in the Assemble options box.

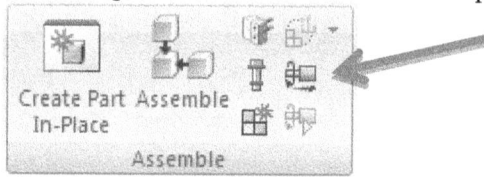

In the Select Moving Part Step click on the Crank.

In the Motor axis Step click on the hole in the center of the Crank to place the direction to rotate.

Click Finish.

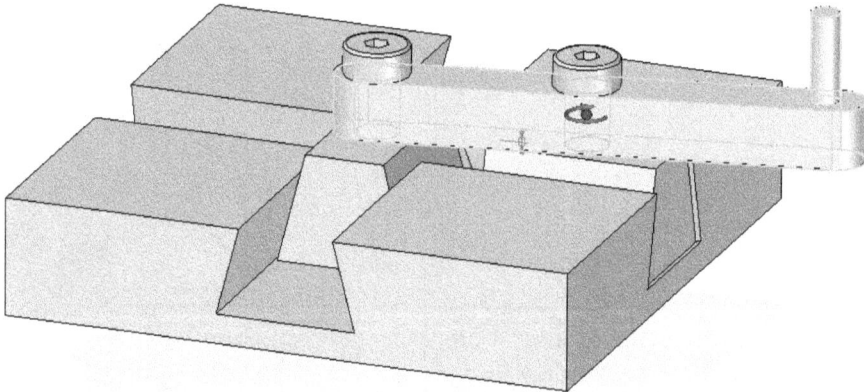

Back in the Motors dropdown menu, click on the Simulate Motor tool click OK when the window pops up.

Go to the Tools tab and click ERA. From there, choose Animation Editor.

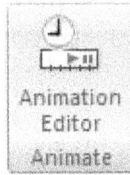

You should get the bar shown below. Right click "Motors" and click "Edit Definition".

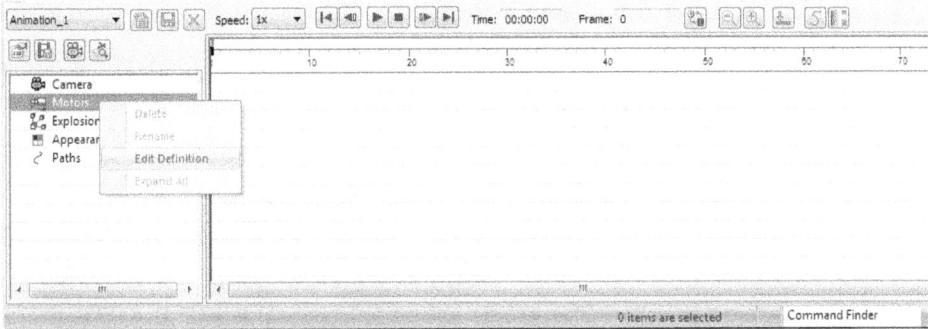

Input the following information into the window that pops up and then click OK.

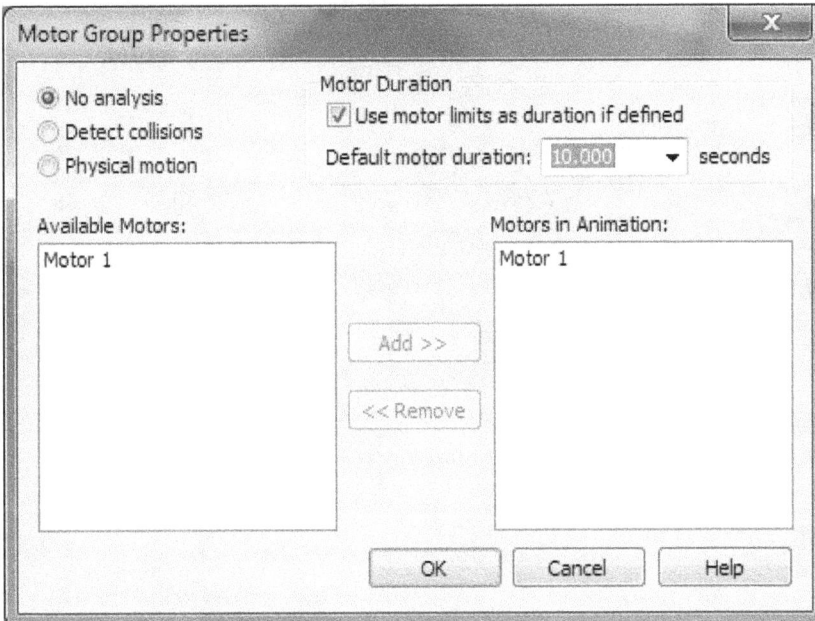

Click the **Play** ▶ button at the top of the **Animation bar** that should be at the bottom of the screen.

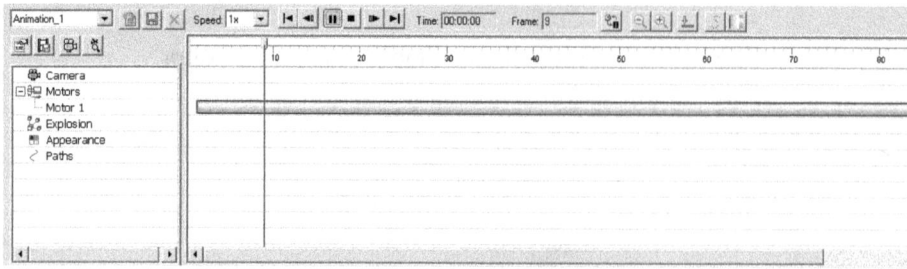

The **Do Nothing** is complete and should be moving on its own!

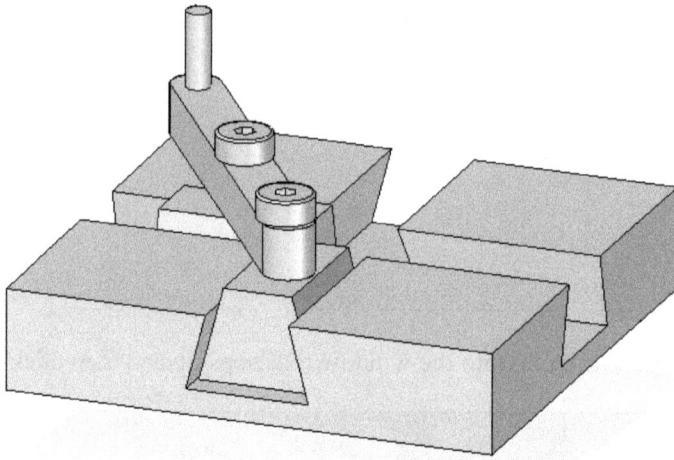

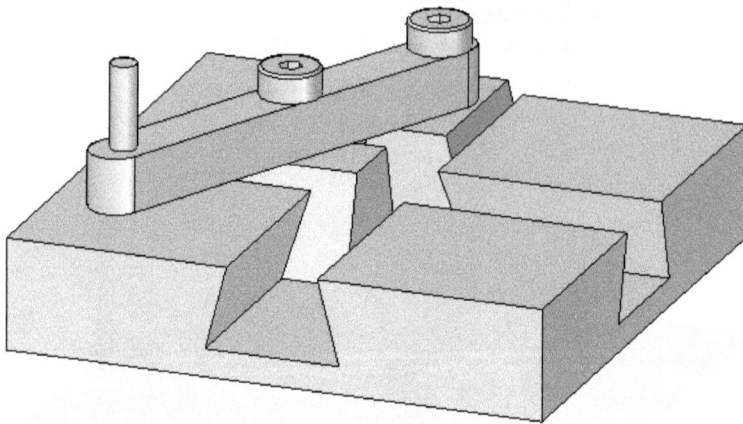

Exercise Complete

Exercise 52: Cam tool "Hurdy Gurdy"

Create a solid model that looks like the part shown below and Save it as **Back.par.**

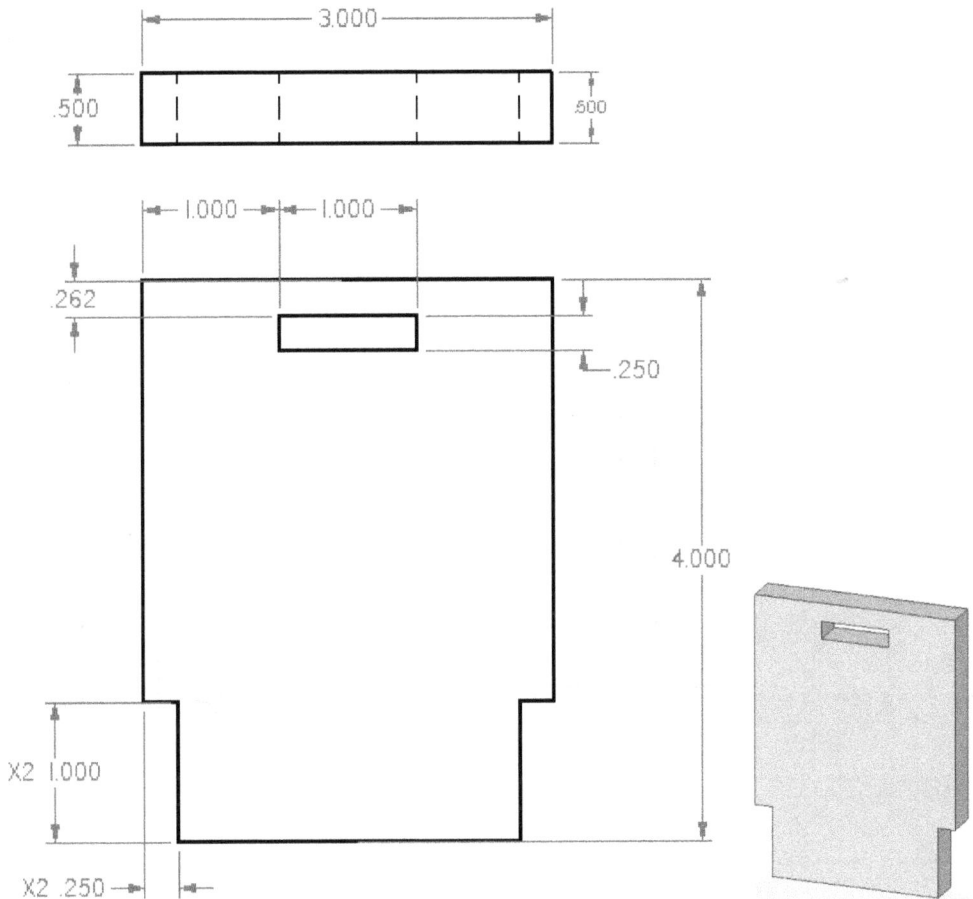

Create a part that looks like the figure shown below and Save it as **Cam Reader.par**

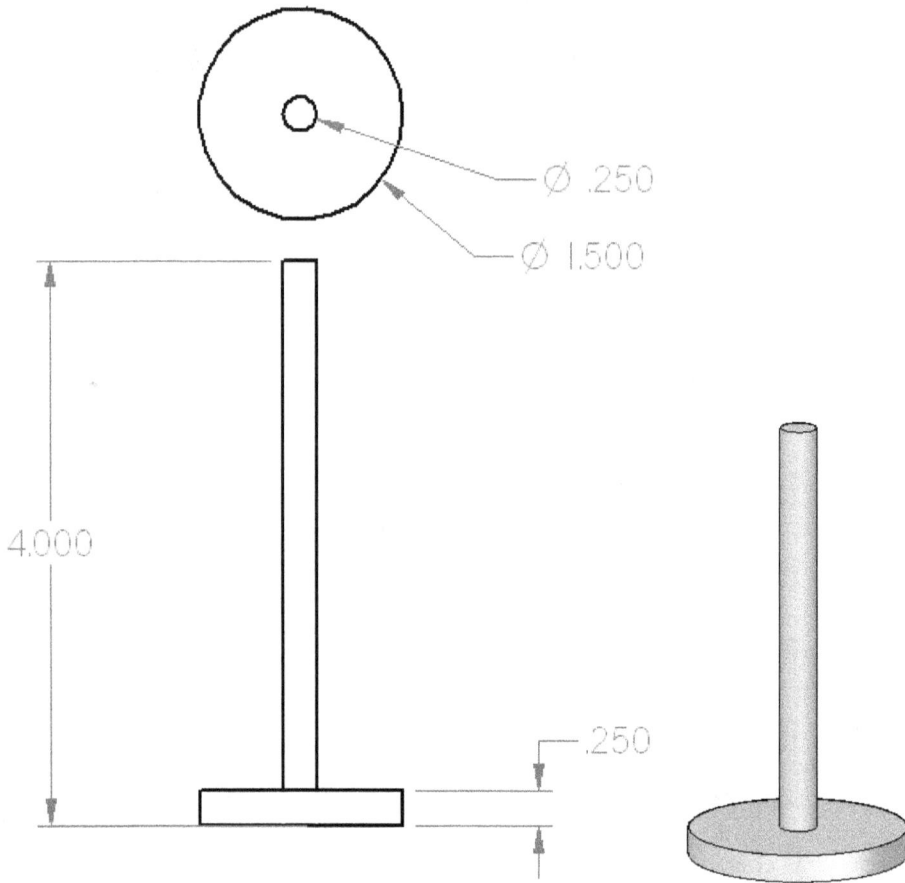

Ø .250

Ø 1.500

4.000

.250

Create a part using an arc and an ellipse that looks like the figure shown below and save it as **Cam Rod.par**

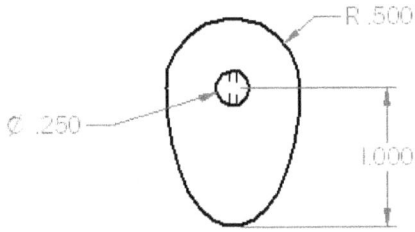

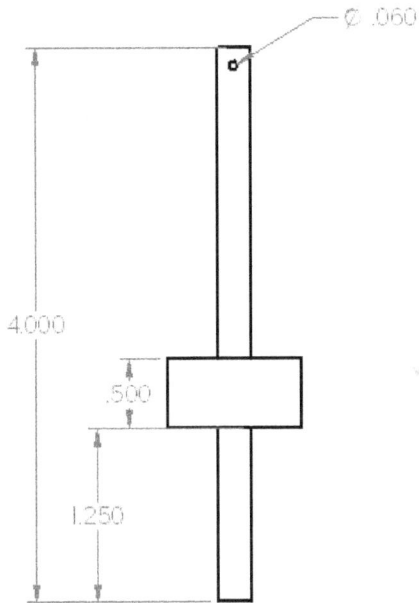

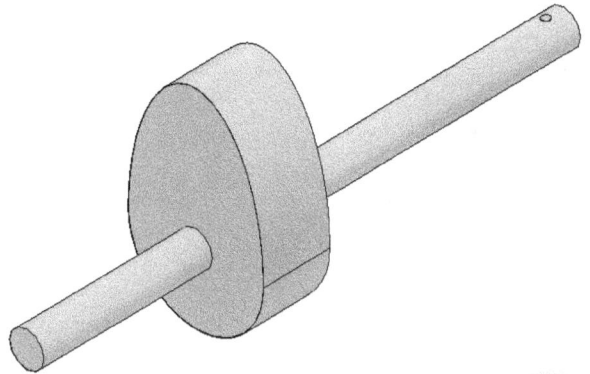

R .500

Ø .250

1.000

Ø .060

4.000

.500

1.250

Create another part that looks like the figure shown below and Save it as **Side.par**

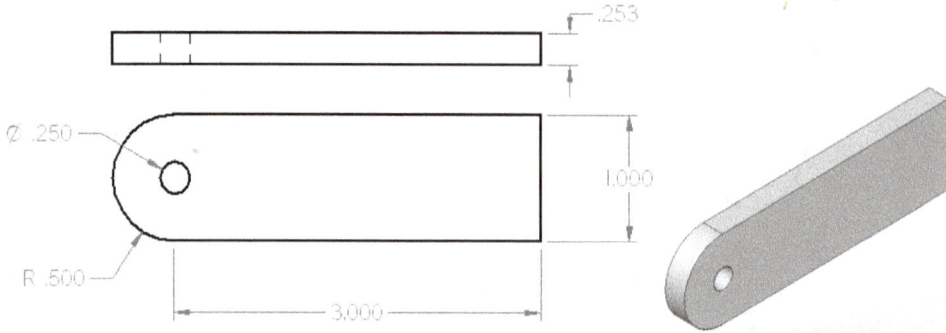

Create another part that looks like the figure shown below and Save it as **Crank.par**

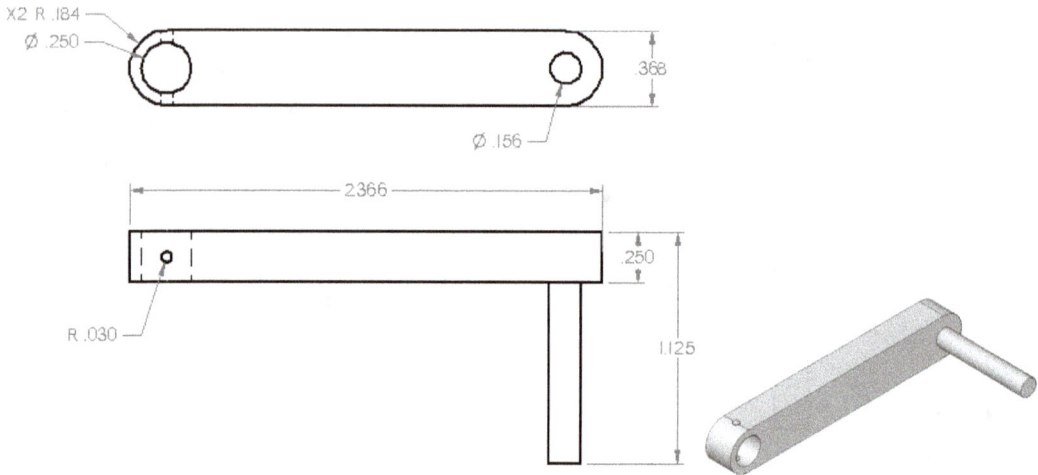

Create another part that looks like the figure shown below and Save it as **Pin.par**

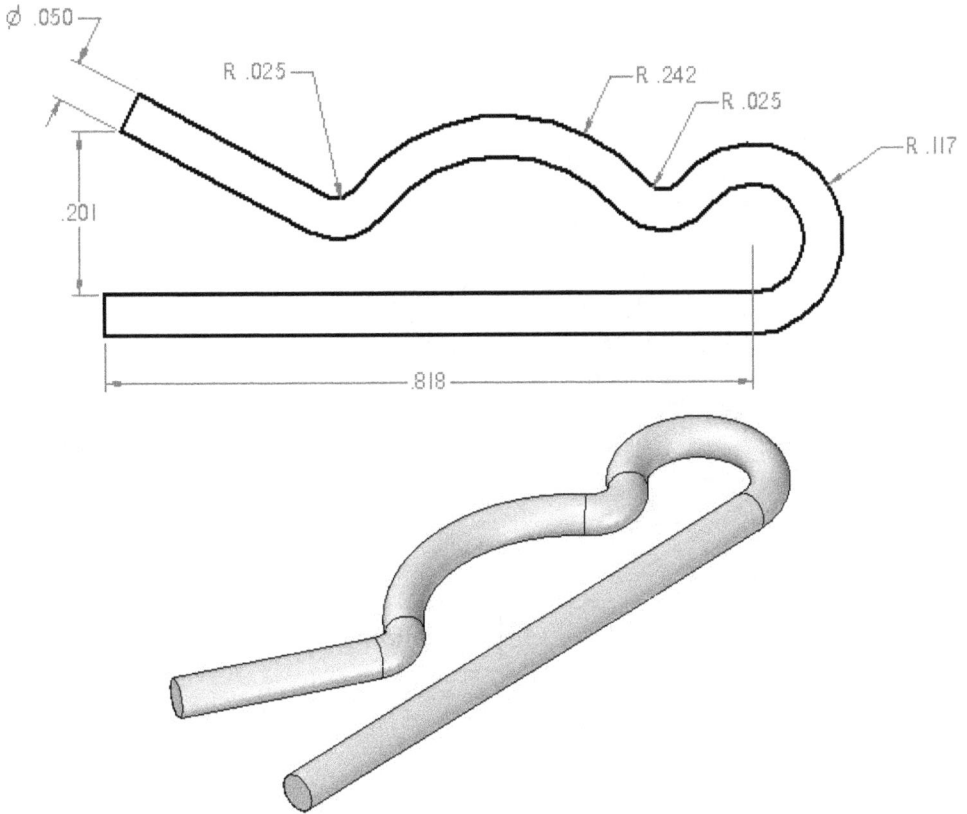

Open a new ANSI Assembly file.

Note: remember to change the units to inches and the style to ANSI.

Drag the Back from the Parts Library into the screen first followed by three Sides.

Position the first Side into the slot of the Back by **mating** the two sides shown in figure 1, and using the planar align tool on the side in both figure 2 and figure3.

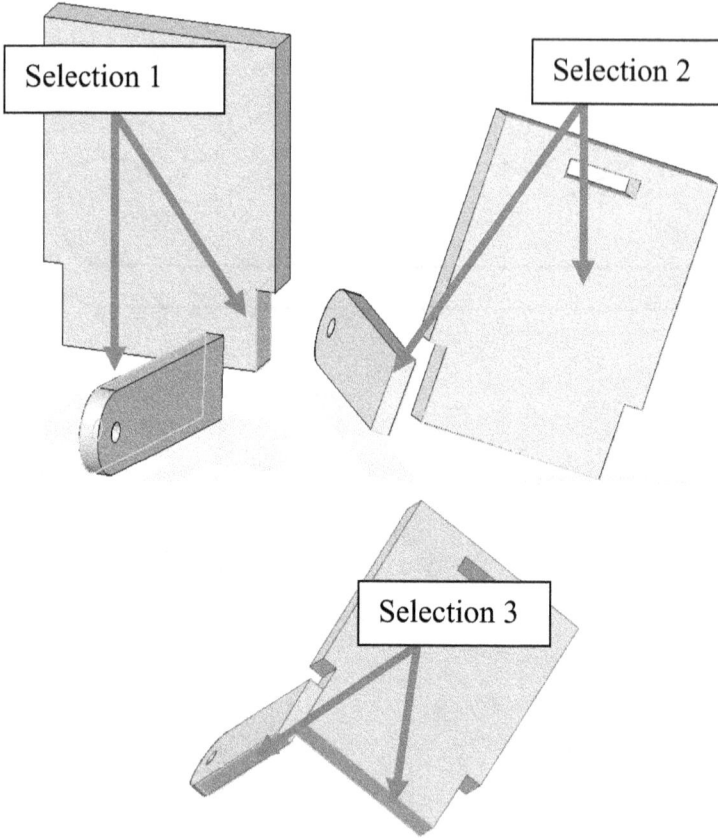

Selection 1

Selection 2

Selection 3

Repeat this process for the second side.

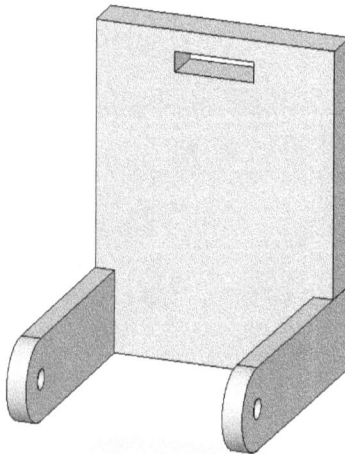

Insert another Side piece into the slot at the top of the Back by **mating** the sides shown in figures 1 and 2 and planar aligning the sides in figure 3.

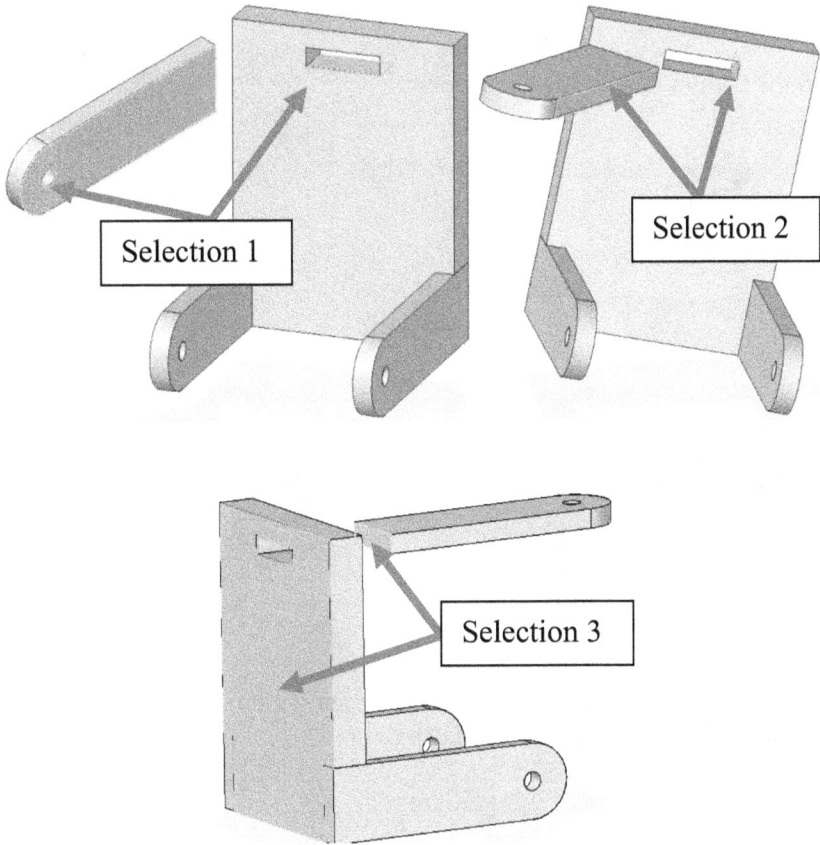

Drag the Cam rod into the screen and **axial align** it into the hole of the first Side. Then planar align the end to the shaft without the small hole to the outside of the second Side. Your model should now look like the following figure.

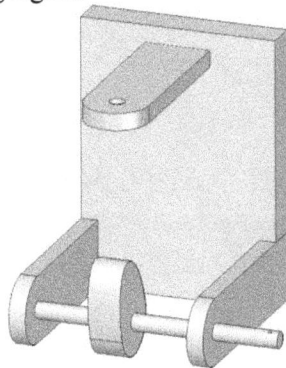

Insert the Cam reader into the screen. **Axial align** the shaft and the hole of the 3rd Side then use the **Cam relate tool** to relate the cam to the reader.

Select the edges of the Cam as shown below on the left and click of the green check mark in the left menu.

Now click on the bottom face of the Cam reader.

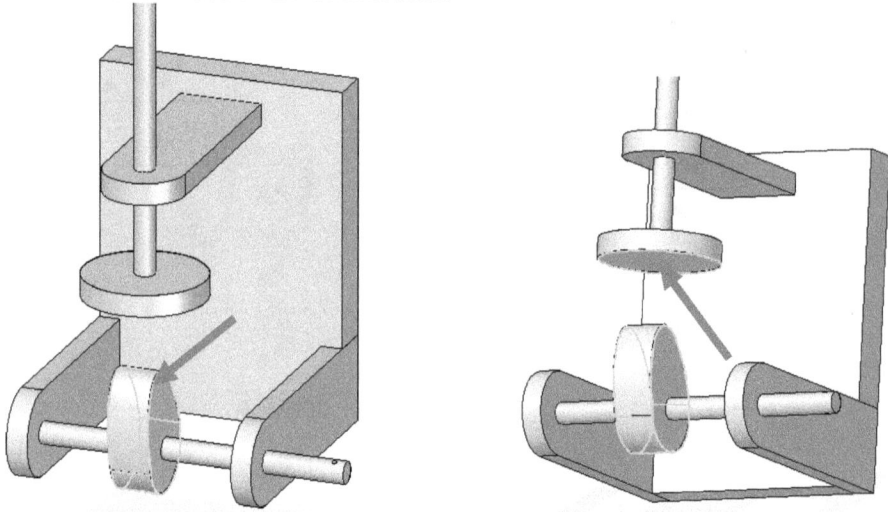

Insert the Crank into the screen and place an **axial align** on the small hole in the end of the Crank with the small hole at the end of the Cam rod.

Next **axial align** its hole to the shaft of the Cam rod.

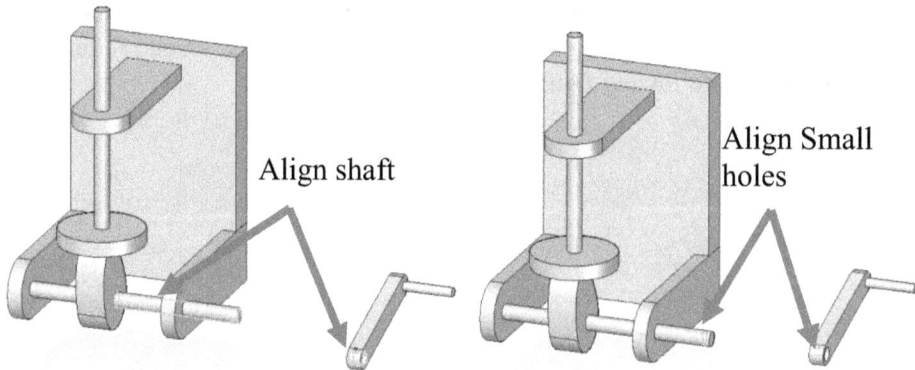

Place a **Motor** about the Cam rod and click **finish** to exit the motor tool.

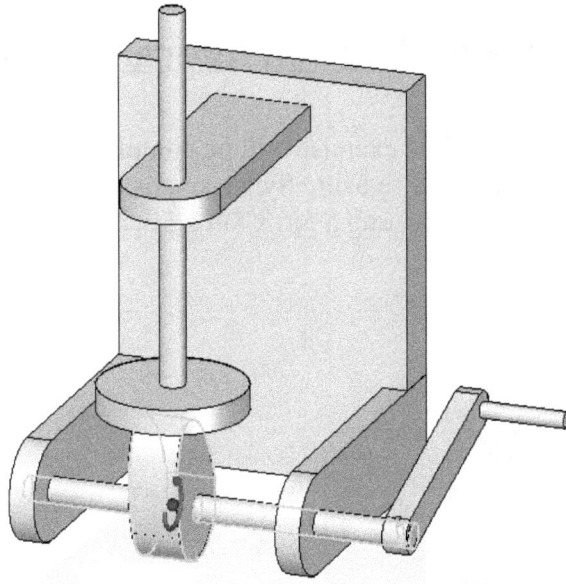

Simulate the motor and watch the cam move.

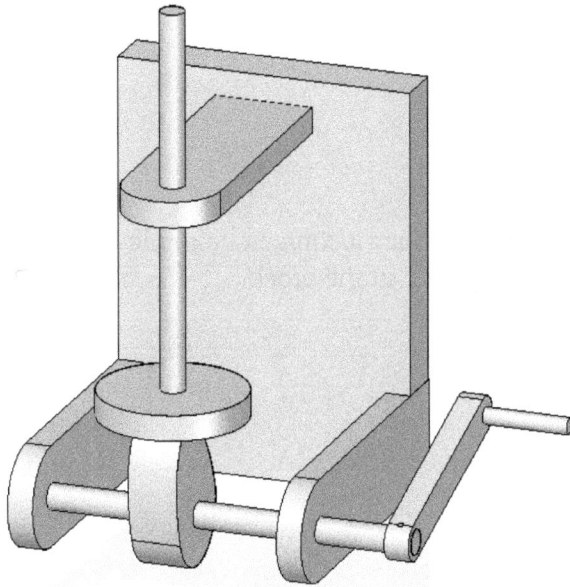

Exercise Complete

Exercise 53: Part Families and Alternate Assemblies

Create a new **ANSI Part** file. This exercise will be completed using Ordered modeling, so switch to that if you're using Synchronous.
Start by clicking **Extrude** and creating a 5in x 5in x 5in cube.

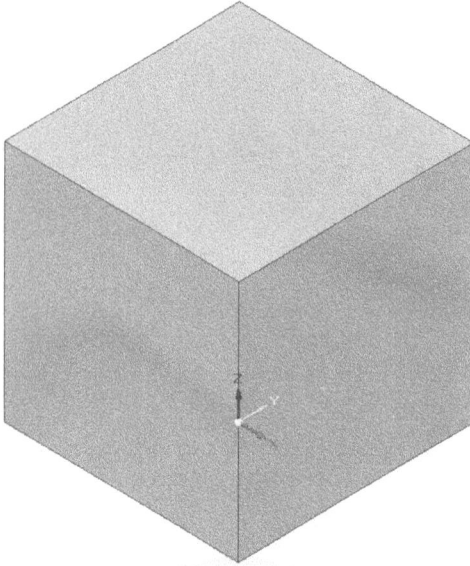

Now click the **Hole** command. Place a countersunk hole with the dimensions shown below in the center of the top face of the block.

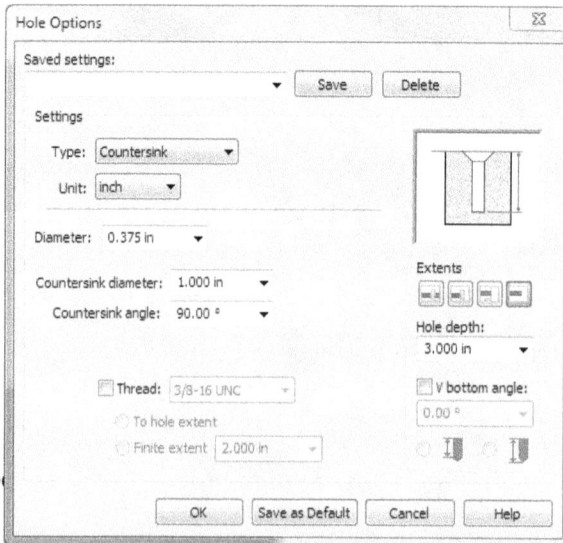

Your blo[...]wireframe view for clarity)

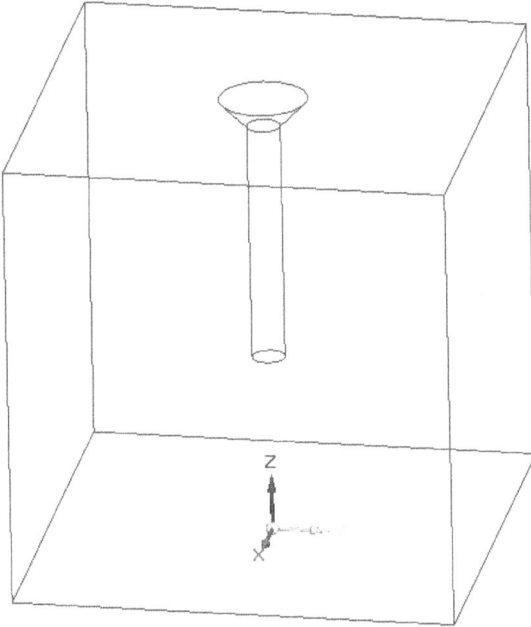

Save this block as **Coutersunk_Block.par**. We will be using it later in this exercise. Open a new **ANSI part** and click **Revolve**. Draw the following sketch. (the bottom line is dashed because after sketching it, I designated it the axis of rotation)

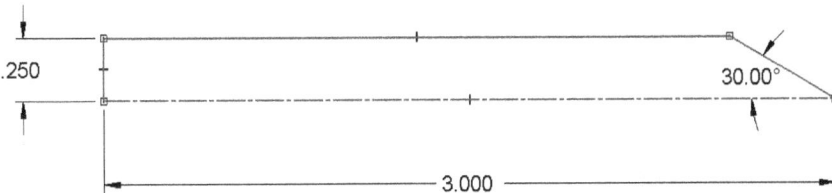

.250

30.00°

3.000

Rotate the sketch a full 360 degrees. Your model should now look like this.

Create a sketch on the **Front Plane**. (it should run straight through the center of your part) Your sketch should look like the following.

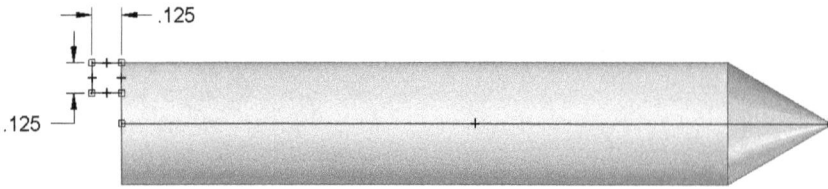

Click **Helical Cut**. First, choose the small square you just sketched as the cross section. Then, choose the line you sketched as the axis of rotation. Choose the end of the line directly below the square as the start point and then enter the information shown.

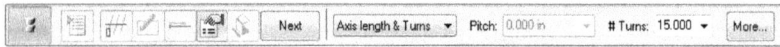

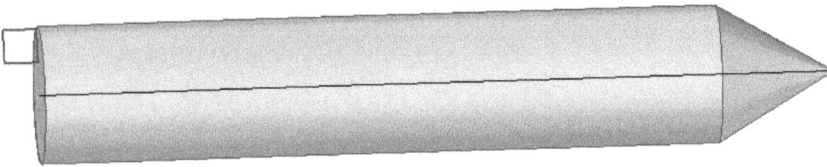

The result should look like the following part.

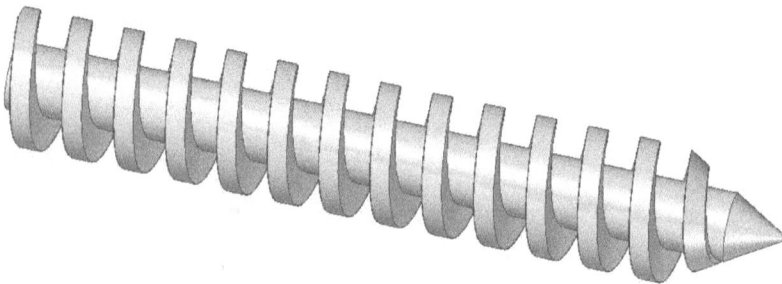

Next, create a circular sketch on the end of your part with a diameter of 0.5in. Then create another circular sketch, except this time offset 0.25in from the previous sketch and with a diameter of 1in.

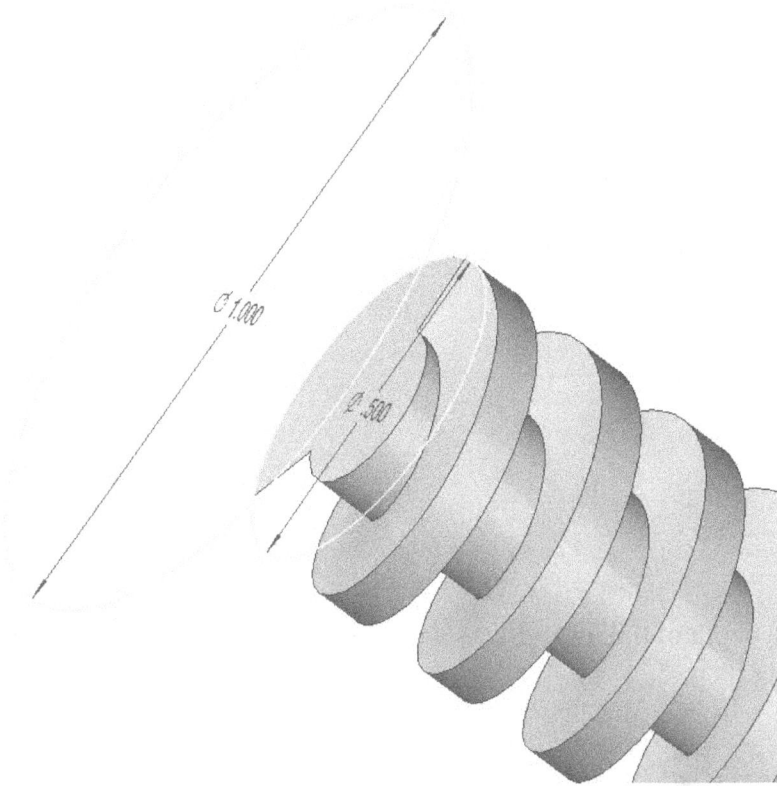

Click **Loft** and then choose both of the circles you just created. The result should be the following part.

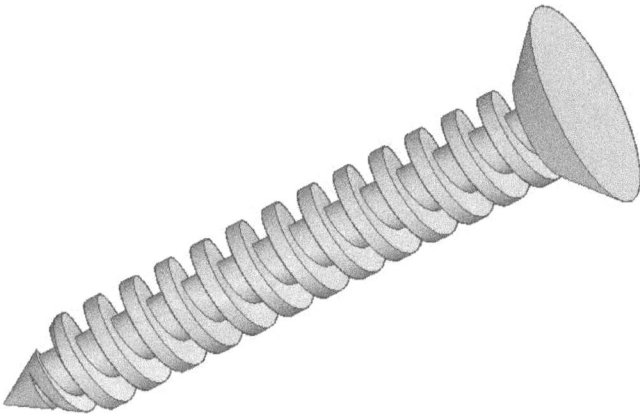

Create the following sketch on top of the screw we created.

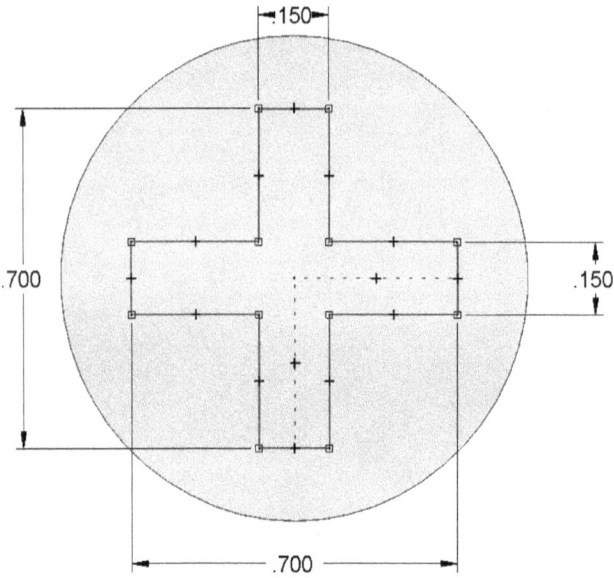

Cut the shape out to a depth of 0.25in and with a draft of 20 degrees inward. It should look like the following image.

Save this part as **Interchangeable_Screws.par**.

Here comes the interesting part: creating a family of parts based around this one. Open the "Family of Parts" tab along the left side of the screen and add a new member called "Phillips Head". Add another member called "Flat Head".

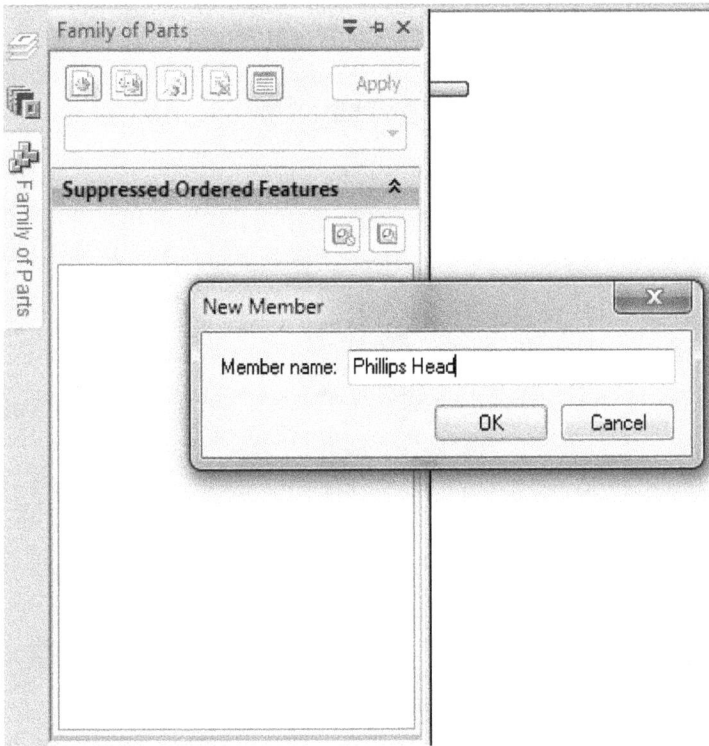

Make sure you are on "Flat Head". Click the **Cutout** feature in the design tree that created the Phillips Head and choose Suppress Feature. Then click apply.

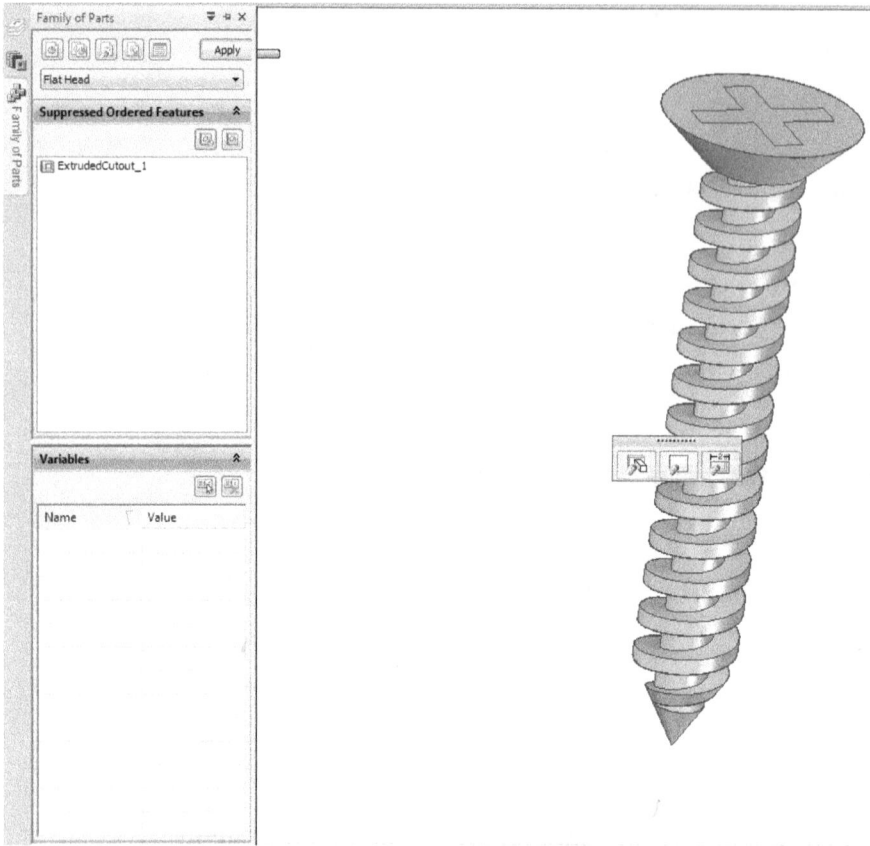

Hide the sketch on the screw head. Now that the head is clear again, create the following sketch on it.

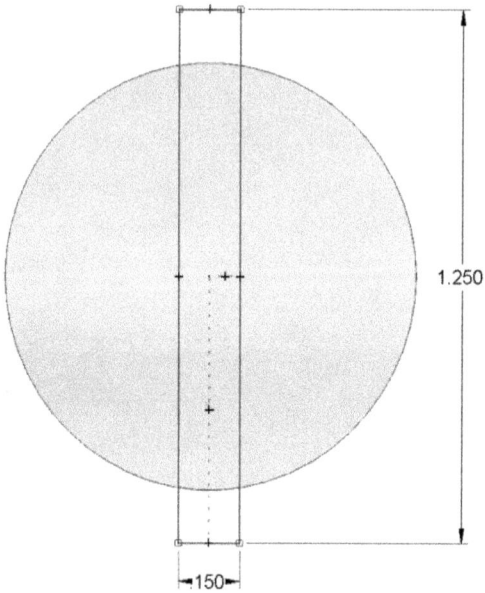

1.250

150

Your part will nc

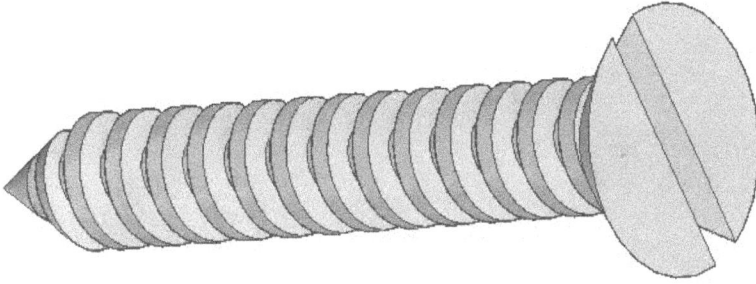

Now go back to your family of parts and switch back to "Phillips Head". Click the **Cutout** feature in the design tree that created the Flat Head and choose Suppress Feature. Then click apply. Now you can freely switch between the two different screws without having to create an entirely new part!

Click the Edit Table button ▤ in the Family of Parts tab. When the window pops up, highlight both screws and click Populate Member(s).

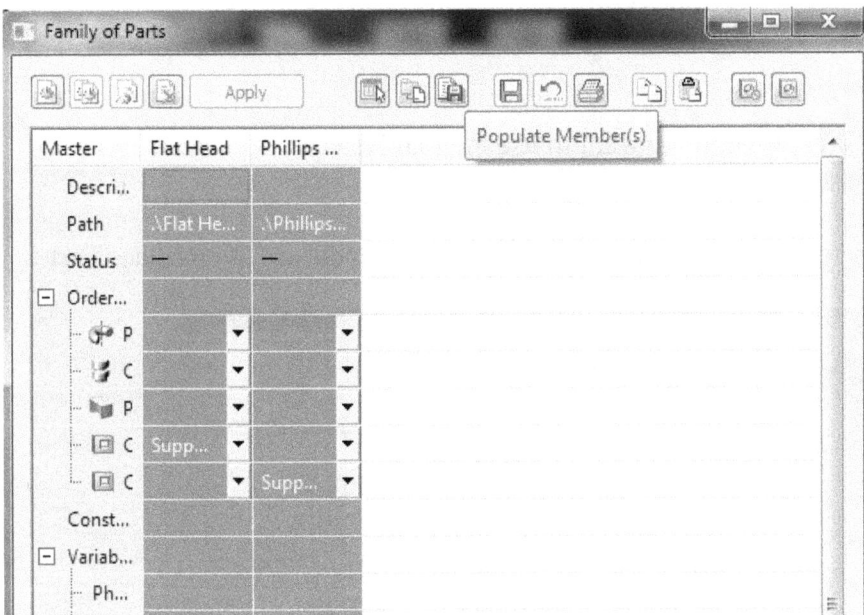

Click OK when a message pops up asking if you want to populate 2 members. Close the window when it finishes populating the members.

Alright, we're on the final stretch of the exercise! Create a new **ANSI Assembly**. Import the block and the Phillips Head screw. Put an Axial Align constraint on the screw and the hole. Put a Planar Align constraint on the screw's head and the top of the block. The screw should now be fully constrained into the block.

"But wait…" you say, "What if we want to use the Flat Head screw instead?" That's where Alternate Assemblies come into play.

First, select your screw and click Define Alternate Components in the Replace Part menu.

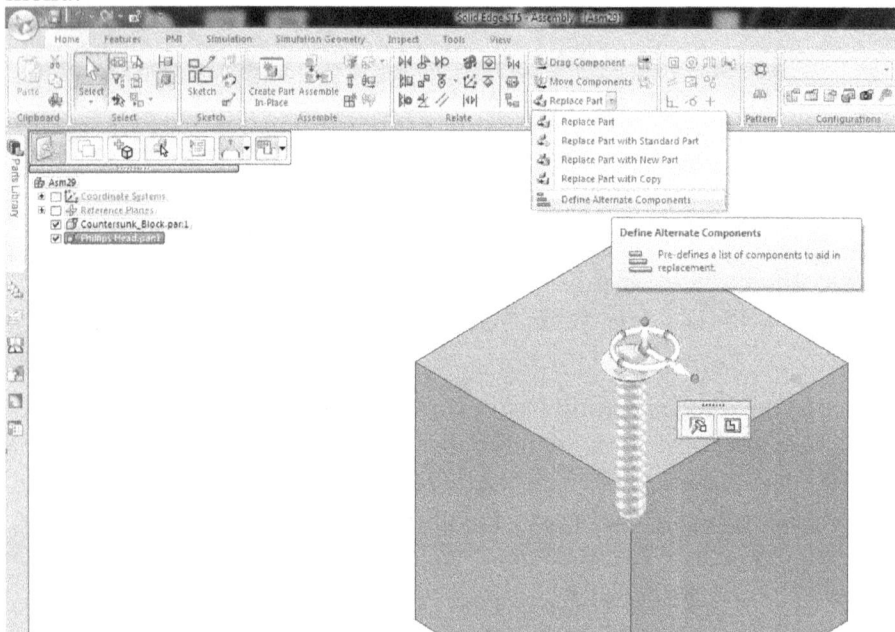

From this menu, add **Flat Head.par** as an alternate component and then click OK.

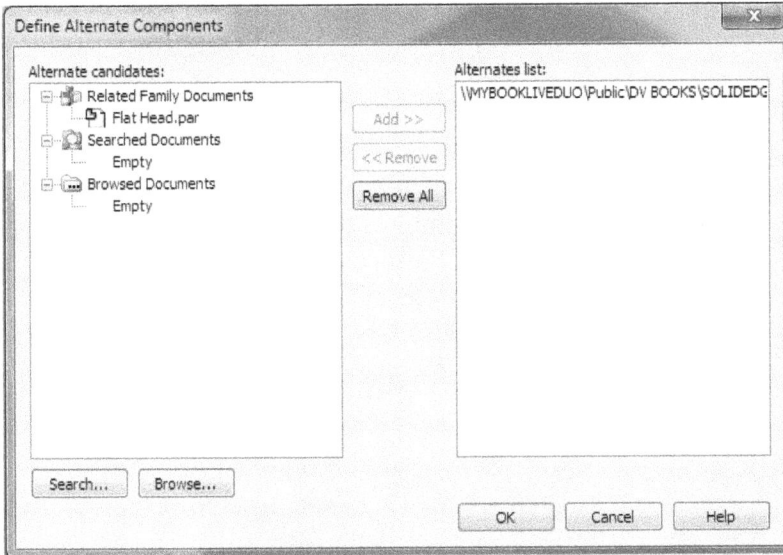

Save the file as **Alternate_Blocks.asm**.

Now go to the Alternate Assemblies menu and click Add New. Input the following information and click OK.

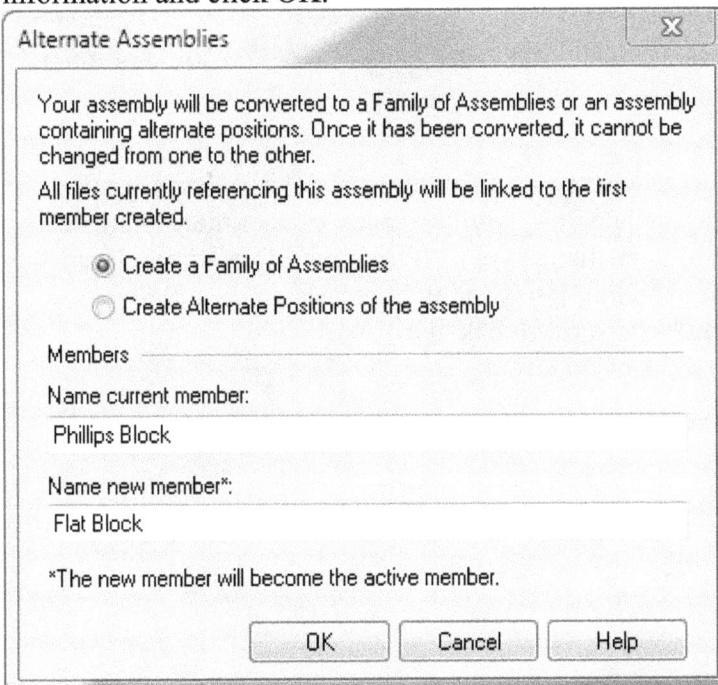

Make sure you are on "Flat Block" before selecting the Phillips screw and clicking Replace Part. When this menu comes up, click OK.

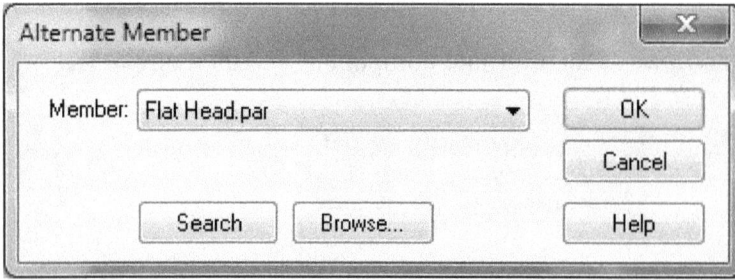

You'll now notice if you go back to Alternate Assemblies that your "Flat Block" has a Flat Head screw while your "Phillips Block" has a Phillips Head screw. Below is a side by side comparison of the alternate assemblies created.

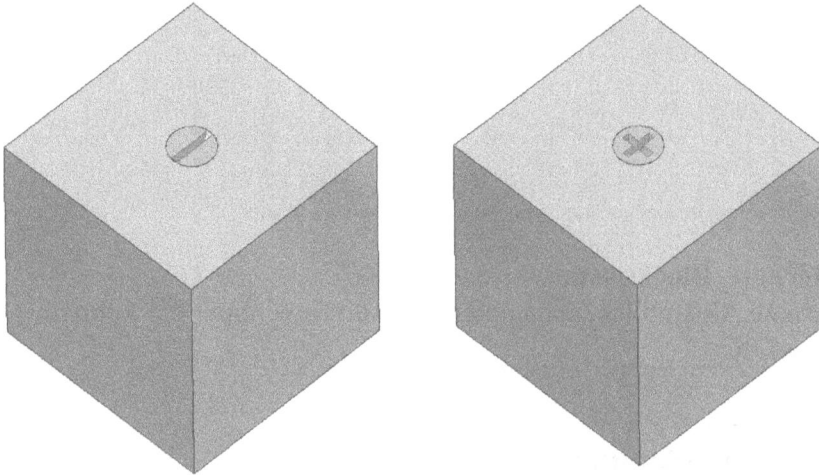

You can have as many members as you please in a family of parts and an alternate assembly. (just remember to repopulate your family of parts every time you create a new member) If you really want to challenge yourself, try creating a Torque Head screw in your family of parts and adding it to your alternate assemblies. Below is a sample image of what a Torque Head screw looks like.

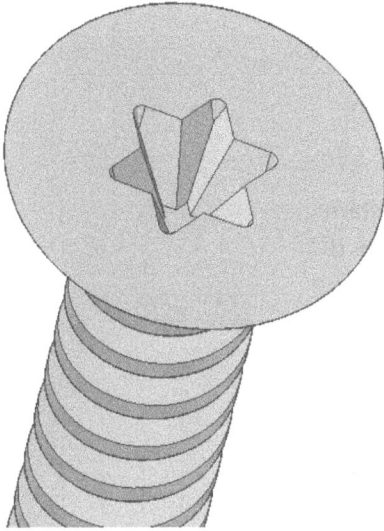

Exercise Complete

Exercise 54: Annotations

Create a **Sketch** that looks like the figure shown below and **Extrude** symmetrically about the sketch plane a distance of 6.25 in.

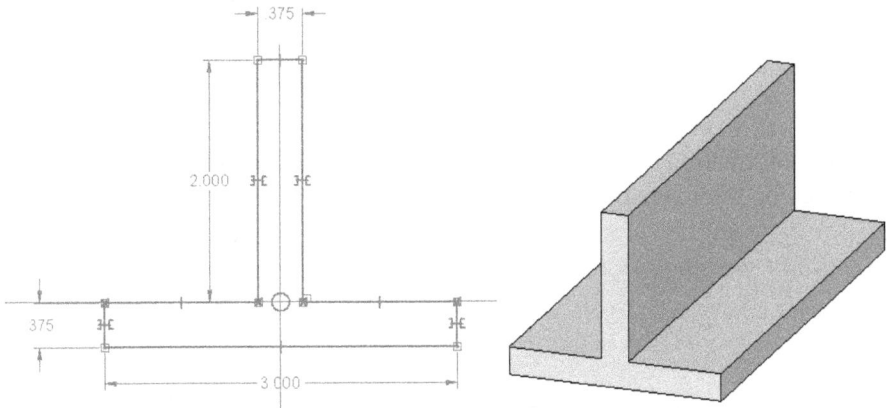

Under the **PMI** tab click on the **Weld Symbol tool** in the **Annotation options** box.

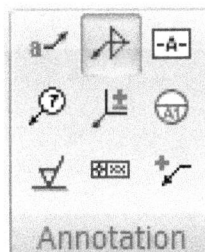

An options window, shown below, will appear.

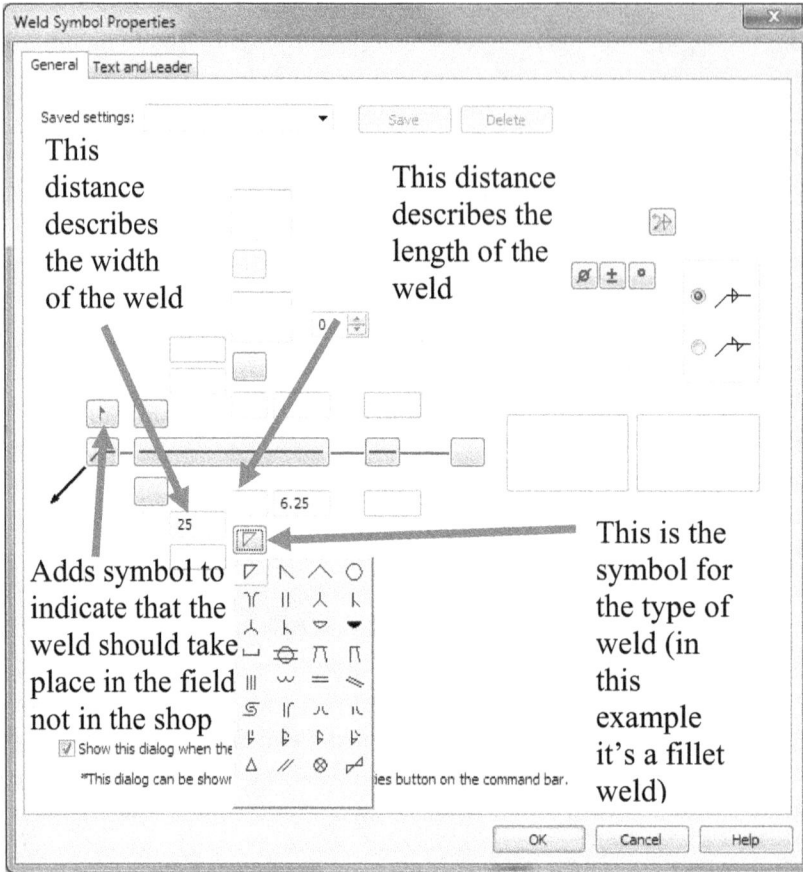

When all of the information shown above is entered click OK. Click the Set Dimension Plane button and select the front face of the part.

Note: This specifies the plane all dimensions will be parallel to.

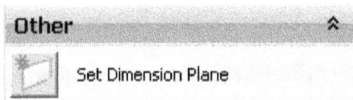

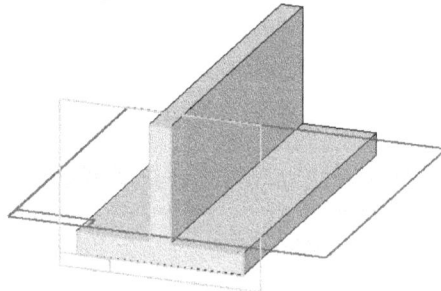

Select the **edge** to be welded as shown below and click to place the **Annotation.**

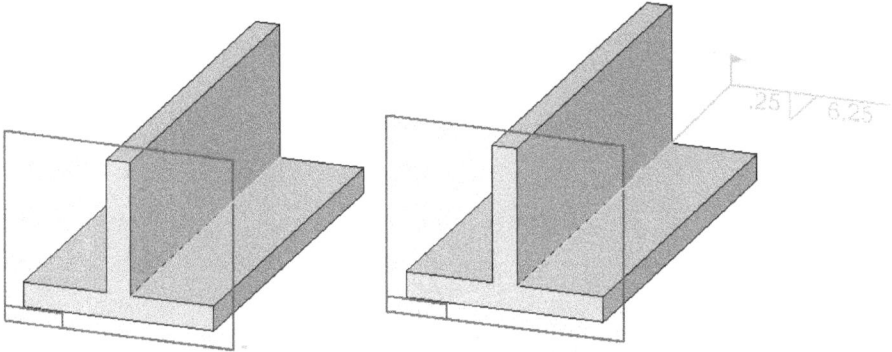

Now that we have placed a Weld symbol let's add a Datum Frame.

Click on the **Datum Frame tool** in the Annotation options box.

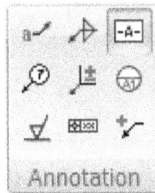

Type "A" in the Text box in the left menu.

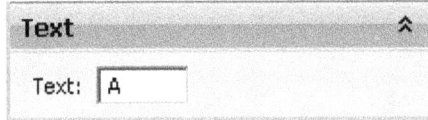

Click on the face shown below and place the Datum Frame.

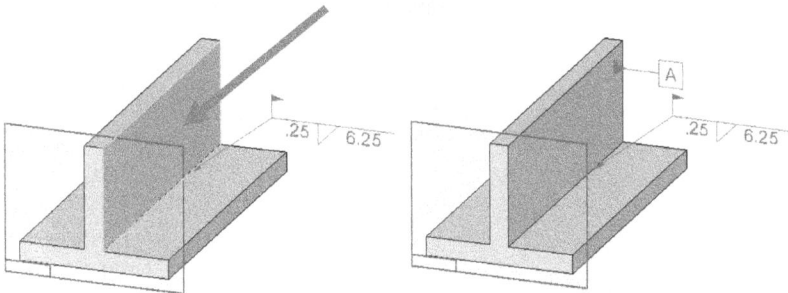

Now identify the **Datum Frame** to place a **Feature Control Frame** that references it.

Click on the **Feature Control Frame** tool in the Annotation options box.

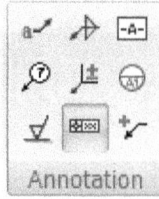

Click on the button with the hand on it to get the options window shown below.

Click on the **perpendicular** button ⊥, the divider │, type in .005, click on the Maximum Material Conditions (MMC) button Ⓜ, the divider │, then type in "A" and click OK.

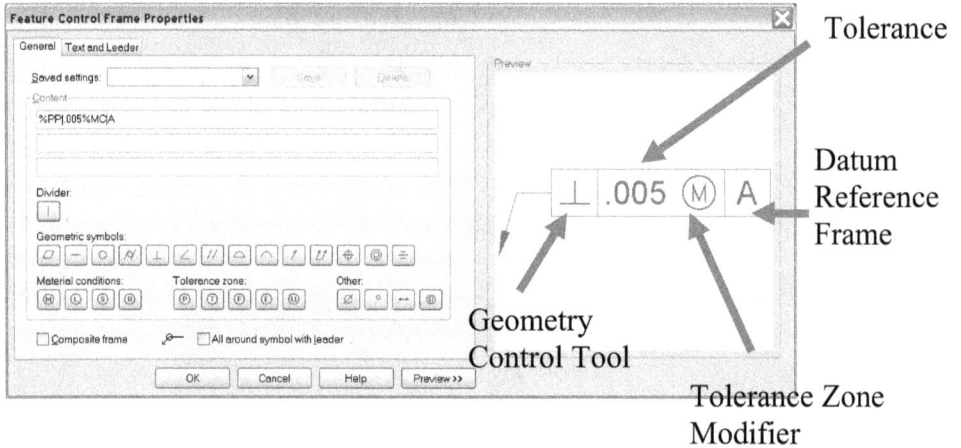

Place the **Feature Control Frame** on the plane perpendicular to the "A" side.

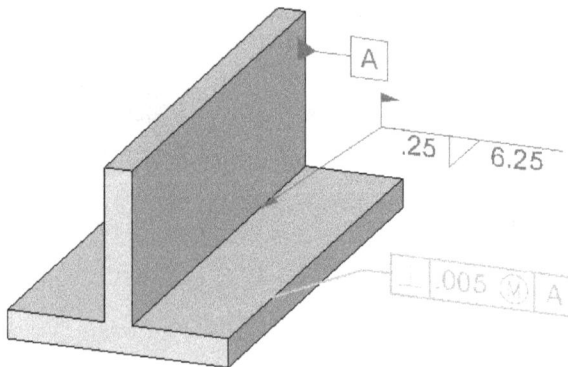

Now let's specify the **Surface Texture** of a side.

Click on the Surface **Texture Symbol** in the Annotation options box.

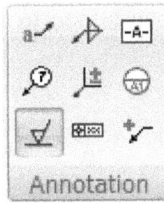

Click on the button with the hand on it to get the options window shown below.

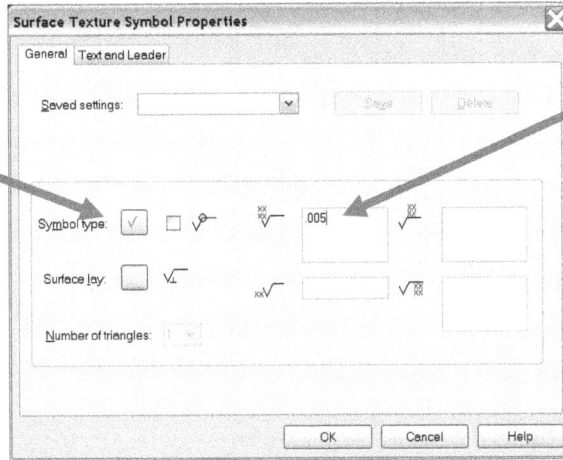

This symbol shows that we want a Basic Texture

This number represents the tolerance of the surface

Click **OK** and place the Surface texture symbol on the plane shown below.

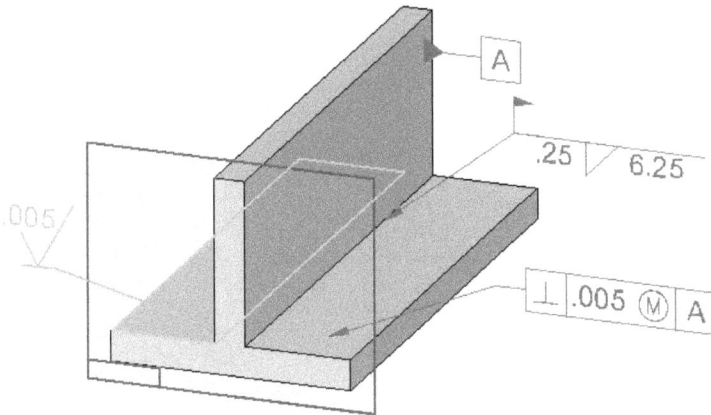

Some basic **GD&T** annotations have been added to the solid model to define the allowable tolerance for manufacturing purposes.

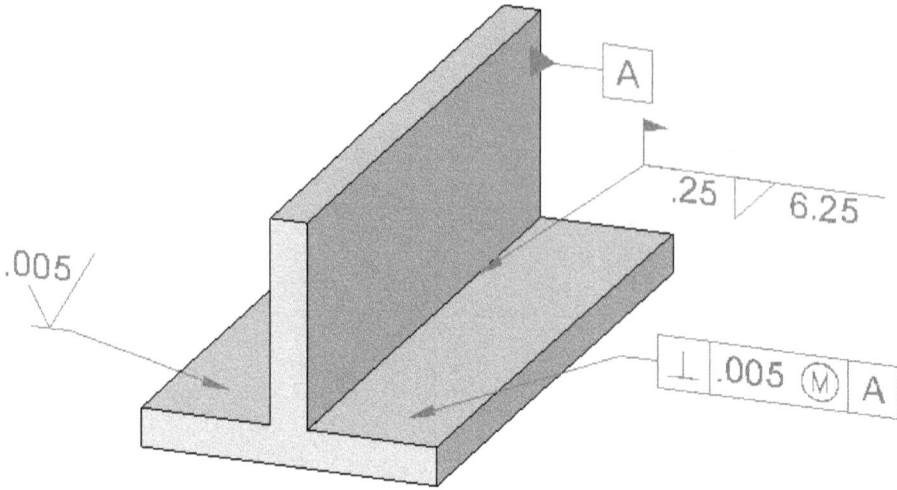

Exercise Complete

Exercise 55: Drafting

Create a solid model that looks like the figure below and Save it as **Drafting Part.par**

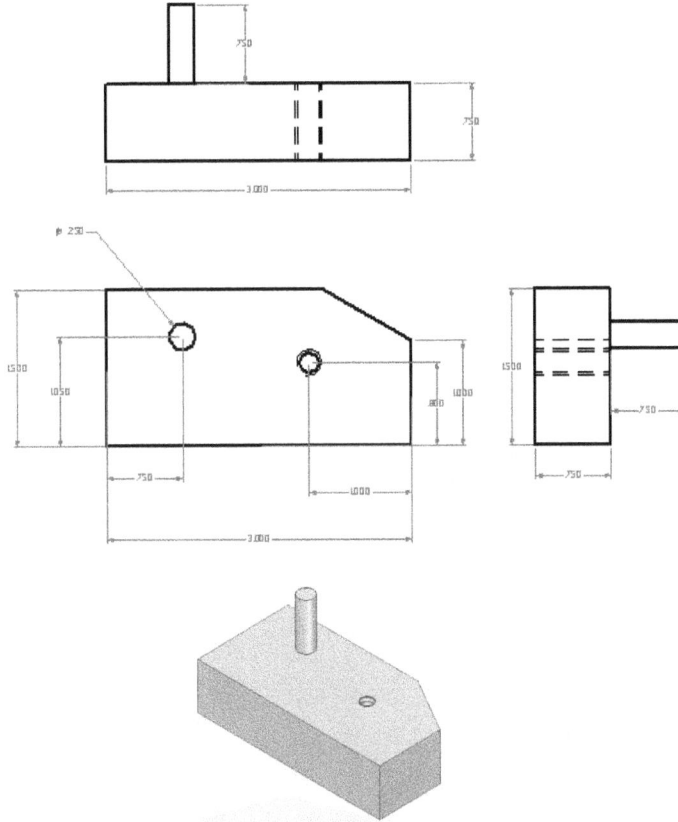

Under the **Solid Edge Symbol** in the top left corner of the window, click on Create Drawing in the New dropdown menu.

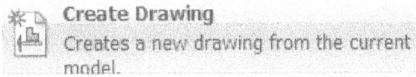
Create Drawing
Creates a new drawing from the current model.

*Note: the **Create Drawing** option is only available when you have the part file of part you want to draft open.*

First Solid Edge will ask for a drawing template. Click **OK** if the default shown below is the template that is normally used. *Note: Most companies have their own custom templates.*

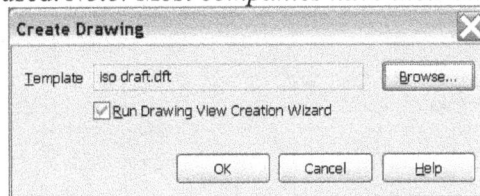

A **Drawing file** should appear, with an options box that looks like the figure below. **Click Next.**

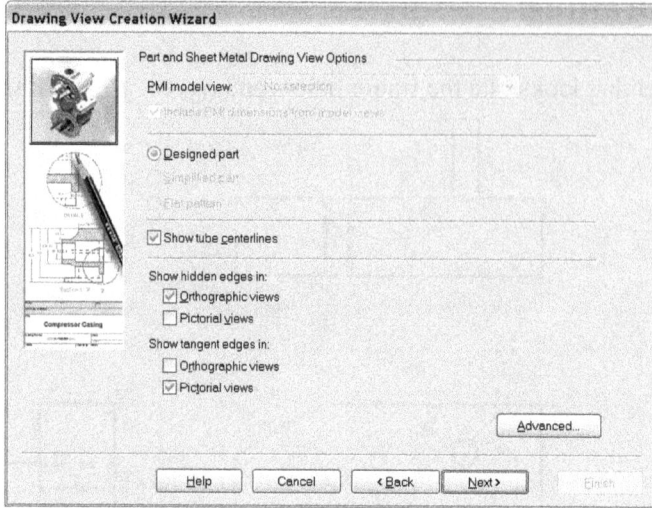

In the **Drawing View Orientation**, pick which view will be the base view. Chose top, and click **Next.**

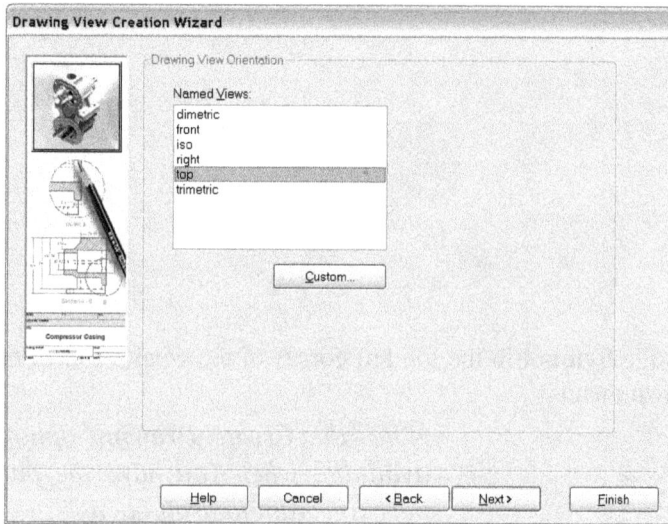

Note: If the part was not created so that the top view is the actual top of the part, use the custom button to choose the top of the part.

The next step is to choose **the Drawing View Layout** click the right side view and the bottom view as shown below, click **Finish** to complete.

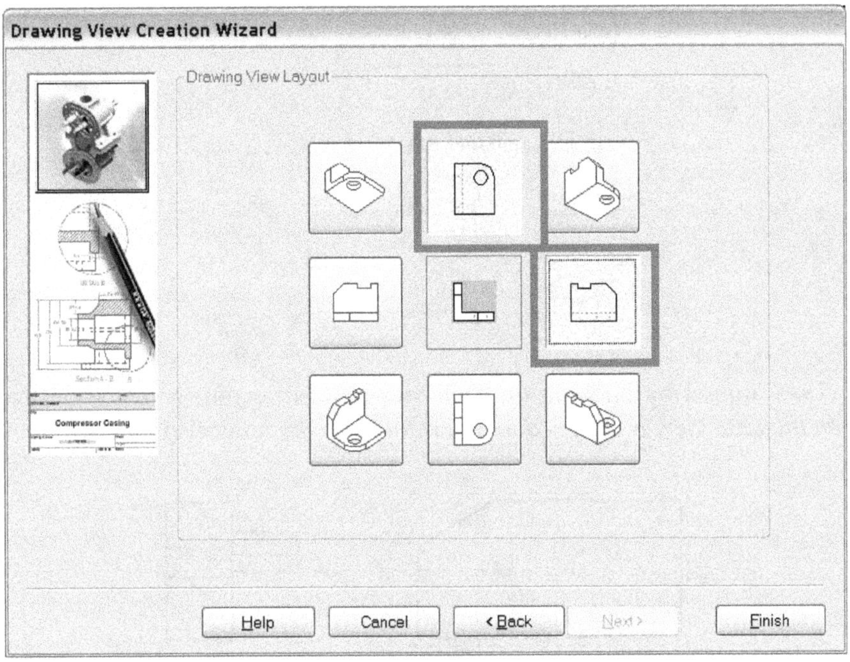

A red rectangle will appear. Click and approximate place for the three views that were generated.

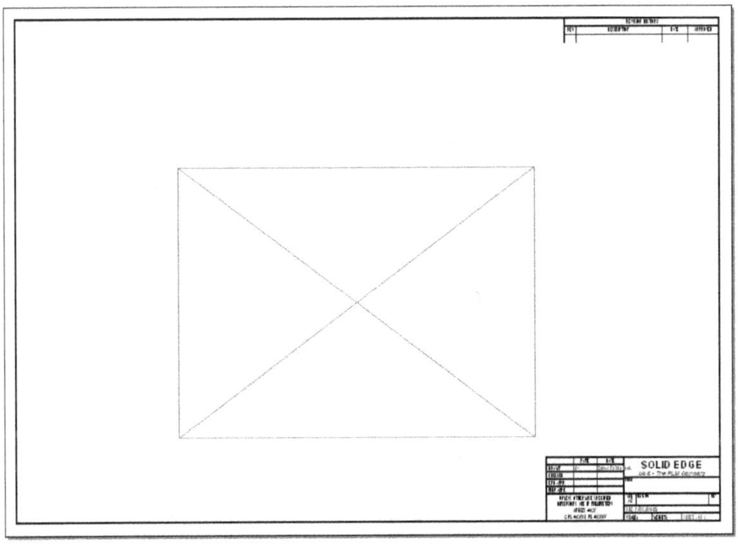

To change the scale, click on the base view and either change the **Scale Value** or choose a Scale from the menu on the left of the screen.

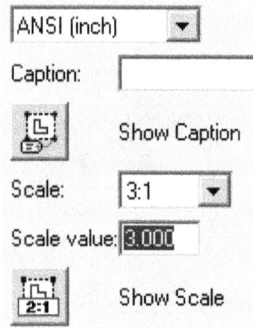

```
ANSI (inch)      ▼

Caption:      [              ]

[icon]        Show Caption

Scale:        [3:1]   ▼

Scale value: [3.000]

[icon 2:1]    Show Scale
```

Move the views around by clicking on each view, and dragging it to the desired position. *Notice: when the base view is dragged to a new location the other two view will move with it.*

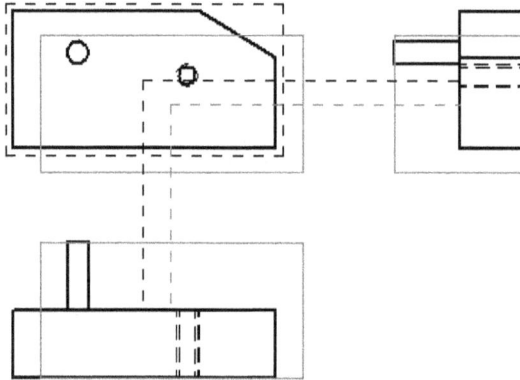

Click on **View Wizard** in the Drawing Views options box. Choose the part that the drawing is being created for and go through the same procedure to insert a new view.

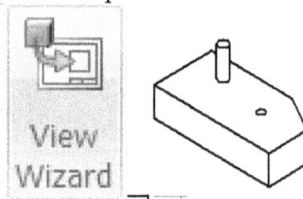

Choose **ISO** from the **Drawing View Orientation** menu, and **click Finish.**

Place and Scale the **ISO** view to fit in the upper right corner of the page.

Click on the **Cutting Plane tool** in the Drawing Views options box.

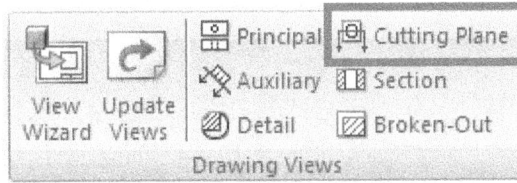

Click on the base view and use the **line tool** to sketch the cutting plane as shown below.

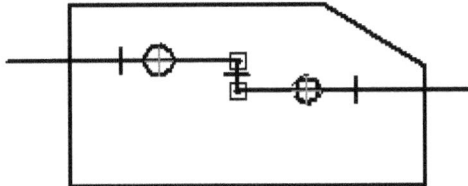

Click **Close Cutting Plane** using the same steps used to close a sketch.

Click below the view to select the direction of the plane.

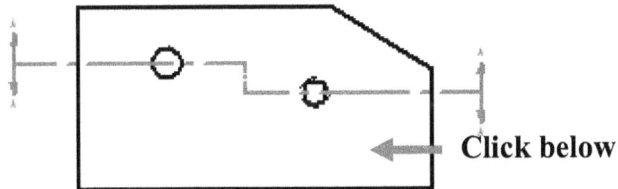

Click below

Click on the **Section tool** in the **Drawing Views** options box and click on the Cutting. Place the section view above the base view as shown below.

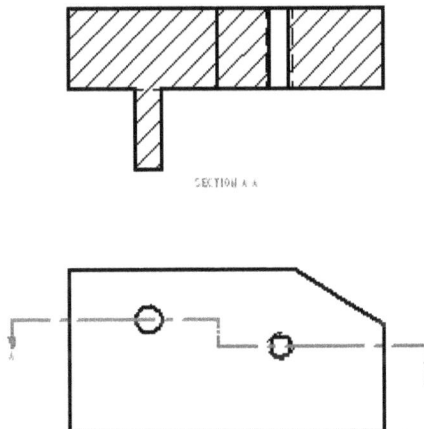

SECTION A-A

Click on the **Detail tool** in the Drawing Views options box.

Click on the top of the cylinder on the bottom view then click again to size the circle. Doing this creates a 'zoomed-in' view of the top of the cylinder that maybe relocated to a new area.

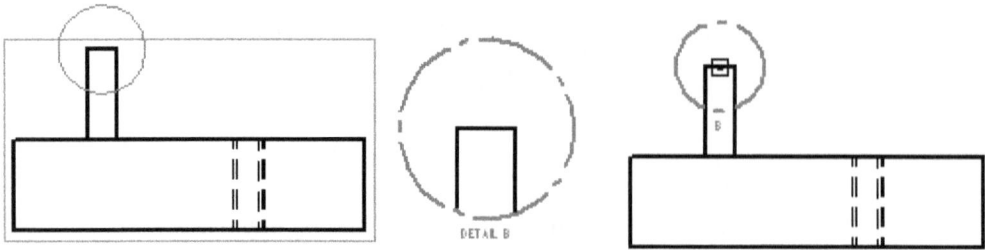

DETAIL B

Using the **Smart Dimension tool** and the **Angle Between tool** in the Dimensions options box add in the dimensions shown below.

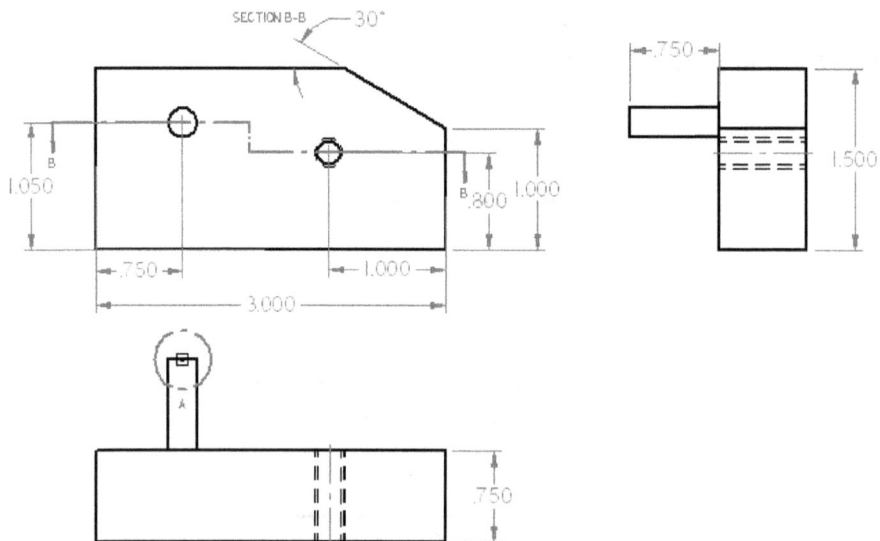

SECTION B-B   30°

1.050

B

.750

3.000

1.000

B .800   1.000

.750

1.500

.750

With the **Smart dimension tool** selected, click on the top of the cylinder in the small detail view circle. Click on the **Dimension Prefix** button under the Tolerance menu on the left.

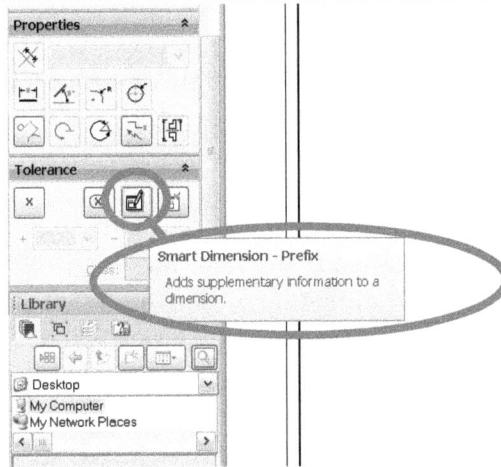

Place the **Diameter** symbol as a Prefix to the dimension by clicking on the prefix box and then clicking the diameter symbol .

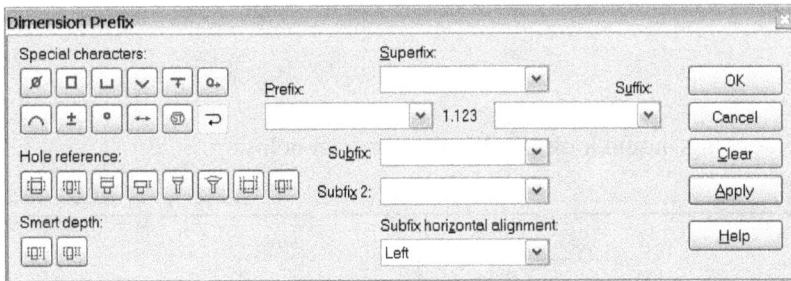

Click **Apply** then **OK.**
Now click on the threaded hole in the base view and use the same procedure to add ¼- 20 UNC as a Suffix and remove the diameter symbol prefix. Click **Apply** then **OK.**

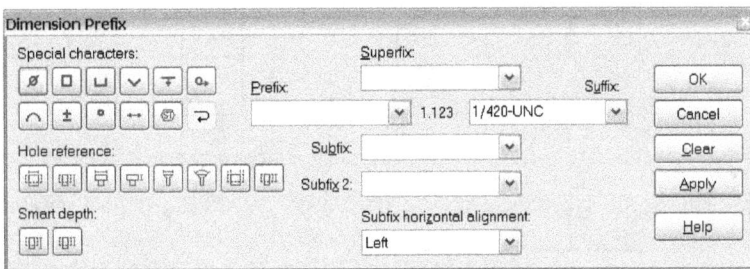

Notice: the diameter is given before the thread code. To hide the diameter, click the X in the Tolerance menu on the left and select the gray X from the dropdown menu.

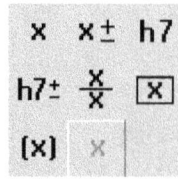

Now place the ¼-20 UNC dimension as shown below.

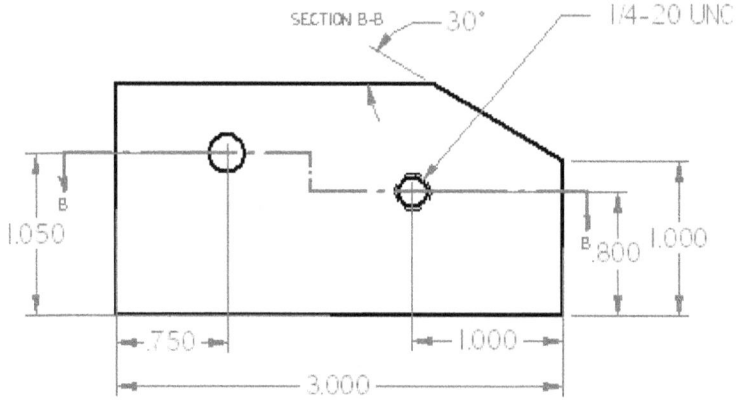

The Finished Drawing should look like the page shown below.

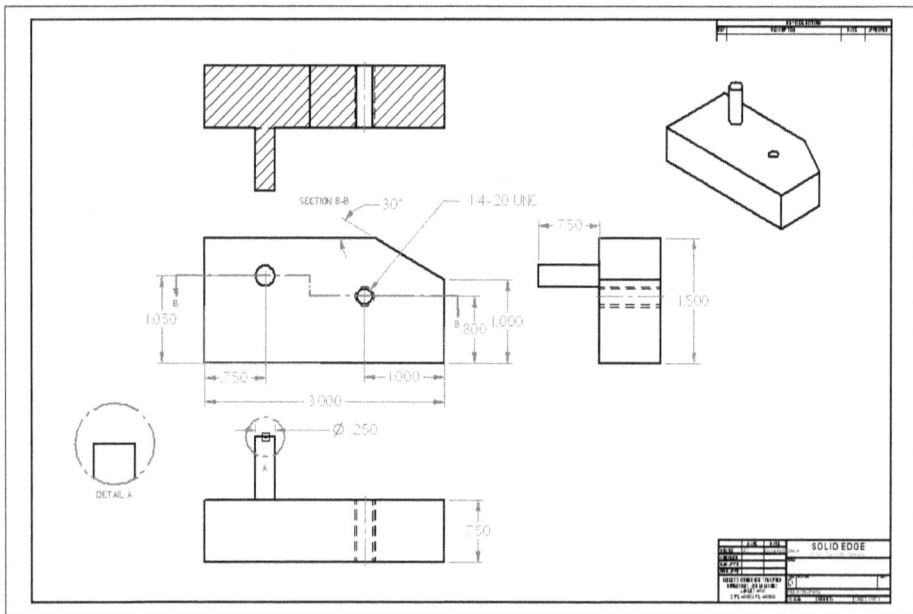

Exercise 56: Drafting (Broken View)

Create a solid model that looks like the part below.

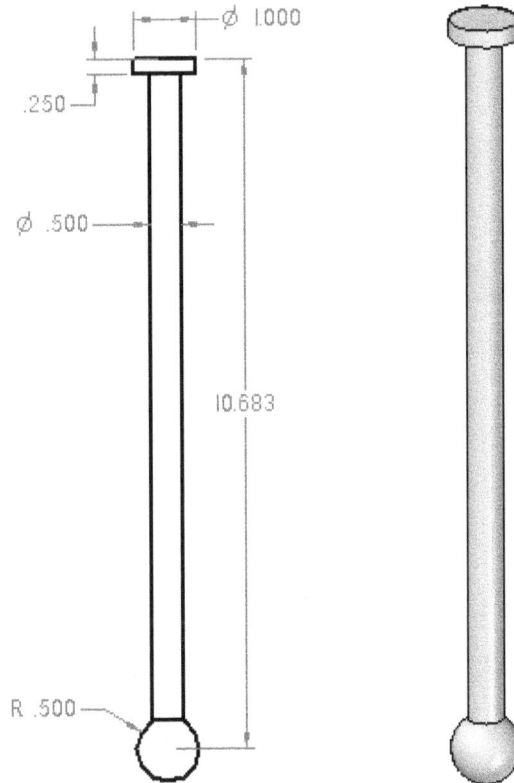

With the part open, select **Create Drawing** from the New dropdown under the **Solid Edge Symbol** in the top left corner.

Click **OK** to the ANSI draft.dft template pop-up. If using a millimeter part file choose **ISO draft.dft.**

Click **Next** when the **Drawing View Creation Wizard** box appears. Then select Front as the **Drawing View Orientation** and click **Next**.

Click on the top view as well then, Next.

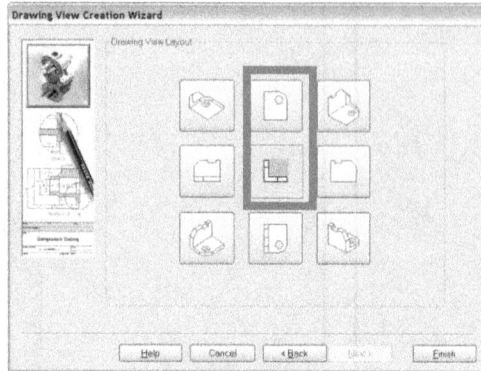

Click **Finish.**

Place the views then set the Scale to 2:1.

Notice: the part is too long to fit on the sheet.

Right Click on the long base view, and select **Add Break Lines** from the shortcut menu.

Click to place the first break line, and then click to place the second.

Click **Finish** in the menu on the left menu or right click outside of the menu.

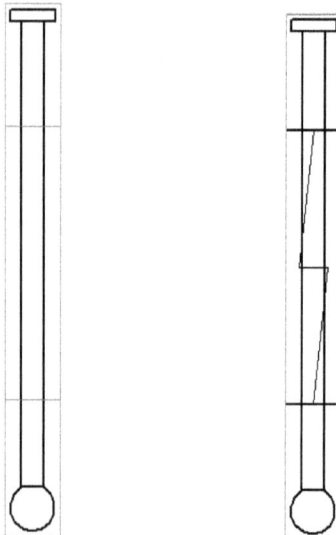

The part that was previously too long will now fit nicely on the page. *Notice: it is possible to add dimensions to the drawing and the break will not impact the actual length of the shaft.*

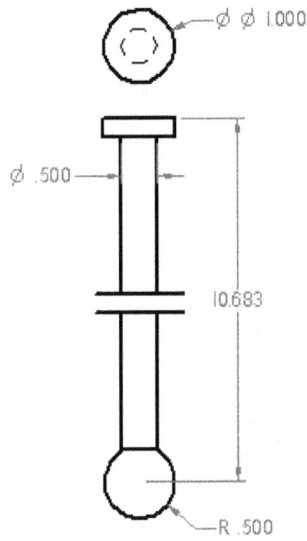

Exercise Complete

Exercise 57: Drafting Assemblies

Create a solid model of each of the parts below.

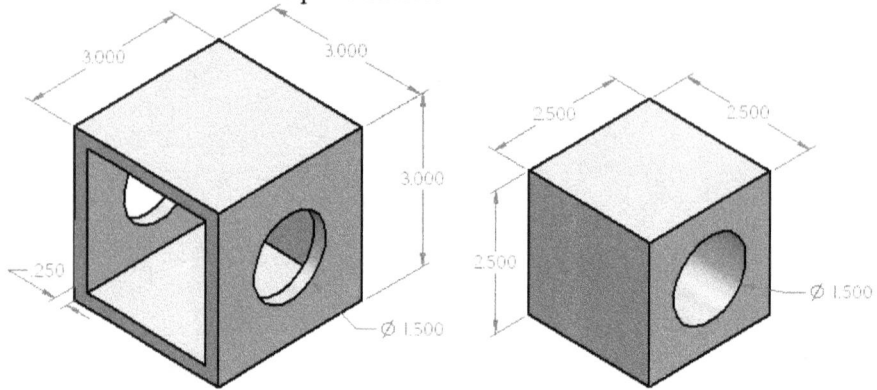

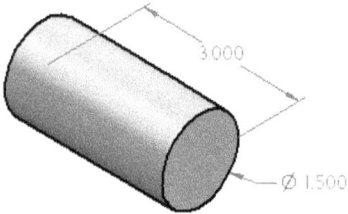

Create a **ANSI Assembly file** and assemble the parts like the figure below.

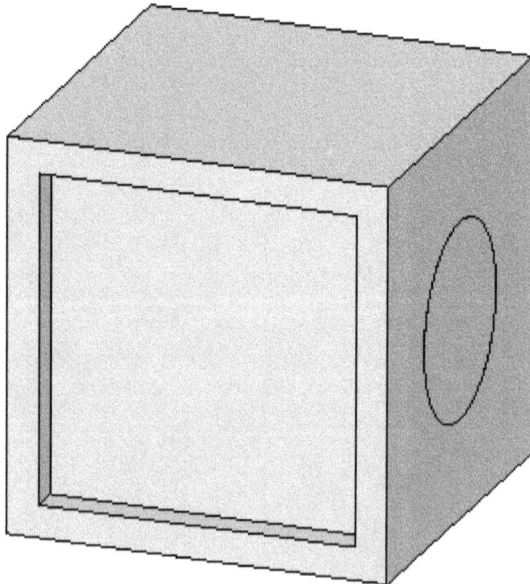

In the **Tools** tab click on the **ERA tool** in the Environs options box.

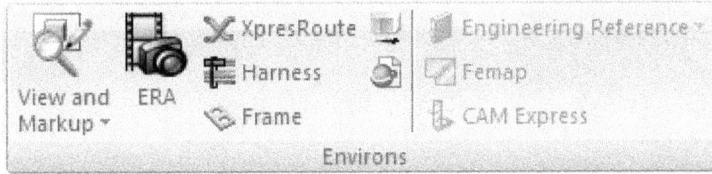

Click on the **Explode tool** in the Explode options box.

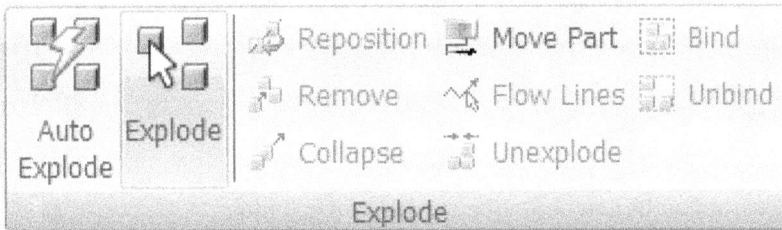

In the **Select Parts Step** click on the cylinder then, **Accept.**

Select the hollow Base cube as the Base feature in the Base Step.

In the Face step select the right face of the cube and click again to specify the direction away from the assembly. Click the **Explode** button at the top of the left menu then Finish.

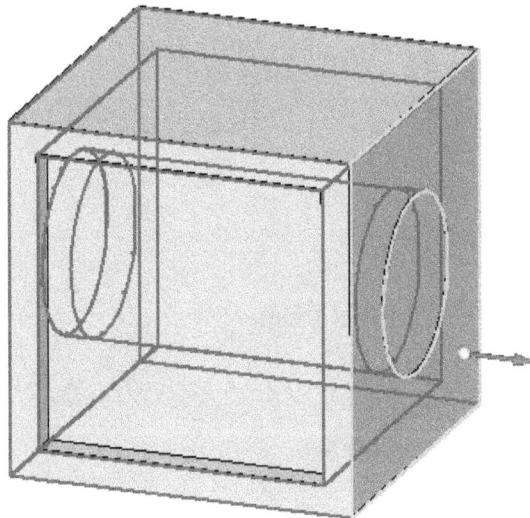

Using the same procedure create another 'Explode' on the smaller cube.

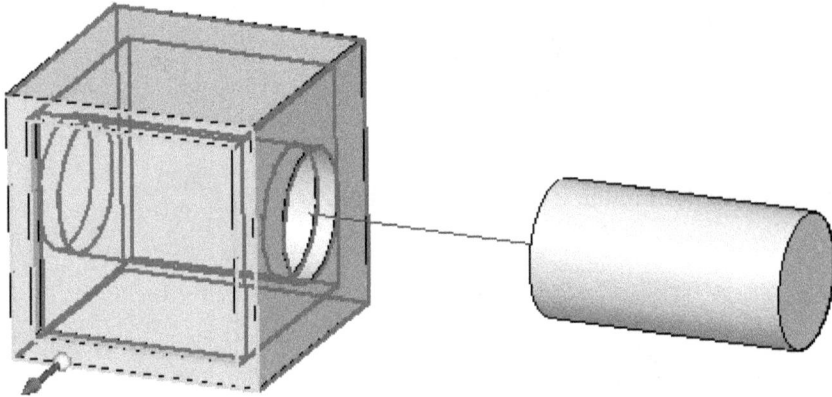

The Exploded view should now look like the following figure.

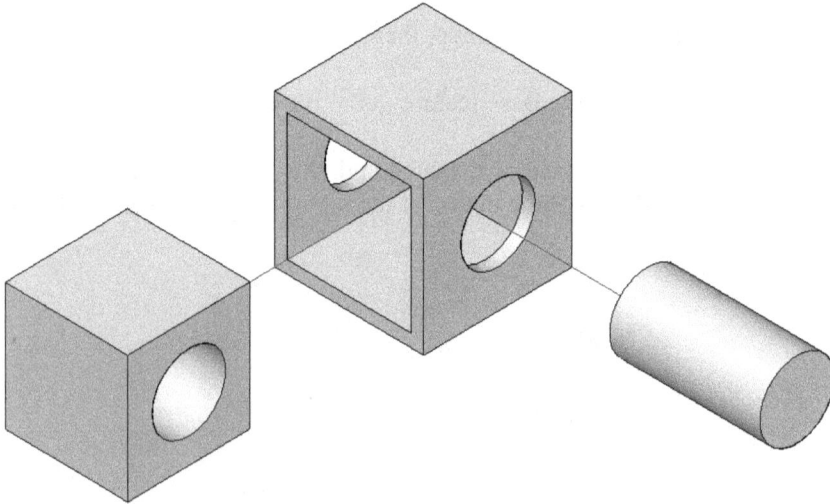

Save the assembly and click the **Close ERA** button.

Under the Solid Edge symbol in the New dropdown menu click **ANSI Draft.**

Back under the Solid Edge symbol select Sheet Set Up.

In the Size tab select a Standard Size of B Wide (17 X 11 in) and in the Background tab select a Background sheet of B Sheet.

Click **OK.**

Click on the View Wizard tool in the Drawing views options box and open the Assembly above. In the Drawing View Creation Wizard choose the **explode.dv** option from the Configuration or PMI model view dropdown menu and click Finish.

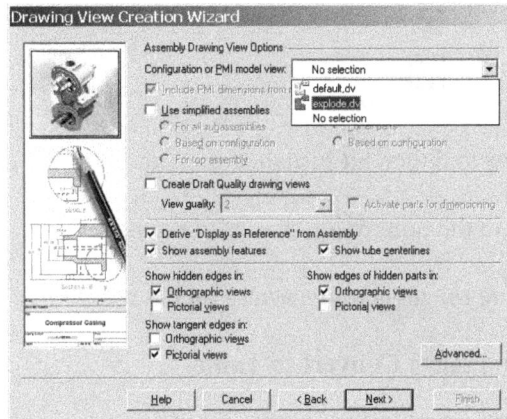

Place the **exploded view** on the page, and scale it to fit. *Notice: It will appear as it was last oriented in ERA mode.*

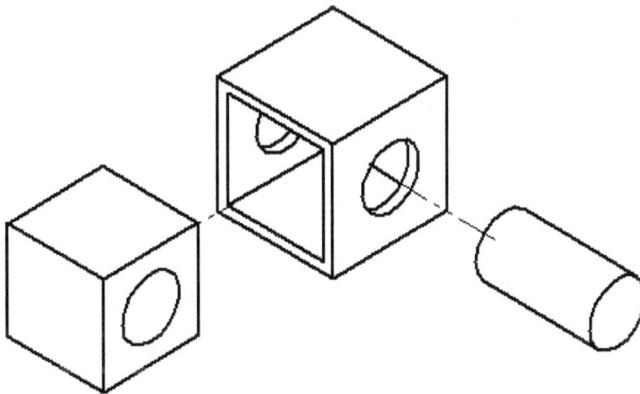

Click on the **Parts List tool** in the tables options box.

Click on the **exploded view** and then hit **Finish** at the bottom of the left menu.

A Parts list should appear in the bottom left corner of the page.

Notice: it is possible to drag the table to anywhere on the screen.

Right click on the table, and select **Convert to Table**.

Right click the Table again and click "Properties". The following box should appear.

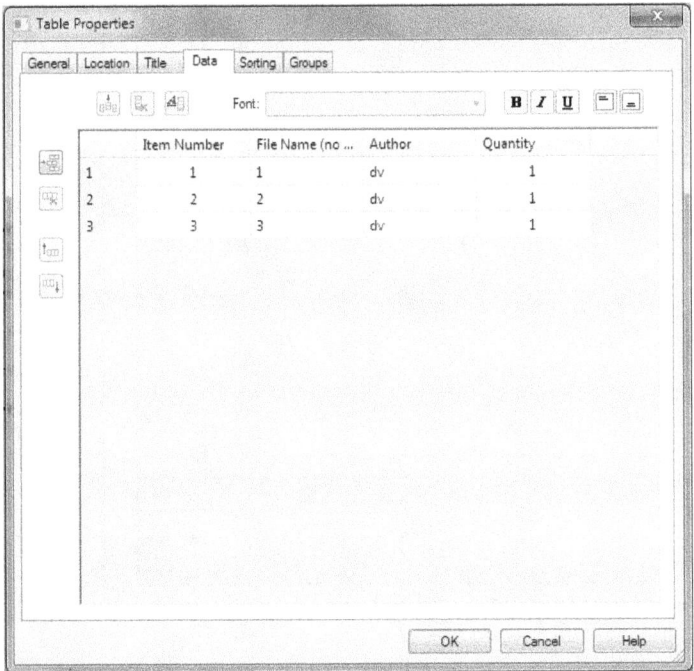

Select each individual cell to add content to the table. *Note: The table headers can be changed or more can be added until the table meets your needs. You can access these features by right clicking one of the existing headers.*

Change the content of the Block by selecting the B-sheet tab at the bottom of the page.

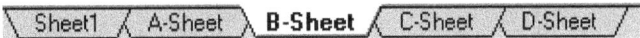

Click on the text in the block to edit it.

Notice: I changed the SOLID EDGE logo to a Design Visionaries logo.

The finished drawing with an **Exploded View** and a **Parts List** should look like the figure below.

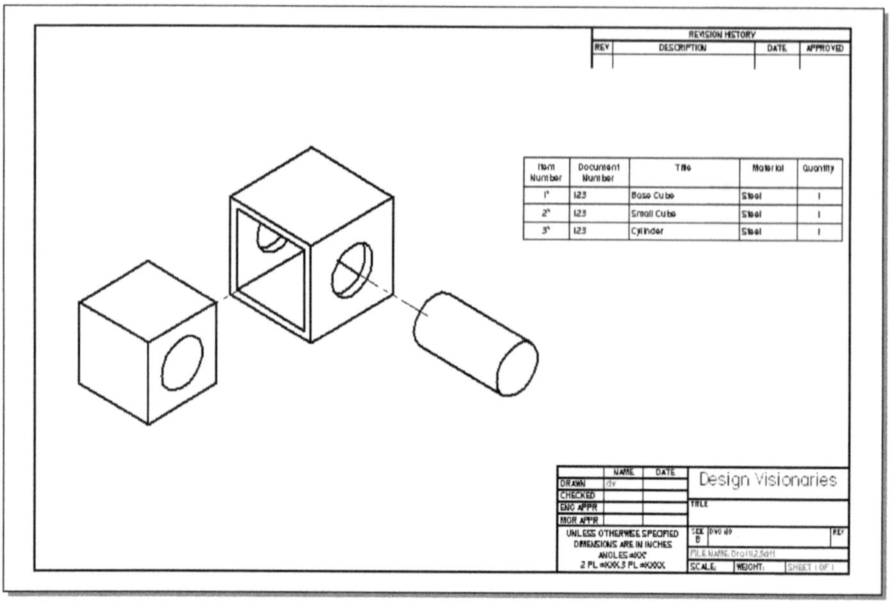

REVISION HISTORY
REV
DESCRIPTION
DATE
APPROVED
Item Number
Document Number
Title
Material
Quantity
1
123
Base Cube
Steel
1
2
123
Small Cube
Steel
1
3
123
Cylinder
Steel
1
NAME
DATE
DRAWN
dv
CHECKED
ENG APPR
MGR APPR
UNLESS OTHERWISE SPECIFIED
DIMENSIONS ARE IN INCHES
Design Visionaries
TITLE
SCALE:
WEIGHT:

Exercise Complete

Introduction to Sheet Metal

Sheet Metal files can be created by clicking the Solid Edge logo in the top left corner of your screen and then selecting **ANSI Sheet Metal** in the "New" menu.

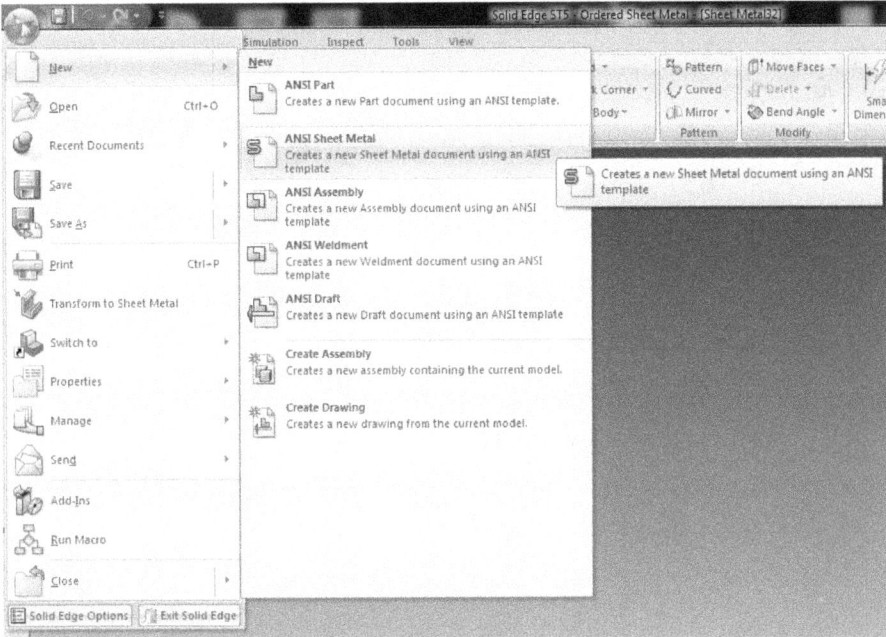

You'll notice immediately that the features available in a Sheet Metal file are different from those of a regular part file. Features such as Tab, Flange, Contour Flange, and Dimple are all specially tailored for use with Sheet Metal models.

In the exercises that follow, we'll cover these features and more as we go over the creation of Sheet Metal models.

Exercise 58: Tabs and Flanges

Create a new **ANSI Sheet Metal** file and click **Tab**. Tab can be thought of as the Extrude of the Sheet Metal world. It's most basic way of creating a sheet metal body onto which features can be added later. When prompted to do so, create the sketch shown below.

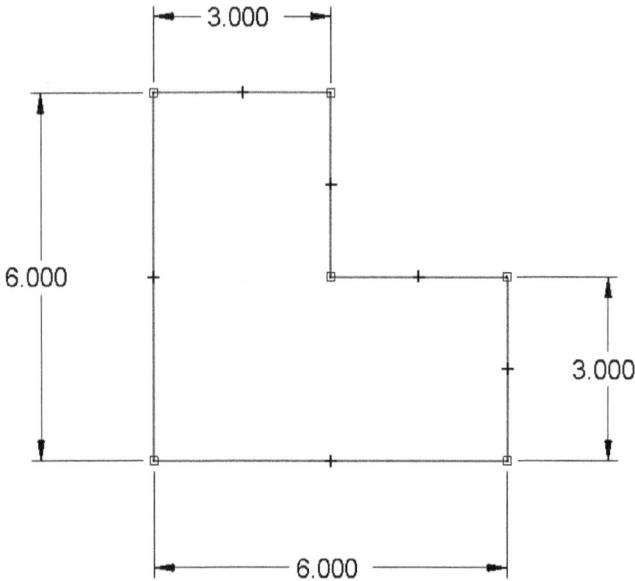

Input a depth of 0.0359 in (20 gauge steel thickness). Your model should now look like the following image.

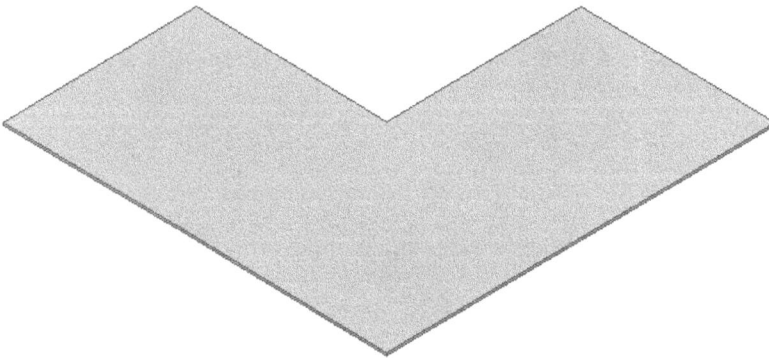

Now click **Flange**. Flange is a command to add metal to your base Tab at a chosen angle. Click the options button and input the following data. Once you finish, click OK.

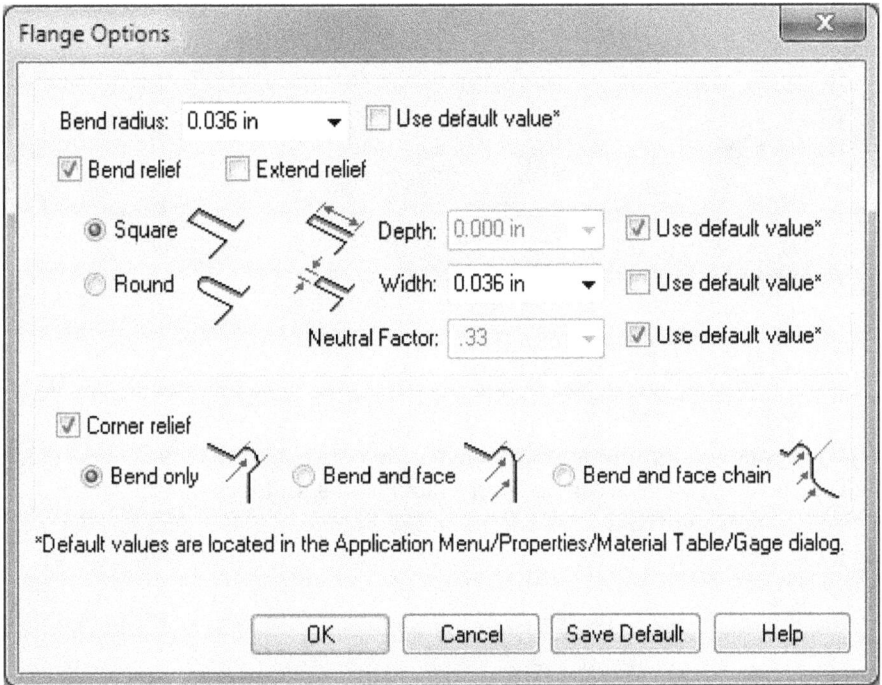

Make the edge selection shown below and input the given values.

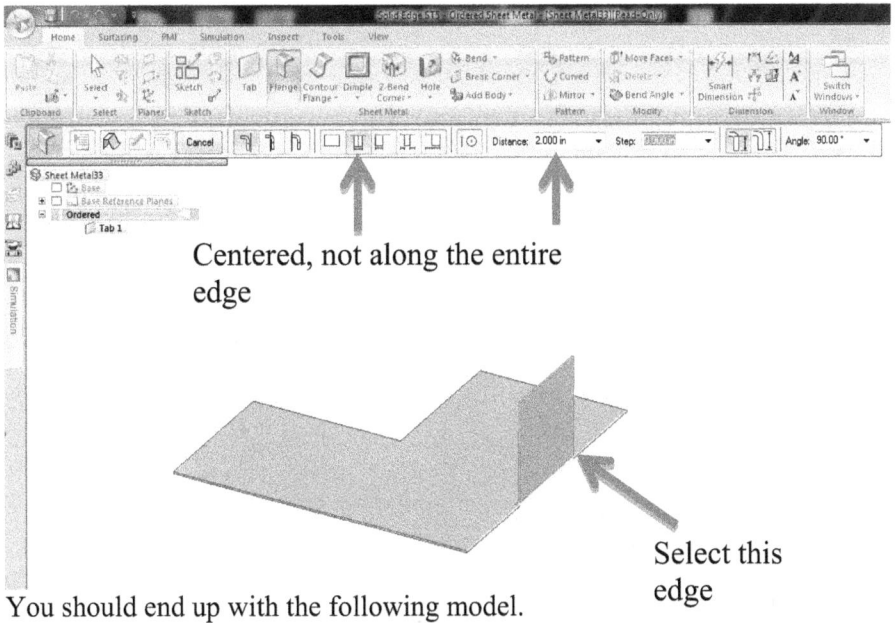

Centered, not along the entire edge

Select this edge

You should end up with the following model.

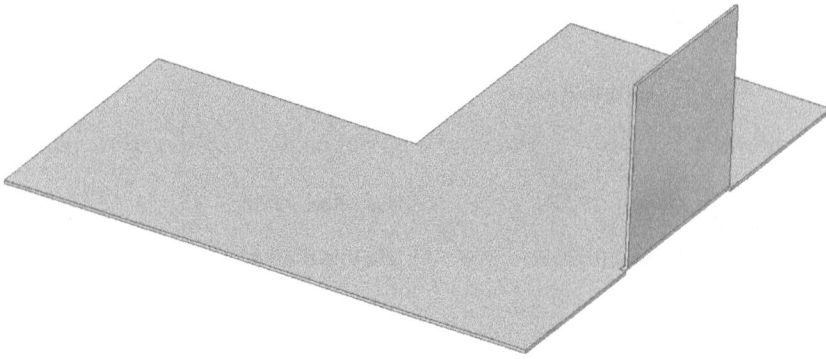

Notice how the flange is contained within the parameters of the original Tab. This is called an Inset Flange. Below is a comparison between an Inset Flange and a regular Flange. Under those images are the selections required to create that kind of Flange. (these selections can be made while creating a Flange)

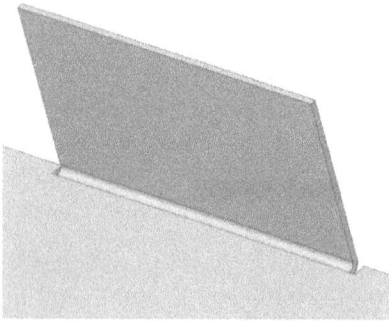

Inset Flange

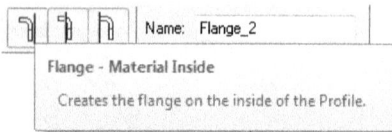

Regular Flange

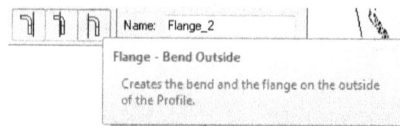

			Name: Flange_2
Flange - Material Inside			
Creates the flange on the inside of the Profile.			

			Name: Flange_2
Flange - Bend Outside			
Creates the bend and the flange on the outside of the Profile.			

Exercise Complete

Exercise 59: Contoured Flanges

Create a new **ANSI Sheet Metal** file and click **Tab**. Create a Tab that is 10in x 5in x 0.038in. When finished, click **Contour Flange** and create a sketch plane normal to one of the 10in sides of the Tab.

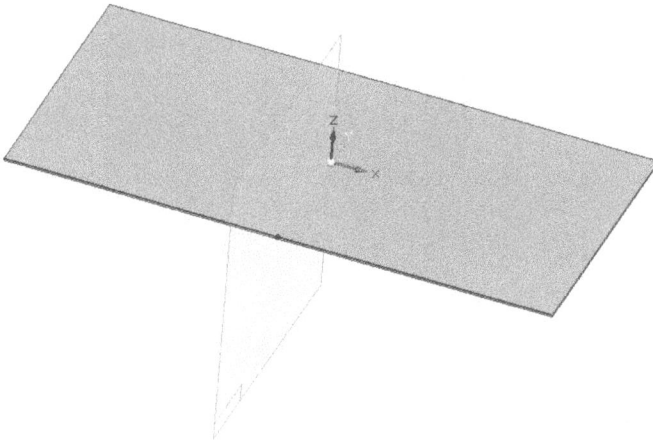

Create the following sketch on the plane you just created. The sketch should touch the top edge of the Tab. Close the sketch when finished.

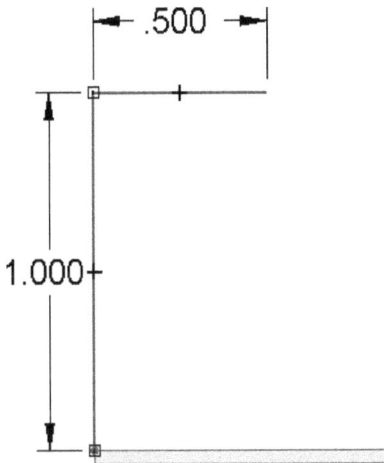

When prompted to choose an extent, choose chain. After choosing chain, select the top edge of the Tab. Click Accept and then Finish.

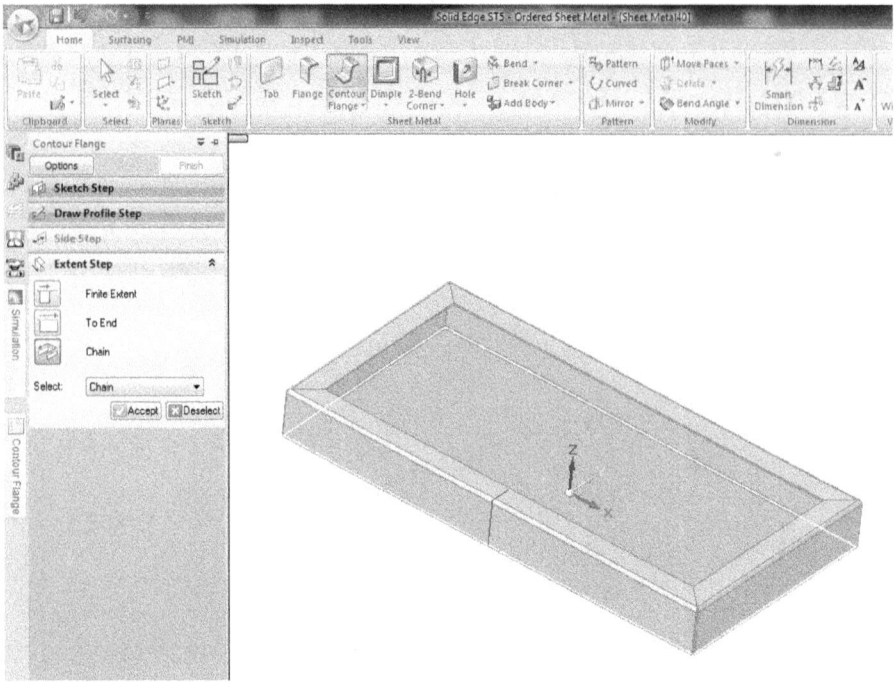

You should now have the following model.

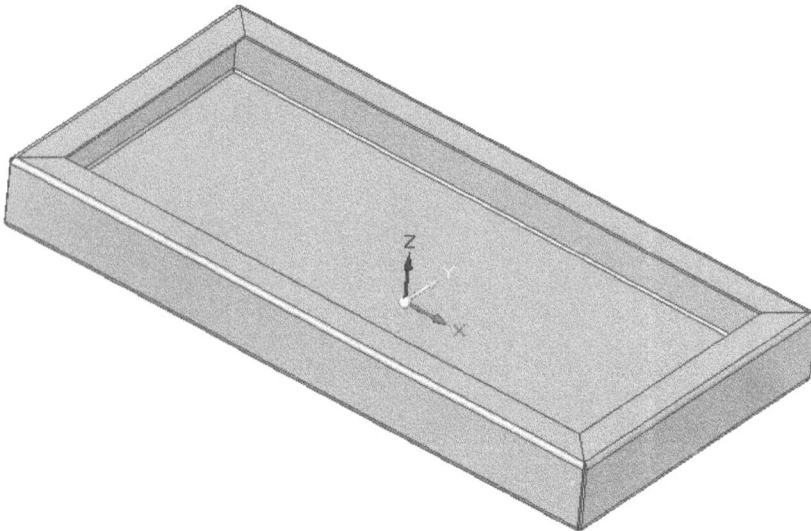

Exercise Complete

Exercise 60: Bending Sheet Metal

Create a new **ANSI Sheet Metal** file and click **Tab**. Create a Tab that is 5in x 5in x 0.038in.

Draw the following sketch on top of the tab. To do this, create another 5in x 5in square on top of the existing one, and then input an angle of 45 deg. Then create another 5in x 5in square, but this time with a 0 deg. angle so that it exactly lines up with the Tab you've already created. Then trim everything leaving only the 4 lines shown below.

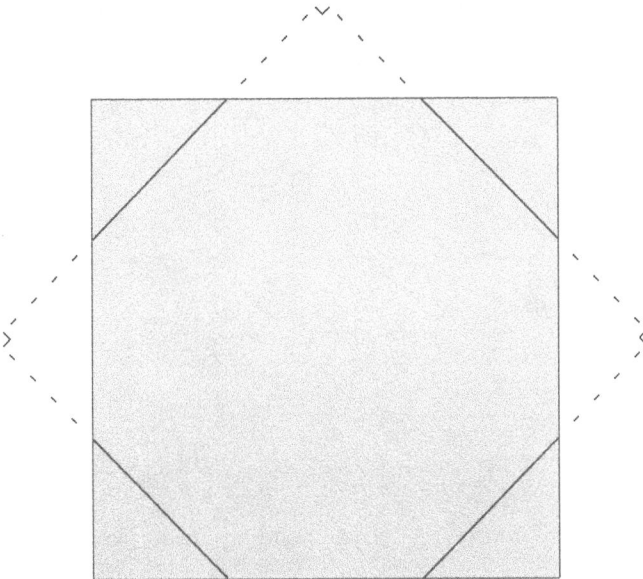

Now click **Bend**.

Select one of the lines you just drew on the 5in x 5in Tab. Click Accept and then enter the information shown below.

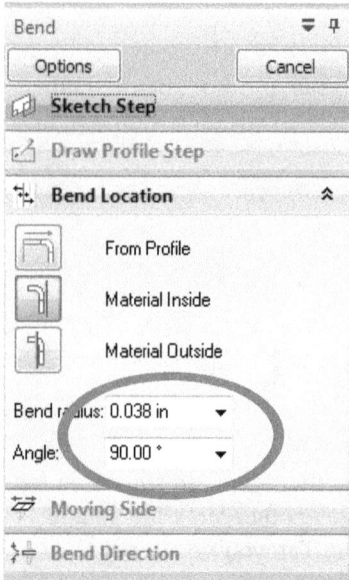

Click the emblem next to "Material Inside" once you've finished. Make the "Moving Side" arrow point outward so that the body of the model stays stationary and only the corner bends. Then select a bend direction, either one is fine. Finally, click Finish. You should now have the following model.

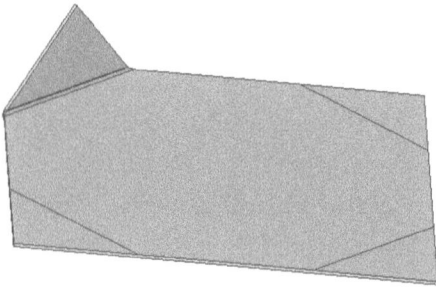

Now do the same thing with the remaining 3 corners. Alternate bending the corners up and down until you have the following model.

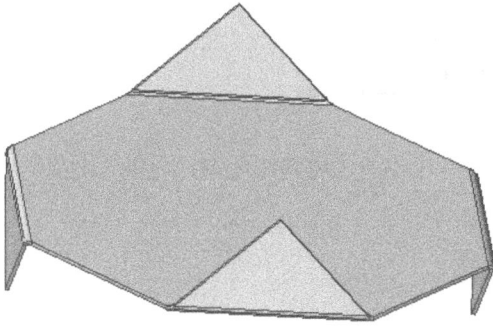

As a finishing touch, use the **Cut** command (located in the "Hole" drop down menu) to cut a 2in diameter circle out of the center of the model.

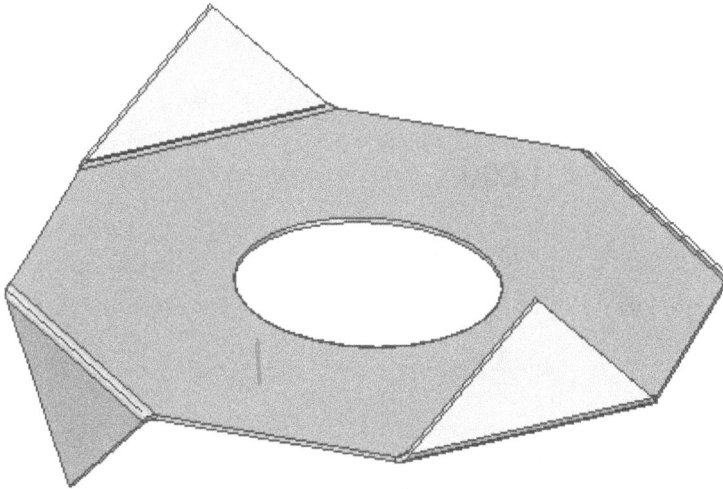

Exercise Complete

Exercise 61: Unbending and Re-Bending

Create a new **ANSI Sheet Metal** file and click **Tab**. Create a Tab that is 2in x 2in x 0.038in.

Click **Contoured Flange** and create the following sketch on one of your Tab's edges. Close the sketch once you've finished

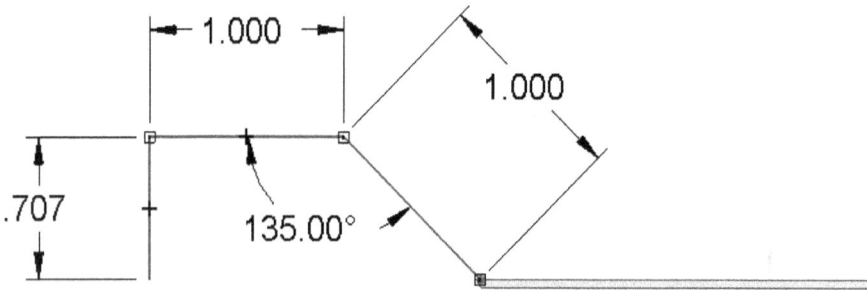

Click "To End" for the extent. You should now have the following model.

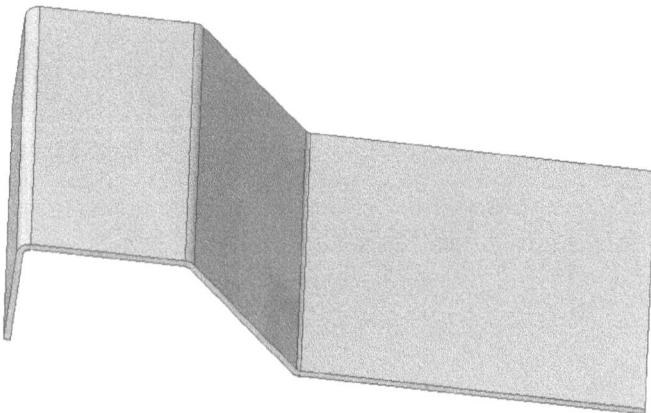

Click **Unbend** in the "Bend" drop down menu.

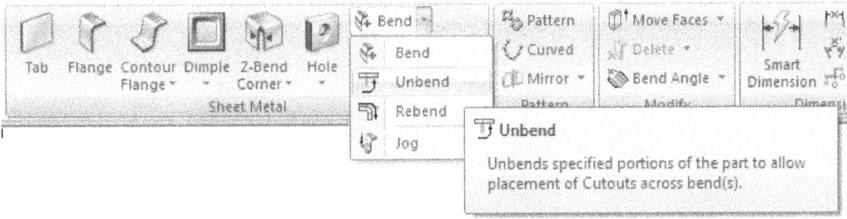

Choose your original 2in x 2in Tab as the fixed face and then select the 3 bends created by the Contoured Flange.

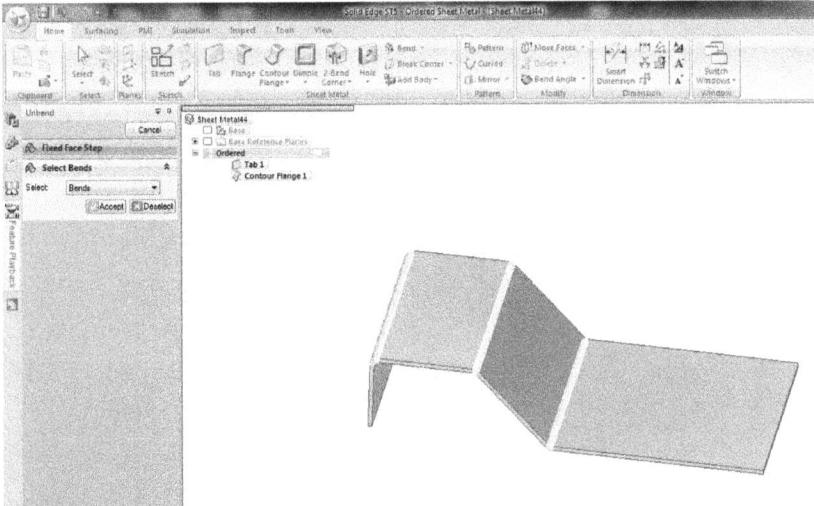

Click Accept, Preview, and then Finish. You should now have a flat sheet of metal. There is a neat little feature included in this operation though: It stores all of that bend data so that if you want to re-bend it later, (as we will be doing) you can.

In the "Hole" menu, click **Cut**. Create the following sketch on your sheet. Its dimensions aren't important, so get creative with it! Just be sure to make it cross all 3 bend marks in the sheet.

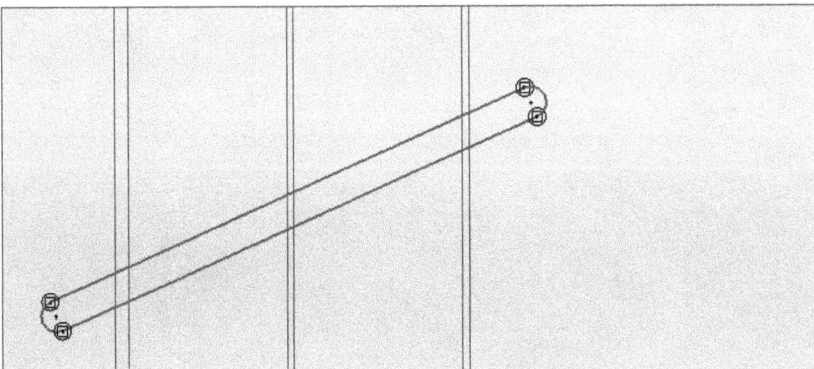

Close the sketch once you've finished. When prompted, choose "Through All" for the cut depth. You should now have the following model.

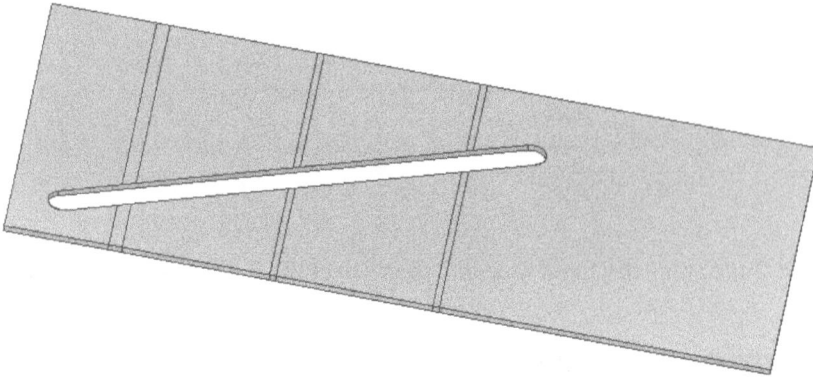

Now go to the "Bend" drop down again, but this time select **Rebend**.

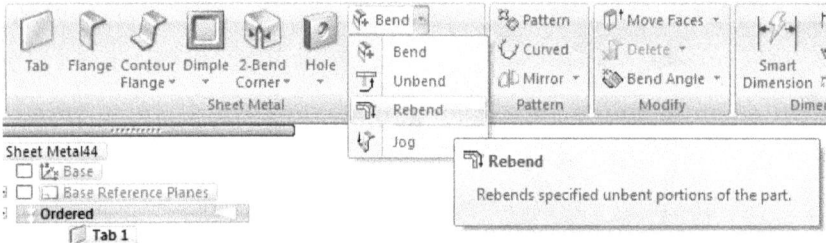

Since Solid Edge remembers how your model was originally bent, all you have to do is select the 3 bend marks.

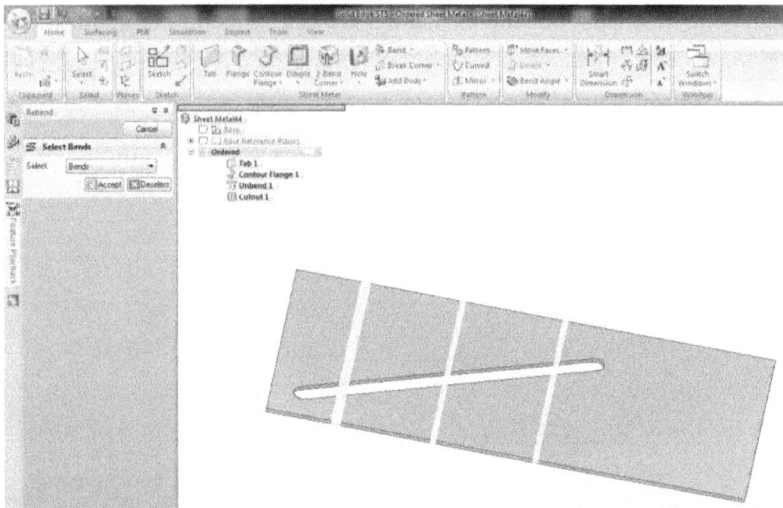

Click Accept, Preview, and the Finish. Your model should now resemble the following image.

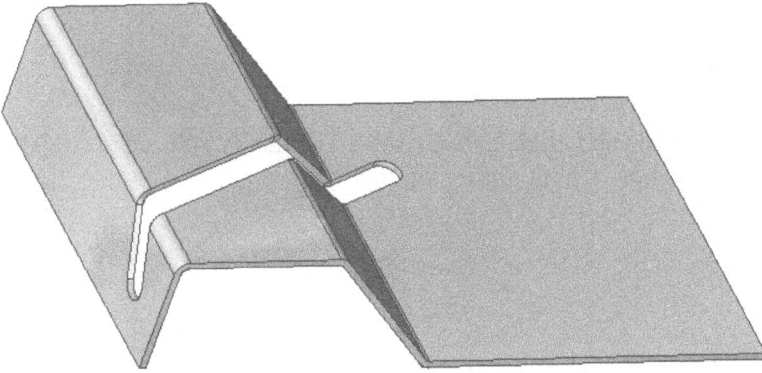

Save this as **Unbending_Exercise.psm**. We will be using this model in a later exercise.

Exercise Complete

Exercise 62: 2-Bend Corners

Create a new **ANSI Sheet Metal** file and click **Tab**. Create a Tab that is 10in x 5in x 0.038in. Apply 1in Flanges to all sides as shown in the following figure.

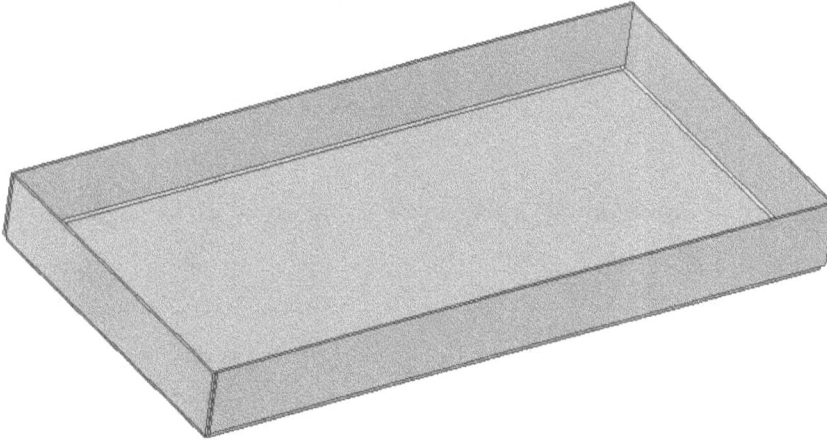

Click **2-Bend Corner**. You'll notice there are quite a few different corner treatments available for your choosing. For this exercise, we will be using a "Closed" treatment with no gap. (after this exercise, feel free to experiment with all of the different corner treatments)

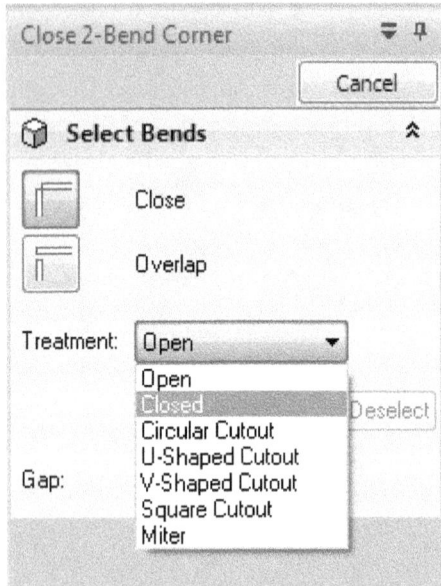

After choosing a "Closed" treatment, select two bends that meet at a corner as shown below.

Click Accept. Below is a comparison between the original corner and the new one.

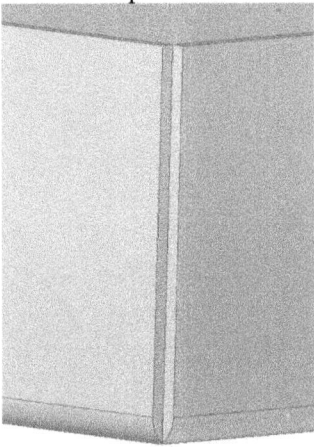

New Corner

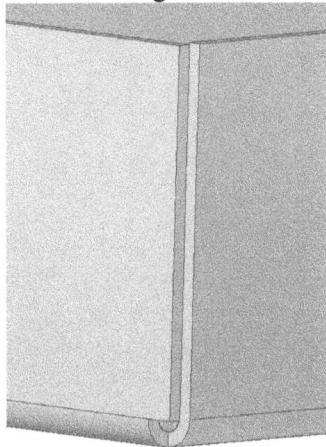

Original Corner

Exercise Complete

Exercise 63: Dimples, Drawn Cutouts, and Etching

Create a new **ANSI Sheet Metal** file and click **Tab**. Create a hexagonal Tab with the following dimensions.

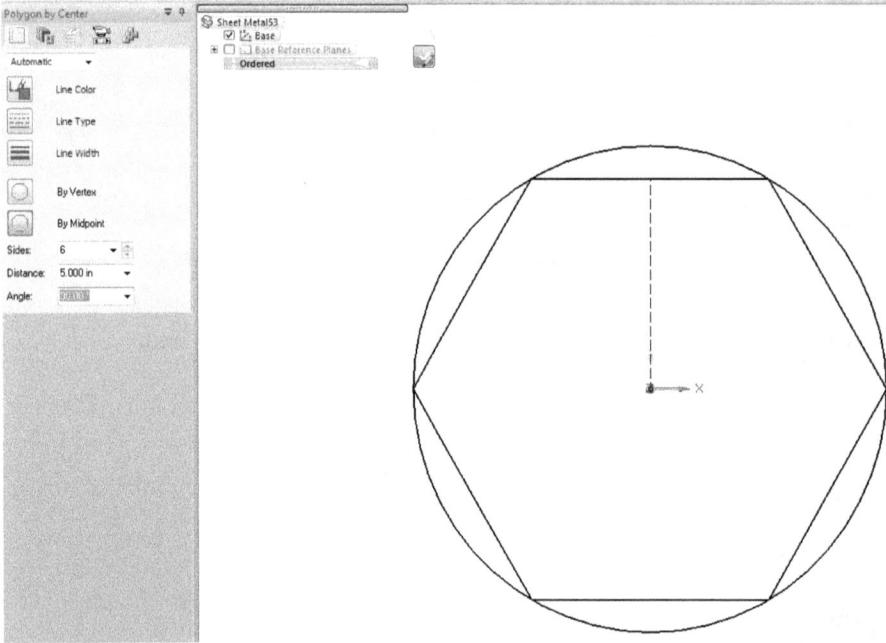

Give it a thickness of 0.038in. You should now have the following model.

Create a sketch on top of the Tab. Click the "Include" command shown below.

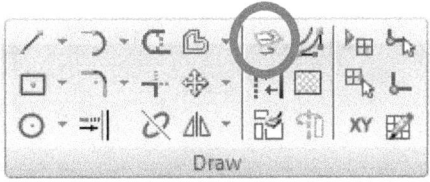

Choose "Include with offset" and then click OK. Select all 6 edges and then click Accept. Enter a distance of 0.5in and offset it inward.

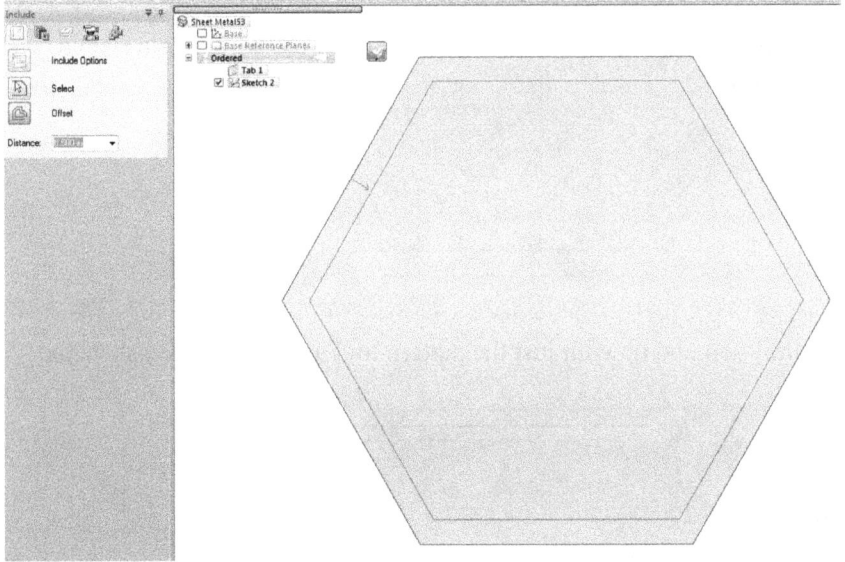

Create the following sketch after offsetting the edge of the hexagonal Tab.

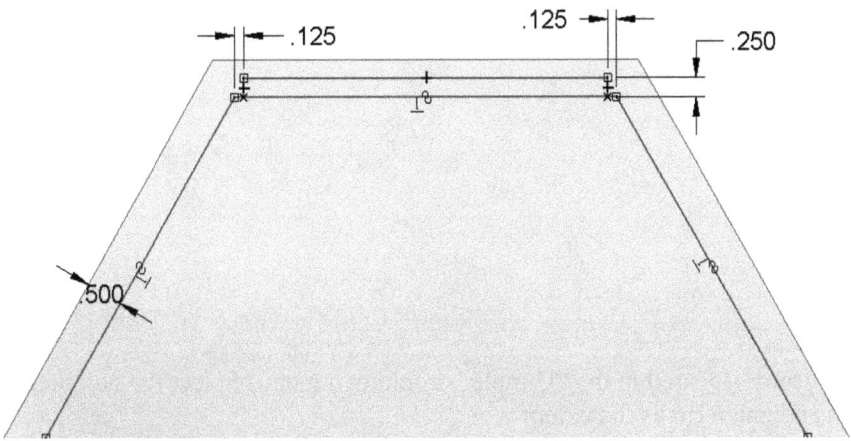

Click **Circular Pattern** and create the circular pattern shown below. It has 6 instances and is inscribed within the hexagonal sketch.

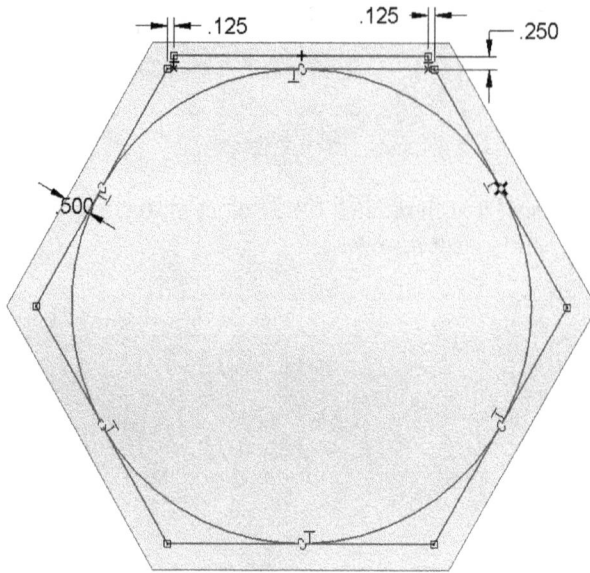

Lastly, use the trim tool until you end up with just the pattern tool and the box you sketched.

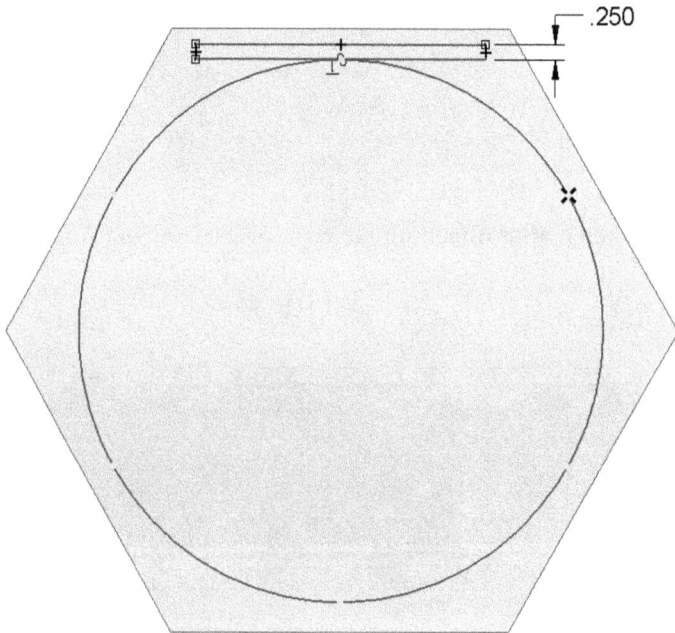

Now click **Drawn Cutout**. (located in the "Dimple" drop down menu) Select the box you've drawn as the profile and enter 1in as the extent.

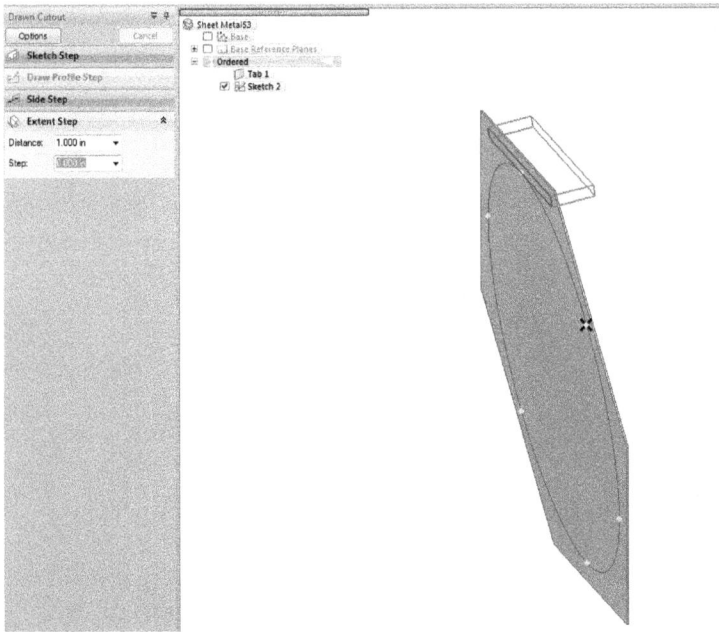

Hit Finish. Now click pattern. Select the **Drawn Cutout** as the feature to pattern and the circular pattern we sketched earlier as the profile around which to pattern. Click Finish and you should end up with the following model.

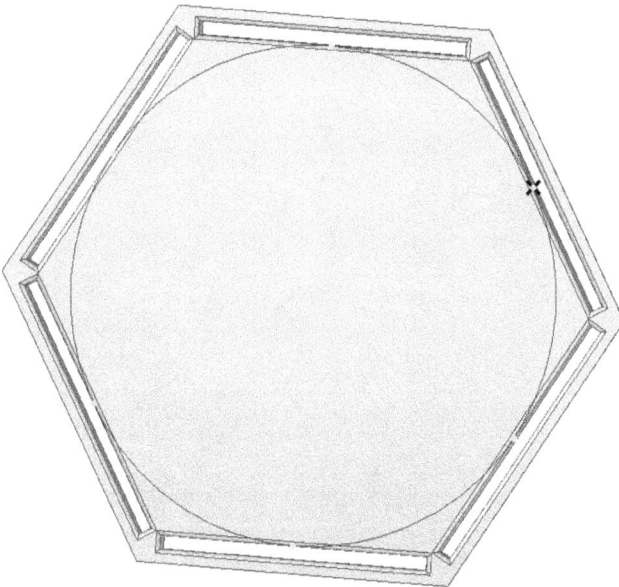

Now click **Dimple** and create the following sketch.

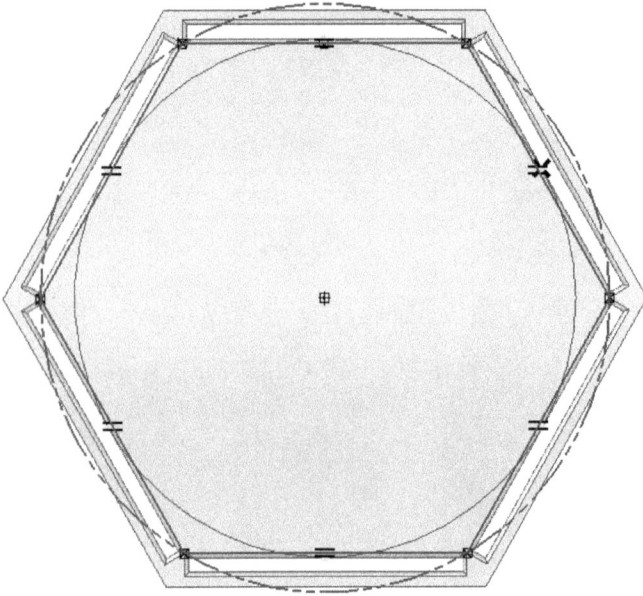

Close the sketch and input an extent of 1in again, but this time in the opposite direction of the cutouts. Click Finish and you'll get the following model.

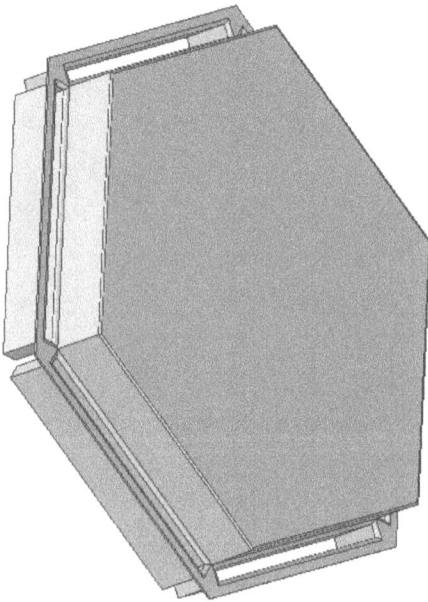

Create a sketch on top of the **Dimple** we just created. Go to Tools and under "Insert" click Text. The actual text isn't all that important, so you can get creative with it. Just make sure it fits on the model. An example is shown below.

In the "Dimple" drop down menu, click **Etch**. Select your text and click OK. (you can change the color and thickness of the etch using the "Etch options" box) Your model should now resemble the example below.

Exercise Complete

Exercise 64: Hems and Jogs

Create a new **ANSI Sheet Metal** file and click **Tab**. Create a Tab that is 6in x 3in x 0.038in. It should look like the following model.

Click the **Hem** command located under "Contour Flange". Select one of the bottom edges of one of the 3in sides. The result is a **Hem** that bends under the model.

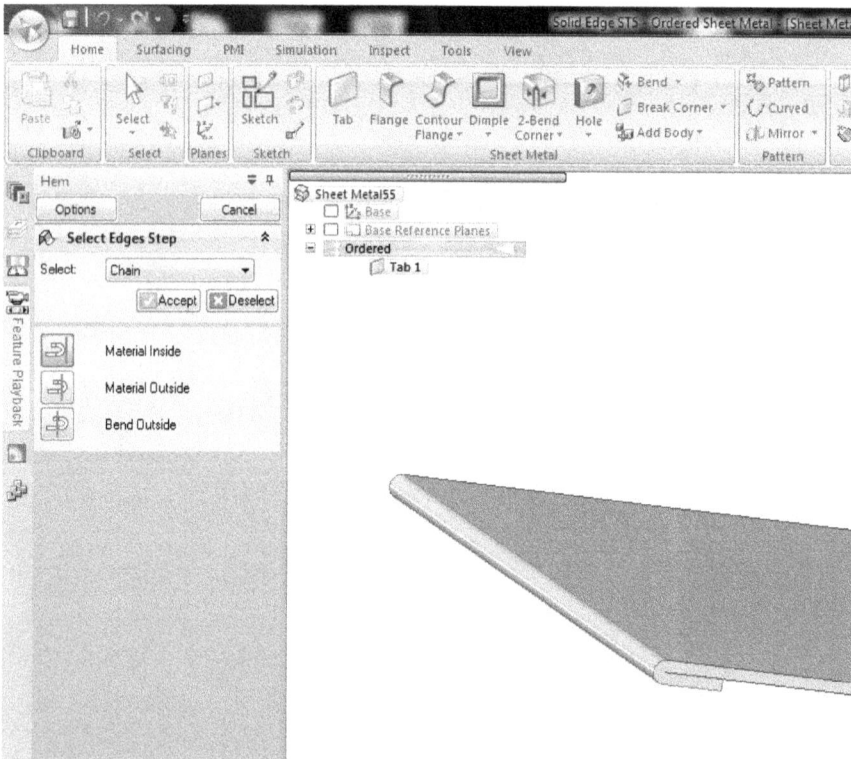

Click Accept and then Finish. Now click the **Jog** tool located in the "Bend" drop down menu. Draw the following sketch on the top face of the model. Its dimensions aren't particularly important.

When prompted to, choose the larger of the two sides as the side to **Jog**. Then input the following information in the "Extent" step.

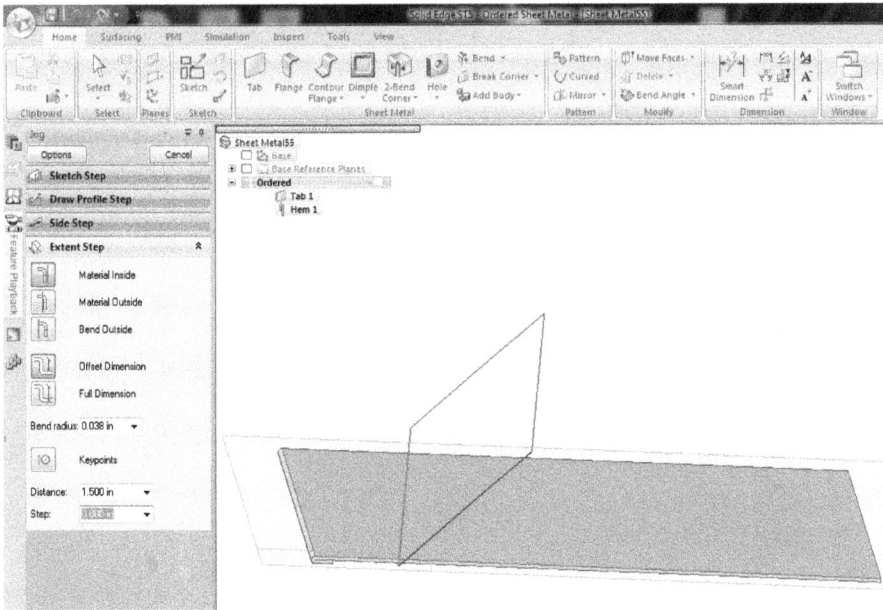

Click Finish. You should now have the following model.

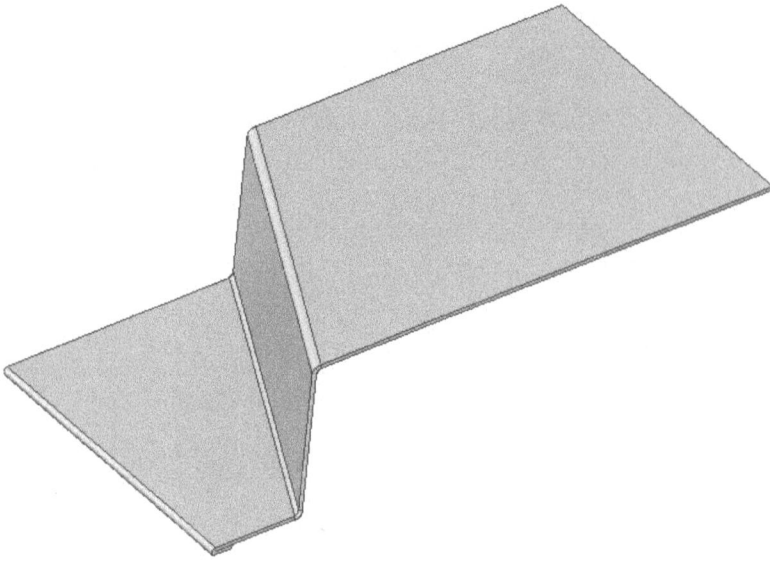

Exercise Complete

Exercise 65: Gussets

Create a new **ANSI Sheet Metal** file and click **Tab**. Create a Tab that is 5in x 5in x 0.038in. Add a 5in flange to one side of the Tab. You should have the following model.

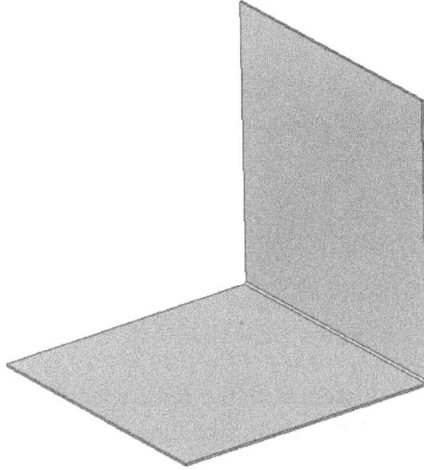

Now click the **Gusset** command located in the "Dimple" drop down menu. First, click options. Input the following information into the options box.

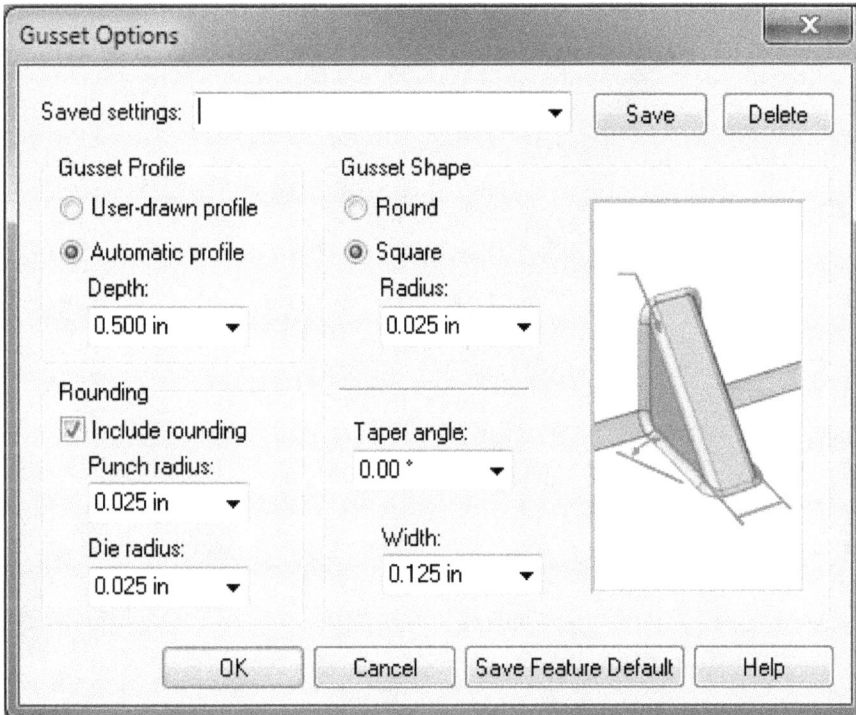

Click Save, as this will help us create more of these without having to input all of the specifications again, then click OK. Select the inner bend between the Tab and Flange. Place the gusset 1.25in from the end of the Tab as shown.

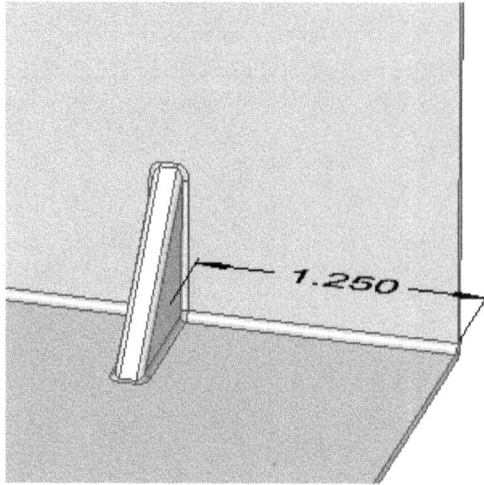

Click Finish, then create two more gussets with the same specifications. Place each one 1.25in from the previous one. Your model should look like the following image.

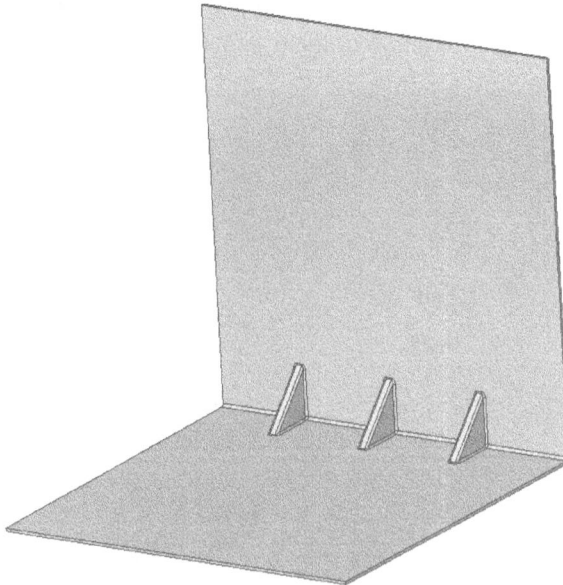

Exercise Complete

Exercise 66: Louvers

Create a new **ANSI Sheet Metal** file and click **Tab**. Create a Tab that is 12in x 12in x 0.038in. It should look like the following model.

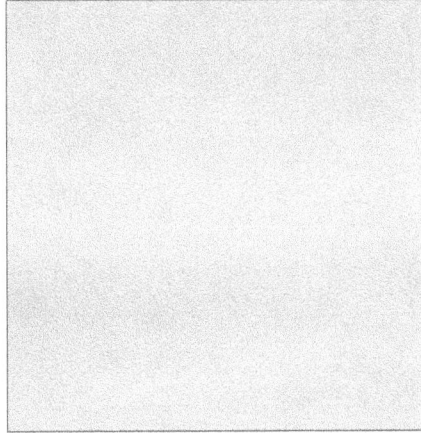

Click **Contoured Flange** and create the following sketch on one of the Tab's edges.

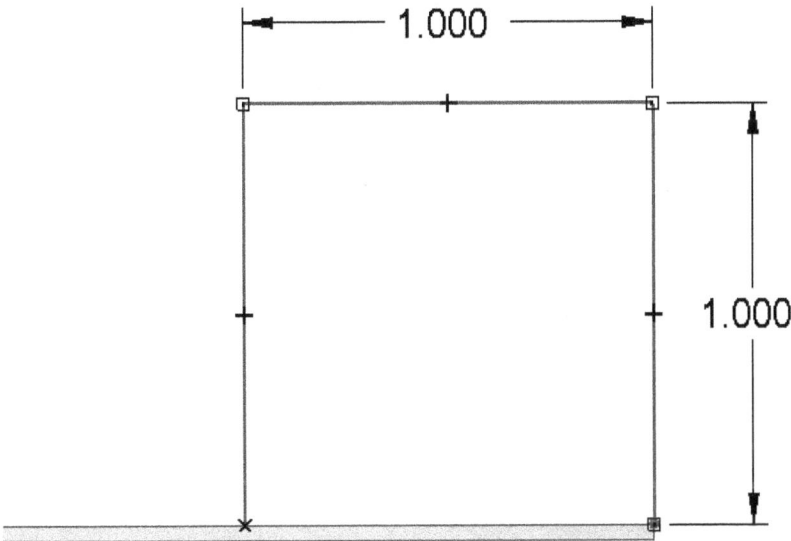

Close the sketch and choose "Chain" for the extent. Select the edge of the Tab as the path. Click Accept and then Finish. You should get the following model.

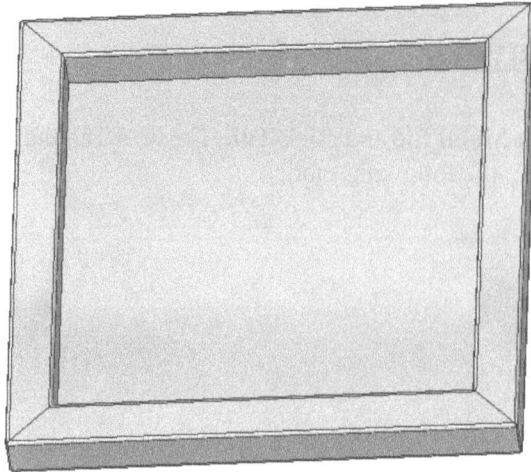

Under "Dimple", click **Louver**.

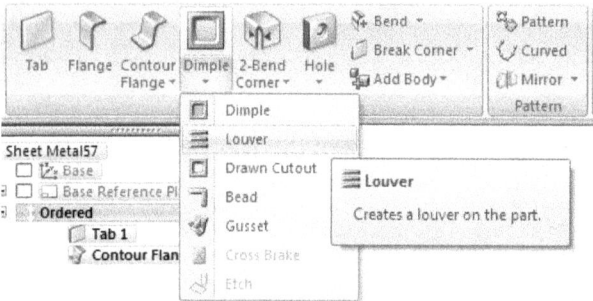

Select your original tab as the sketch plane. Create the following sketch and then click close sketch.

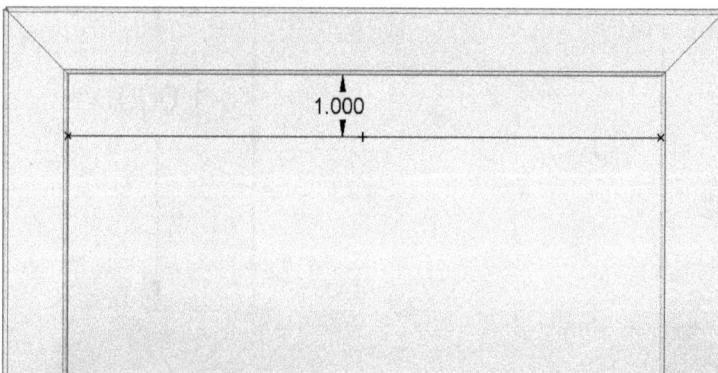

Select 1in for the depth and orient it such that the red outline of the louver lines up perfectly with the inner edge of the contoured flange.

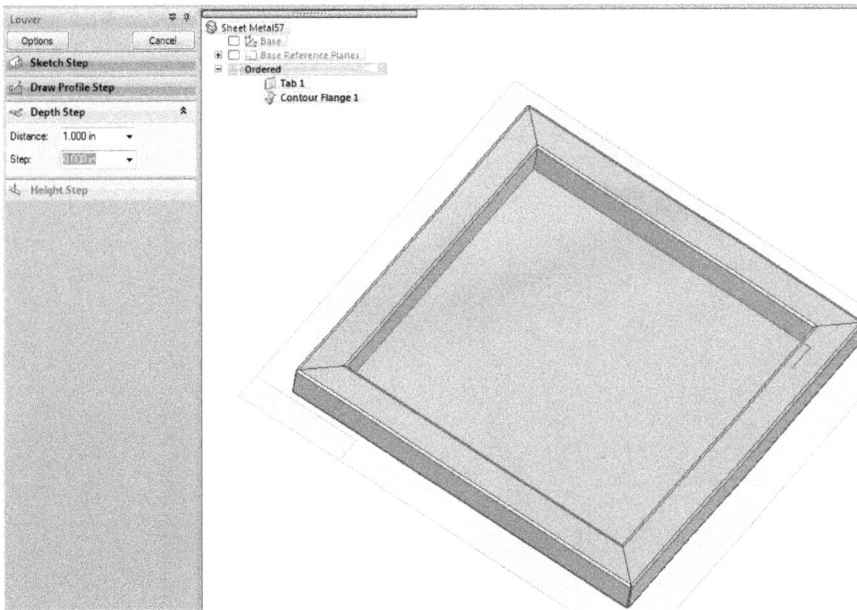

Input a height of 0.962in and orient the red outline outward as shown in the image below.

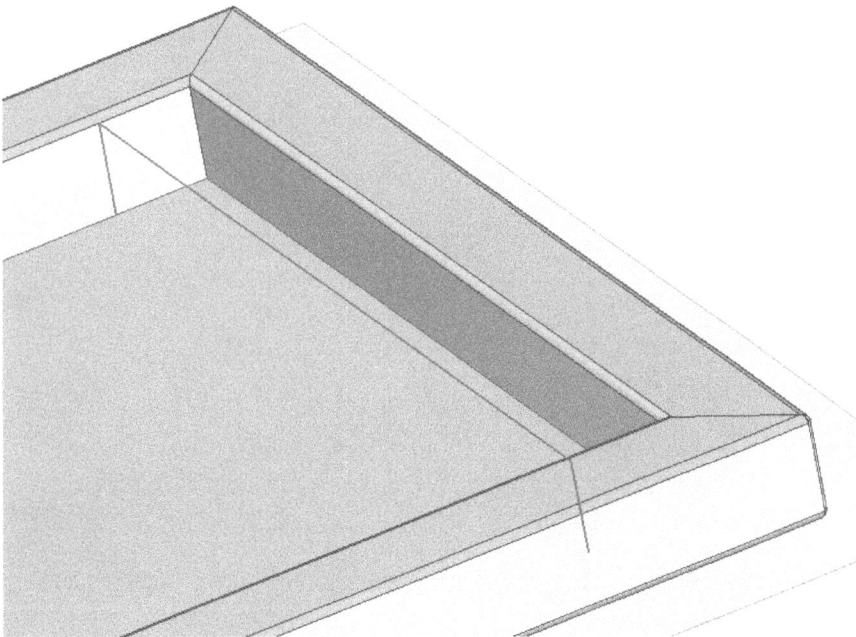

Click Finish. You should now have the following result.

Now create the following sketch on your original Tab. Close the sketch once you finish.

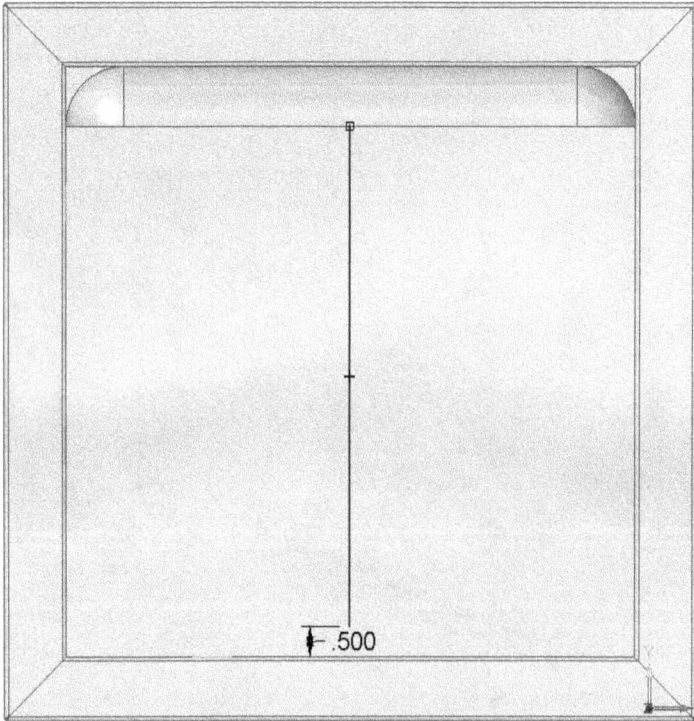

.500

Now click the **Curved Pattern** tool. Select the louver as the feature to pattern and the line as the curve to pattern it with. Input the following information on the "Select Curve" step. Click Accept and then choose the end of the line touching the louver as the anchor point. Orient the direction of the arrow such that the patterned louvers are on your original tab.

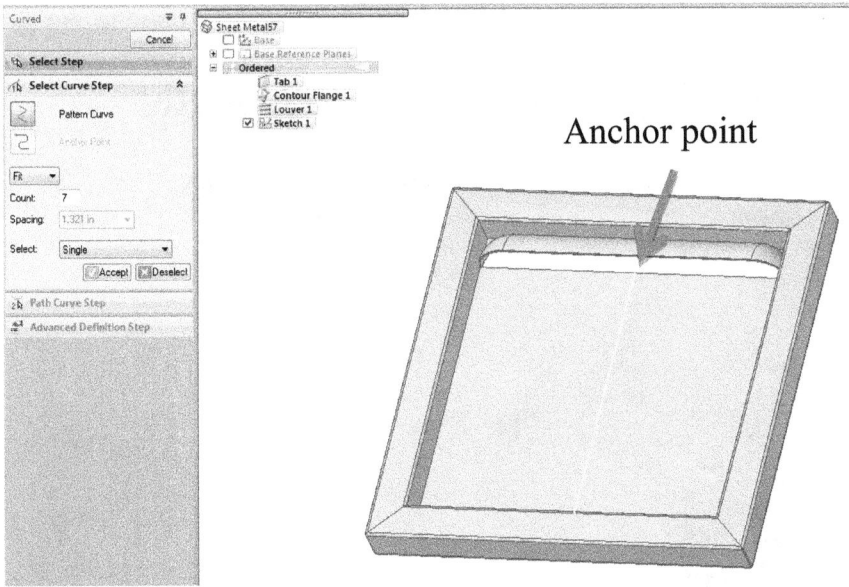

Anchor point

Click Preview and then Finish. You should now have the following model.

Exercise Complete

Exercise 67: Converting Solid Parts to Sheet Metal

Open a new ANSI part file. Click **Extrude** and create a 10in x 10in x 10in cube. Then click **Extrude** again and create a 5in x 5in x 5in cube on top of your first cube. (make sure it is centered)

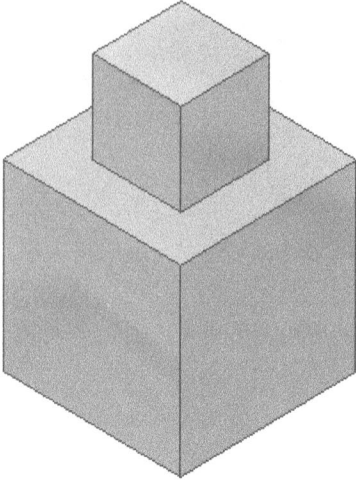

Now click **Thin Wall** and choose a thickness of 0.038in. Choose the top of the 5in x 5in x 5in cube as the face to remove. Click Accept, Preview, and then Finish. You should get the following part.

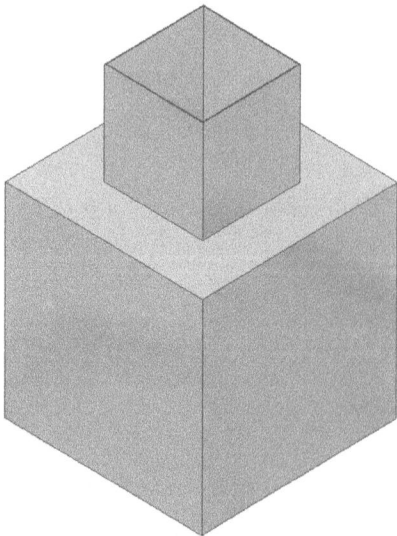

Create the following sketch on top of the 10in x 10in x 10in cube. These lines will be used as rip lines when we convert it to Sheet Metal.

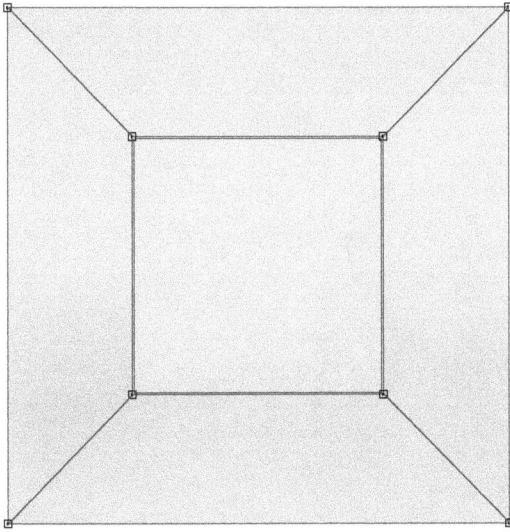

Now click the Solid Edge logo in the top left corner of the scree and under "Switch to" select Sheet Metal.

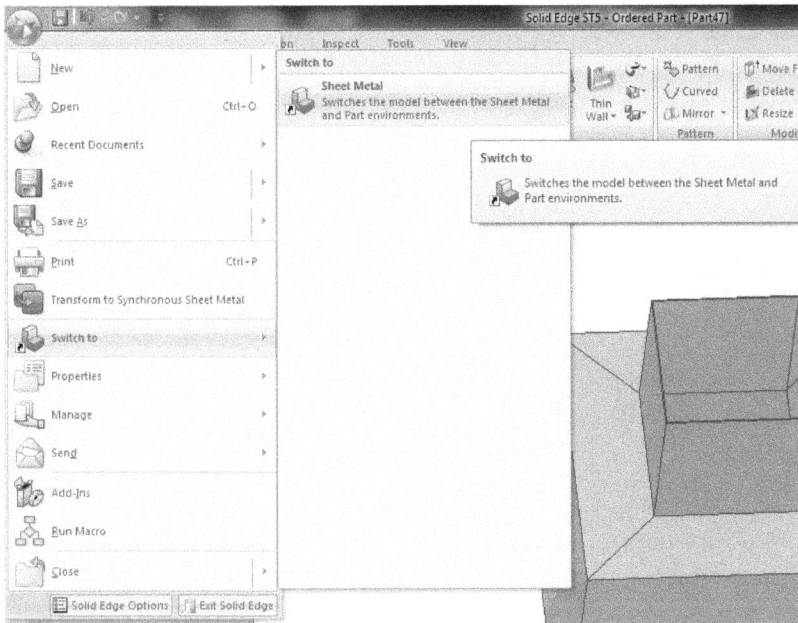

Under "2-Bend Corner", click **Rip Corner**. Select the vertical edges of both the 5in cube and the 10in cube and the sketch lines created earlier in the exercise. *Note: Do **not** select any horizontal edges.* When you have made the appropriate selections, click Accept, Preview, and then Finish.

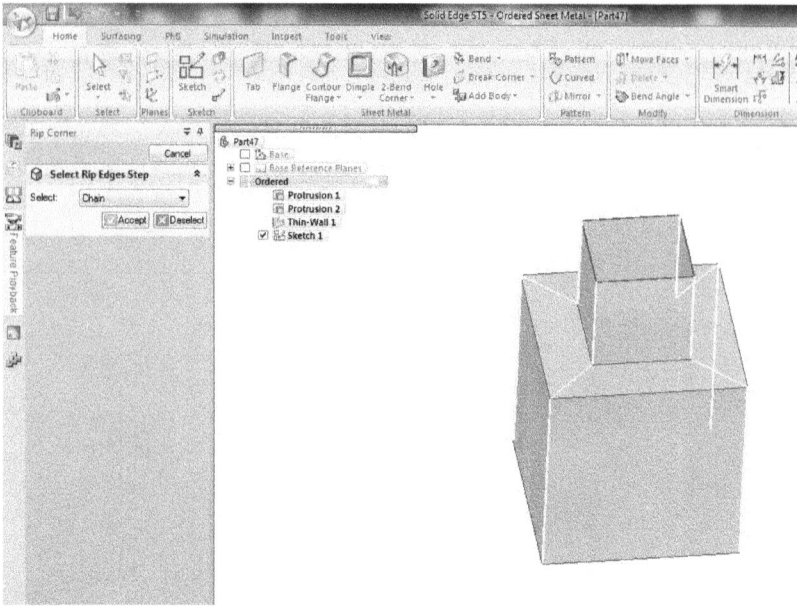

Now click the Solid Edge logo again and choose "Transform to Sheet Metal". Select the bottom of the 10in cube as the Face Step and then click Finish.

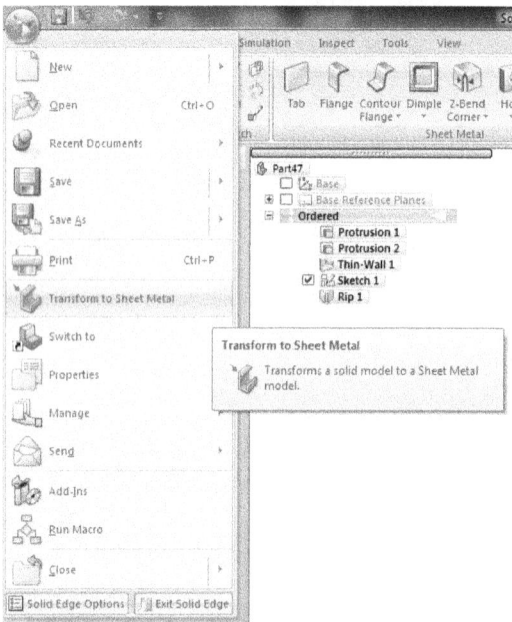

The only noticeable change will be that the edges show the deformation that would occur in bent metal.

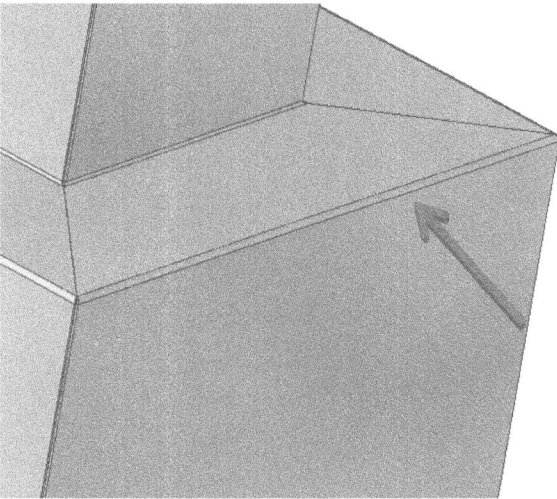

Since this is so discrete, we will **Unbend** it to prove that it is in fact Sheet Metal. Click **Unbend** and then select the bottom of the 10in cube as the Fixed Face. When prompted to, select every bend on the part.

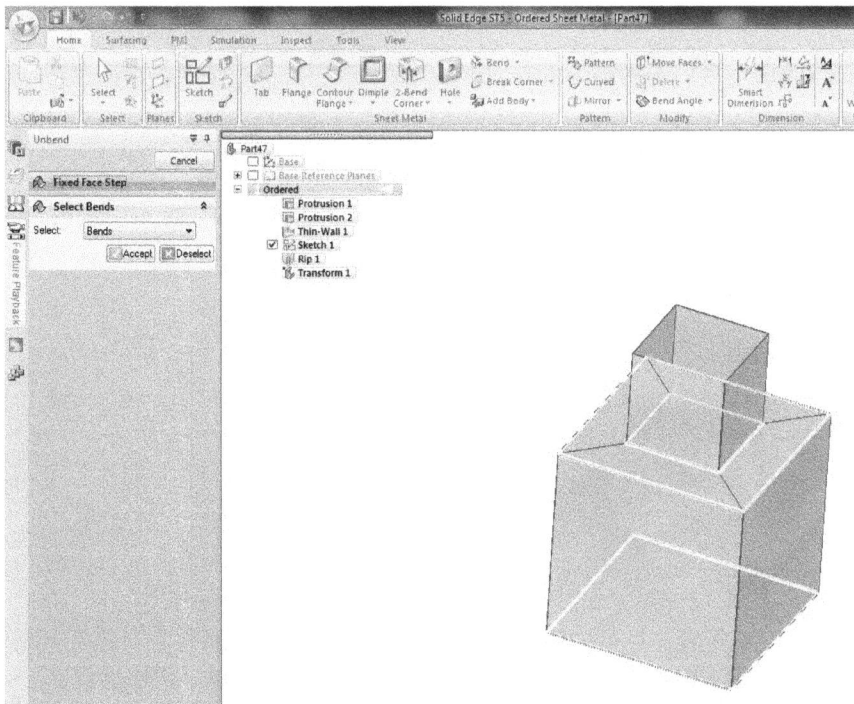

Click Accept, Preview, and then Finish. Hide the sketches used to rip the part. You should now have the following Sheet Metal model.

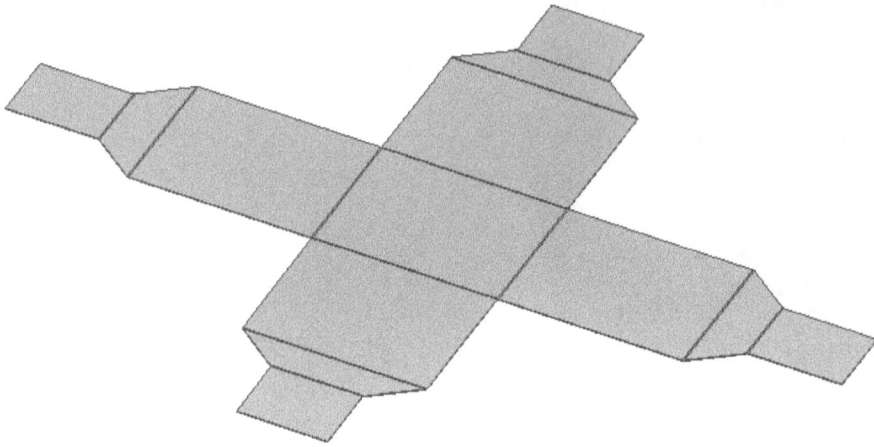

Exercise Complete

Exercise 68: Modifying Sheet Metal Models

Open **Unbending_Exercise.psm**, the Sheet Metal model created during the Unbending and Re-Bending exercise.

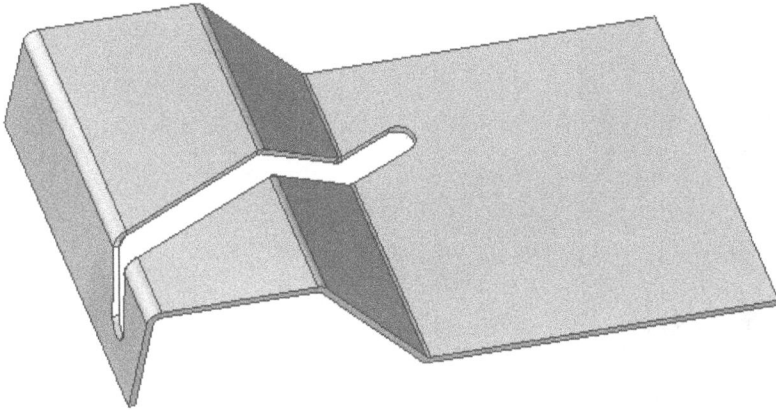

Click **Bend Angle** located in the "Modify" box. Select the bend highlighted in yellow below as the bend to modify and the face highlighted in red as the fixed face.

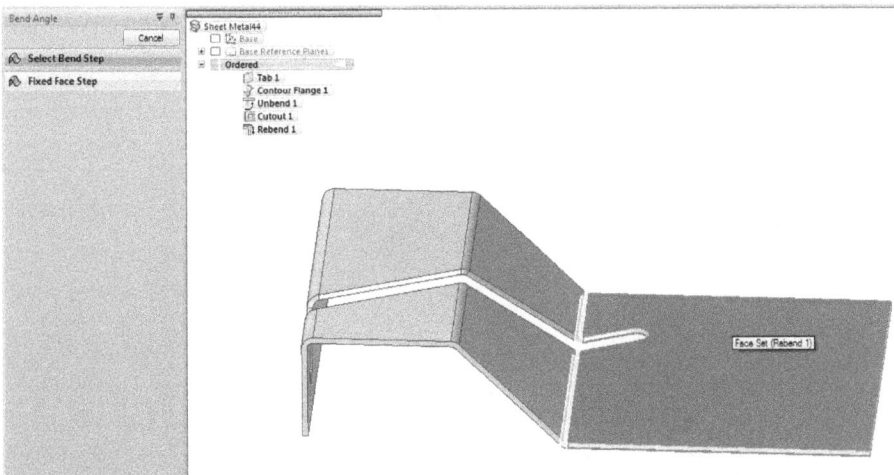

Enter 90 deg. and click Preview and then Finish. You should now have the following model.

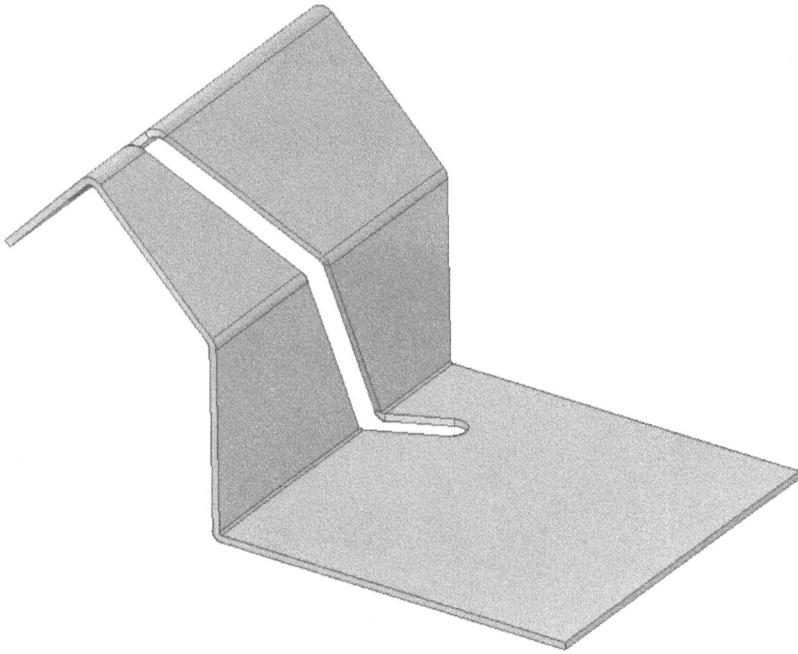

Now click **Bend Radius** (located in the "Bend Angle" drop down menu) and select the bend shown below.

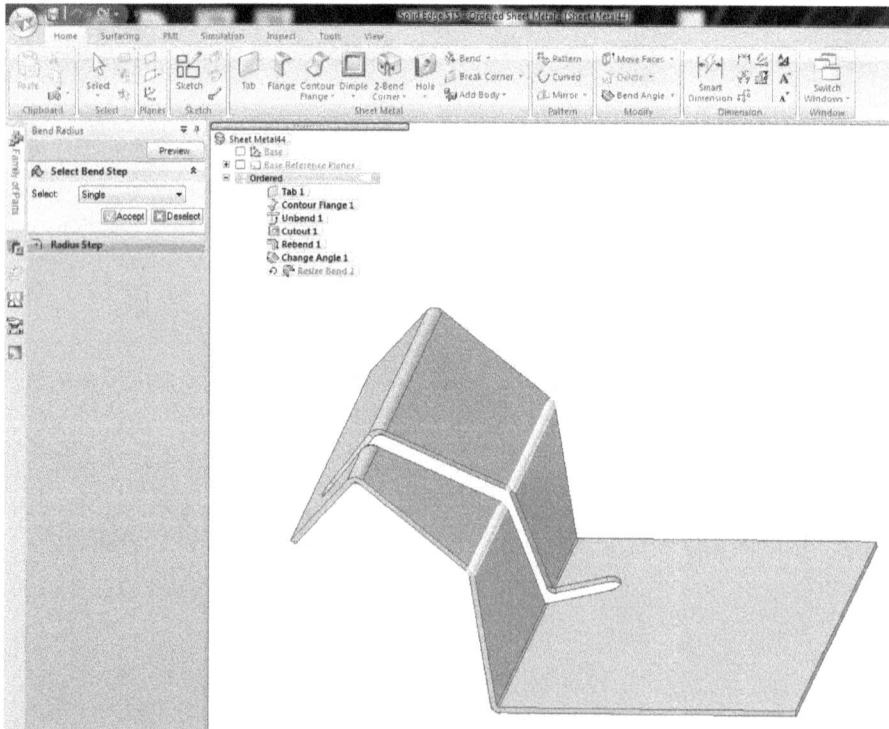

Change the radius to 0.25in and click Preview, and then Finish. You should now have the following model.

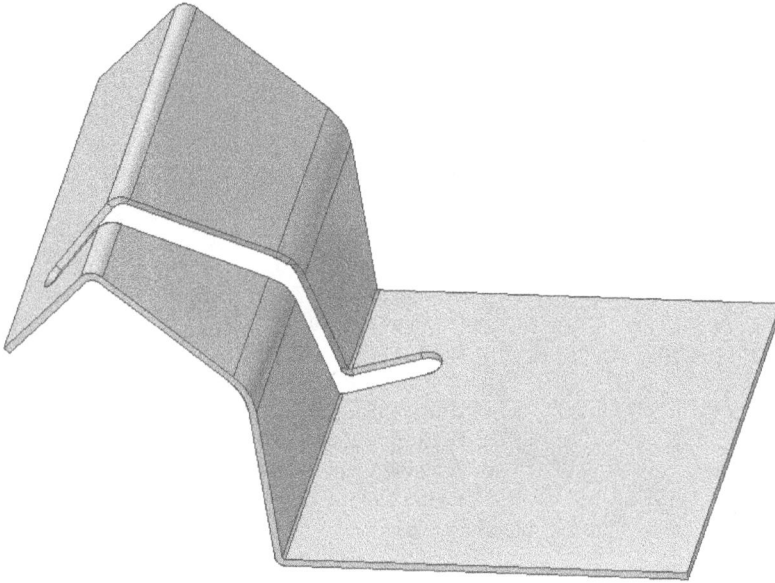

Other Modify operations such as **Offset Face**, **Move Face**, and **Rotate Face** can also be applied to Sheet Metal models. They have the same functionality when they are applied to Sheet Metal as when they are applied to Solid Parts. For instruction on these operations, see "Using Synchronous Modeling in an Ordered Part File".

Exercise Complete

INDEX

www.ingramcontent.com/pod-product-compliance
Lightning Source LLC
Chambersburg PA
CBHW080653220326
41598CB00033B/5194